IUTAM SYMPOSIUM ON
VARIABLE DENSITY LOW-SPEED TURBULENT FLOWS

FLUID MECHANICS AND ITS APPLICATIONS
Volume 41

Series Editor: **R. MOREAU**
MADYLAM
Ecole Nationale Supérieure d'Hydraulique de Grenoble
Boîte Postale 95
38402 Saint Martin d'Hères Cedex, France

Aims and Scope of the Series

The purpose of this series is to focus on subjects in which fluid mechanics plays a fundamental role.

As well as the more traditional applications of aeronautics, hydraulics, heat and mass transfer etc., books will be published dealing with topics which are currently in a state of rapid development, such as turbulence, suspensions and multiphase fluids, super and hypersonic flows and numerical modelling techniques.

It is a widely held view that it is the interdisciplinary subjects that will receive intense scientific attention, bringing them to the forefront of technological advancement. Fluids have the ability to transport matter and its properties as well as transmit force, therefore fluid mechanics is a subject that is particulary open to cross fertilisation with other sciences and disciplines of engineering. The subject of fluid mechanics will be highly relevant in domains such as chemical, metallurgical, biological and ecological engineering. This series is particularly open to such new multidisciplinary domains.

The median level of presentation is the first year graduate student. Some texts are monographs defining the current state of a field; others are accessible to final year undergraduates; but essentially the emphasis is on readability and clarity.

For a list of related mechanics titles, see final pages.

IUTAM Symposium on

Variable Density Low-Speed Turbulent Flows

Proceedings of the IUTAM Symposium
held in Marseille, France,
8–10 July 1996

Edited by

LOUIS FULACHIER

C.N.R.S., and
Universités d'Aix-Marseille I and II,
Marseille, France

JOHN L. LUMLEY

Sibley School of Mechanical and Aerospace Engineering,
Cornell University,
Ithaca, New York, U.S.A.

and

FABIEN ANSELMET

C.N.R.S., and
Universités d'Aix-Marseille I and II,
Marseille, France

SPRINGER-SCIENCE+BUSINESS MEDIA, B.V.

A C.I.P. Catalogue record for this book is available from the Library of Congress.

ISBN 978-94-010-6302-9   ISBN 978-94-011-5474-1 (eBook)
DOI 10.1007/978-94-011-5474-1

Printed on acid-free paper

Chairmen of the IUTAM Symposium :

L. Fulachier (IRPHE)   J.L. Lumley (Cornell Univ.)

Scientific Committee :

R.W. Bilger (Australia)   W. Kollmann (U.S.A.)
C. Dopazo (Spain)   B.E. Launder (U.K.)
H.E. Fiedler (Germany)   H.K. Moffatt (U.K.)
J.C.R. Hunt (U.K.)   A.N. Secundov (Russia)

Organizing Committee :

M. Amielh (Treasurer, IRPHE-IHT2)   L. Fulachier (IRPHE-IHT2)
F. Anselmet (Chairman, IRPHE-IHT2)   J. Quinard (RPHE-CSR)
P. Dupont (IRPHE-SUP)   R. Schiestel (IRPHE-MOD)

Sponsoring Organizations and Companies :

Association Universitaire de Mécanique (A.U.M.)
Centre National d'Etudes Spatiales (C.N.E.S.)
Centre National de la Recherche Scientifique (C.N.R.S.)
Commissariat à l'Energie Atomique (C.E.A./D.R.N., Cadarache)
Commission of the European Communities
Dantec
Deltalab
Département des Bouches-du-Rhône
Electricité De France (E.D.F./L.N.H., Chatou)
Gaz De France (G.D.F./D.E.T.N., La Plaine Saint Denis)
International Science Foundation (I.S.F., U.S.A.)
International Union of Theoretical and Applied Mechanics
Kluwer Academic Publishers (The Netherlands)
Ministère de la Défense-Direction Générale de l'Armement (D.G.A./D.R.E.T.)
Quantel
Région Provence-Alpes-Côte d'Azur
Société Nationale d'Etudes et de Construction de Moteurs d'Avions
Spectra Physics
Université de Provence (Aix-Marseille I)
Ville de Marseille

Contents

PREFACE

The General Assembly of the International Union of Theoretical and Applied Mechanics in its meeting on August 28, 1994, selected for 1996 only four Mechanics Symposia, of which ours is the only one related to Fluid Mechanics : *Variable Density Low Speed Turbulent Flows*. This IUTAM Symposium, organized by the Institut de Recherche sur les Phénomènes Hors Equilibre (Marseille), is the logical continuation of the meetings previously organized or co-organized - on the French or European level, such as Euromech 237, Marseille, 1988 - by the same research group of Marseille.

This meeting focused specifically on the structure of turbulent flows in which density varies strongly : the effect of this variation on the velocity and scalar fields is in no sense negligible. We were mainly concerned with low-speed flows subjected to strong local changes of density as a consequence of heat or mass transfer or of chemical reactions. Compressible turbulent flows - such as supersonic ones - were also considered in order to underline their similarities to and their differences from low-speed variable density flows.

These turbulent flows are of fundamental interest because the conservation equations for thermodynamics, mass and momentum are linked together. Another interesting fundamental aspect is that such flows - in the jet or wake configuration - can develop self-excited oscillations related to absolute instability of the nozzle region. The understanding of these flows is an important problem in its own right but it is also the key to an improved understanding of combusting flows where the coupling between chemical reactions and aerodynamics occurs through local density fluctuations due to chemical heat release and composition changes. In addition, these density variations occur in many practical situations such as in the aerospace industry, in pollution and environmental problems, in engine combustion chambers, and so on.

One of the aims of the Symposium has been to stimulate collaboration between experimentalists, theoreticians and modellers, but also between University scientists and industrial workers. Another goal has been to bring together researchers specializing in turbulence and those involved in combustion.

From about fifty submitted papers, twenty-eight full oral communications and twelve posters were presented. In addition, six invited lectures were given at the beginning of each Symposium session. The present volume contains papers selected by the Scientific Committee of this IUTAM Symposium : they correspond to the six invited lectures - the papers of which are located at the beginning of each section of this volume -, the twenty-eight oral communications, seven posters, and the concluding remarks. In general, the number of pages allocated to each of these papers depends on its scientific impact.

The six sections developed in this book are focused on the following topics concerning variable density turbulent flows :
• Instabilities
• Modelling and experiments
• Modelling and experiments. Buoyancy effects
• Modelling and experiments. Industrial applications
• Experiments and measurement methods
• Compressible flows

From papers presented in this book, various research axes emerge, such as, for instance :
- Development of direct numerical simulations and large eddy simulations in particular in the nozzle region of variable density jets ;
- In the case of strong density variations - such as in combustion, where Favre averaging is intensively used -, modelling must be studied again. In addition, particular attention must be paid to scalar dissipation ;
- Experimental approaches - concerning velocity as well as scalar fields - must be developed to analyze the influence of density variations on turbulence small scales but also on turbulence large scales ;
- The Proper Orthogonal Decomposition (POD) approach must be extended to these flows with variable density as suggested in the concluding remarks.

We hope this book will contribute to the state of the art of variable density turbulent flows : it seems to us there are no other works that deal with this interesting area.

The Editors :

Louis Fulachier
John L. Lumley
Fabien Anselmet

I. Instabilities

INSTABILITIES AND BIFURCATIONS
IN VARIABLE DENSITY FLOWS

P. HUERRE, K. AMRAM & J.M. CHOMAZ
Laboratoire d'Hydrodynamique (LadHyX)
CNRS-Ecole Polytechnique
F-91128 Palaiseau FRANCE

Free shear flows are characterized by the presence of coherent vortices that may be regarded as instability waves evolving on a mean velocity profile. The main objective of the lecture is to review from a hydrodynamic stability perspective some of the recent progress in the modeling of variable density jets and mixing layers.

1. Global Modes in Variable Density Jets.

Very significant advances have been made by Monkewitz and his colleagues in our current understanding of the dynamics of low density jets. Such flows have been shown to undergo a transition from convective to absolute instability as the jet density is sufficiently decreased (Monkewitz & Sohn 1988). The appearance of a finite region of absolute instability is responsible for the onset of self-sustained oscillations, as strikingly demonstrated in the experiments of Sreenivasan et al. (1989) and Monkewitz et al. (1990). These periodic motions obey all the scaling laws that are expected from the Landau amplitude evolution model : jets effectively undergo a supercritical Hopf bifurcation as the jet density is gradually lowered (Raghu & Monkewitz 1991).

These self-sustained oscillations can be modeled as global modes, i.e. extended wavepackets that "live" on the slowly diverging jet flow and beat at a well defined frequency. Asymptotic WKBJ formulations have been proposed to obtain the global frequency (Chomaz et al. 1991, Le Dizès et al. 1996) and the spatial structure of the wavepacket (Monkewitz et al. 1993). Such formulations have been extremely successful in predicting the resonance frequency of wakes behind blunt-edged plates (Hammond & Redekopp 1996) but they remain to be implemented and tested in the case of low-density jets.

Global mode oscillations lead to the formation of intense side-jets that drastically enhance the measured spreading rates of low-density jets (Monkewitz et al. 1989). In recent numerical simulations, Brancher et al (1994) have confirmed the scenario proposed by Monkewitz & Pfizenmaier (1991) : the strong radial ejection of fluid

L. Fulachier et al. (eds.), IUTAM Symposium on Variable Density Low-Speed Turbulent Flows, 3–8.

is not directly related to the deformation of the primary vortex rings but rather to the occurrence of coherent streamwise vortex pairs in the braid region.

2. Bifurcations in non isothermal mixing layers.

The second part of the lecture is devoted to the recent numerical study of temporally evolving two-dimensional compressible mixing layers by Amram (1995). Following Djordjevic & Redekopp (1989), the nondimensional basic velocity and temperature profiles are taken to be

$$U(y) = \tanh y \quad , \quad T(y) = 1 + b \; \text{sech}^{Ma^2} y. \qquad (1a,b)$$

In the above relations the length scale is half of the vorticity thickness, $Ma \equiv \Delta U/(2a_\infty)$ designates the Mach number based on half of the velocity difference ΔU and on the velocity of sound a_∞ at $y = \infty$. Finally, $b \equiv (T_c - T_\infty)/T_\infty$ denotes the temperature ratio, T_c and T_∞ being the temperatures at $y = 0$ and $y = \infty$ respectively. For such a family of basic states, the weakly nonlinear evolution of marginally unstable wavenumbers in the viscous critical layer régime is known to follow the Landau amplitude evolution model (Huerre 1987 , Djordjevic & Redekopp 1989) : the kinetic energy ε of fluctuations integrated over the entire spatial domain of interest is governed by the equation

$$\frac{1}{2} \frac{d\varepsilon}{dt} = \sigma \varepsilon + L \varepsilon^2 , \qquad (2)$$

where σ is the linear temporal growth rate and L the so-called Landau constant.

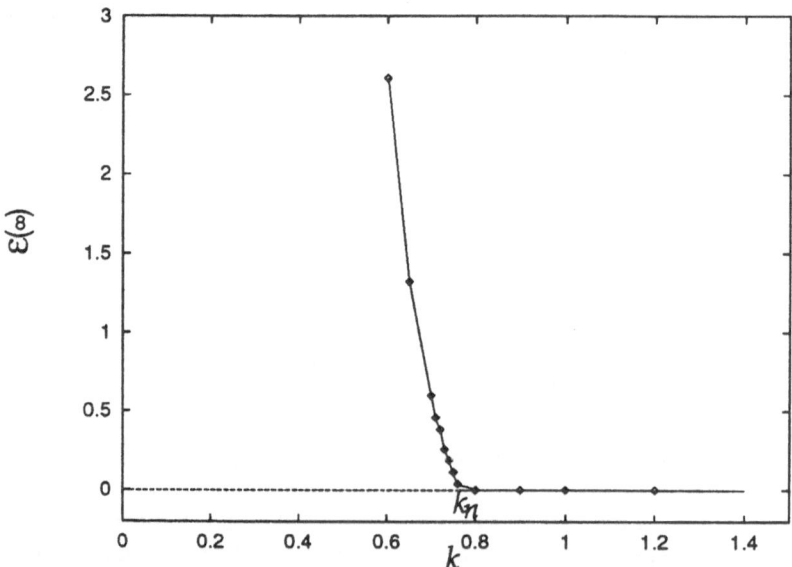

Figure 1 - Supercritical bifurcation diagram $\varepsilon \, (\infty)$ versus k for isothermal mixing layers ; Ma = 0.5, Re = 50, b = 0. Direct numerical simulations.

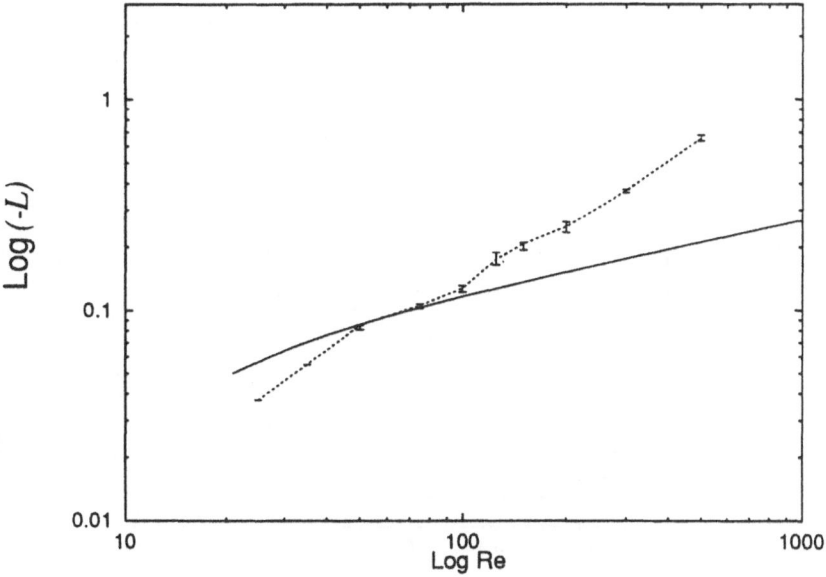

Figure 2 - Numerically determined Landau constant log (– L) versus Reynolds
number log (Re). Solid curve corresponds to weakly nonlinear analysis of
Djordjevic & Redekopp (1989), points to numerical simulations. Ma = 0.5, b = 0.

Direct temporal numerical simulations are used here to test the relevance and
range of validity of this theoretical model in the vicinity of the neutral wavenumber
k_n. Eigenmodes from linear instability theory are chosen as initial conditions and
the two-dimensional compressible Navier-Stokes equations are integrated over
time in a streamwise periodic box of spatial period $2\pi/k$. The wavenumber k then
defines a control parameter of the system, that may be adjusted by varying the
streamwise period of the computational domain.

In the case of isothermal mixing layers (b = 0), the computed long-time dynamics
give rise to a **supercritical** bifurcation (L < 0) as k crosses k_n from above (Figure
1). Furthermore, it is verified that, for k sufficiently close to k_n, the fluctuating
energy ε (t) is effectively governed by the Landau evolution equation (2). As
shown on figure 2, the numerically determined Landau constant L coincides with
the theoretical predictions of Djordjevic & Redekopp (1989) in an intermediate
range of Reynolds numbers 50 < Re < 100, where Re ≡ ΔUδ/(4ν) : L follows the
expected $Re^{1/3}$ scaling law precisely when the critical layer Reynolds number is
sufficiently small.

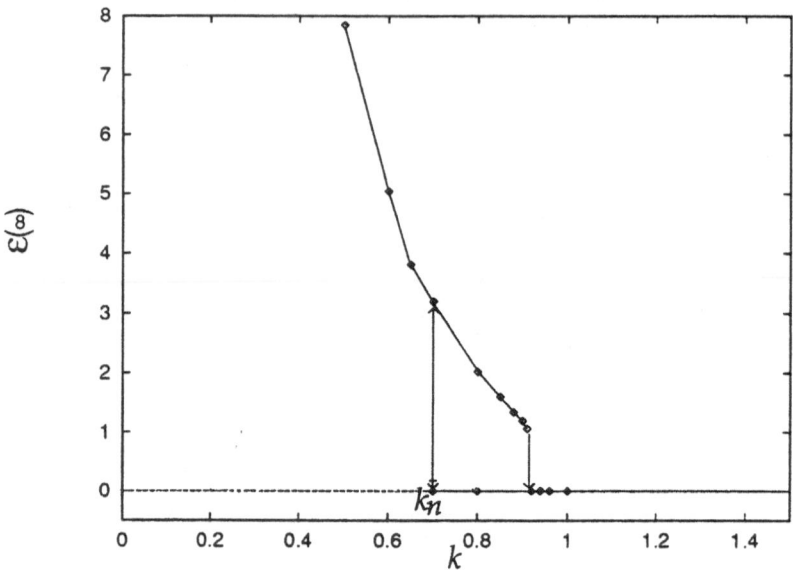

Figure 3 - Subcritical bifurcation diagram ε (∞) versus k for sufficiently cooled mixing layers ; Ma = 0.7, Re = 50, b = - 0.8. Direct numerical simulations.

In the case of non-isothermal mixing layers (b $\neq$ 0), a similar behavior is observed, but as the mixing layer is sufficiently cooled (b < 0), the finite-amplitude régime is found to display hysteresis as the wavenumber k is decreased and increased across the neutral value k_n (Figure 3). The **subcritical** nature of the bifurcation (L > 0) predicted by Djordjevic & Redekopp (1989) is therefore confirmed for symmetrically cooled mixing layers.

This numerical study has therefore established, for the first time in mixing layers, the validity of the Landau nonlinear evolution model, at wavenumbers sufficiently close to the neutral wavenumber k_n.

The financial support (K.A.) of ONERA is gratefully acknowledged.

References.

Amram, K. (1995). Étude numérique des instabilités dans les couches de mélange compressibles. PhD dissertation, Laboratoire d'Hydrodynamique (LadHyX), École Polytechnique, Palaiseau, France.

Brancher, P., Chomaz, J.M. & Huerre, P. (1994). Direct numerical simulations of round jets : vortex induction and side-jets. Phys. Fluids, **6**, 1768-74.

Chomaz, J.M., Huerre, P. & Redekopp, L.G. (1991). A frequency selection criterion in spatially developing flows. Stud. Appl. Maths. **84**, 119-144.

Djordjevic, V.D. & Redekopp, L.G. (1989). Nonlinear stability of subsonic mixing layers with symmetric temperature variations. Proc. R. Soc. Lond. A **426**, 287-330.

Hammond, D. A. & Redekopp, L.G. (1996). Global dynamics of symmetric and asymmetric wakes. J. Fluid Mech. (in press).

Huerre, P. (1987). On the Landau constant in mixing layers. Proc. R. Soc. Lond A **409**, 369-381.

Le Dizès, S. (1994). Modes globaux dans les écoulements faiblement inhomogènes. PhD dissertation, Laboratoire d'Hydrodynamique (LadHyX), École Polytechnique, Palaiseau, France.

Le Dizès, S., Huerre, P., Chomaz, J.M. & Monkewitz, P.A. (1996). Linear global modes in spatially developing media. Phil. Trans. R. Soc. Lond. A., **354**, 169-212.

Monkewitz, P.A., Bechert, D.W., Barsikow, B. & Lehmann, B. (1990). Self-excited oscillations and mixing in a heated round jet. J. Fluid. Mech. **213**, 611-639.

Monkewitz, P.A., Huerre, P. & Chomaz, J.M. (1993). Global linear stability analysis of weakly non-parallel shear flows. J. Fluid. Mech. **251**, 1-20.

Monkewitz, P.A., Lehmann, B., Barsikow, B. & Bechert, D.W. (1989). The spreading of self-excited hot jets by side-jets. Phys. Fluids A1, 446-48.

Monkewitz, P.A., & Pfizenmaier, E. (1991). Mixing by "side-jets" in strongly forced and self-excited round jets. Phys. Fluids A **3**, 1356-61.

Monkewitz, P.A. & Sohn, K.D. (1988). Absolute instability in hot jets. AIAA J. **26**, 911-916.

Raghu, S. & Monkewitz, P.A. (1991). The bifurcation of a hot round jet to limit-cycle oscillations. Phys. Fluids A **3**, 501-503.

Sreenivasan, K.R., Raghu, S. & Kyle, K. (1989). Absolute instability in variable density round jets. Exp. Fluids **7**, 309-17.

GLOBAL SELF-EXCITED OSCILLATIONS IN A TWO-DIMENSIONAL HEATED JET : A NUMERICAL SIMULATION

V.G. CHAPIN, F. SERS AND P. CHASSAING
Ecole Nat. Sup. d'Ing. de Constructions Aéronautiques
1, Place Emile Blouin, 31056 Toulouse Cedex.

1. Introduction

The aim of this work (Sers 1995) was to develop a numerical methodology to gain insight in the low-density jet spatially developing dynamic with a *nonlinear* approach. Numerical simulations are shown to differentiate convective and absolute instability regimes and to capture a self-excited global mode in an open flow : the 2D hot jet.

The first part is devoted to numerical methodology and its validation on this unsteady problem which is known to be noise sensitive. The second part presents numerical results. They confirm theoretical and experimental results on the development of self-excited global oscillations of the jet column when the density ratio is lower than a critical value. The global mode and its associated Hopf bifurcation are identified.

2. Numerical methodology

The time-dependent compressible Euler equations with an hyperviscosity model (bilaplacian numerical dissipation) and boundary conditions, are time marched with a 5-stage 2nd-order Runge-Kutta scheme (Jameson & al. 1981). The spatial discretization is based on 2nd-order central differences with a finite volume method. Equations written in conservative variables are of the following form :

$$\frac{\partial U}{\partial t} + \nabla.F(U) = \nu_4 \nabla^4 U \tag{1}$$

The used calculation domain is 18-D long (with D the exit jet height) and 10-D wide. Mesh refinement is used in the potential core region with one third of the total number of gridpoints. For precision, a typical large scale

9

L. Fulachier et al. (eds.), IUTAM Symposium on Variable Density Low-Speed Turbulent Flows, 9–15.
© 1997 *Kluwer Academic Publishers.*

vortical structure in the finely resolved part of the fine grid (286x213) is approximately described with 40x40 gridpoints. The boundary conditions have been chosen in the following way : slip conditions at the lateral boundaries, a non-reflecting condition at the outlet and a reservoir condition at the inlet with prescribed velocity and density profiles.

The independence of the global mode frequency has been controlled towards grid resolution, outlet boundary condition location and simulation time. Two calculations were made with a coarse mesh (143x107) and a domain length of 18-D and 32-D. Frequencies were identical within the frequency resolution of 3%. Then, simulations were performed for a long enough time integration to check the statistically stationnarity of the frequency within 1% (the resolution of the Fourier analysis).

3. Numerical experiments

Our laminar jet configuration (figure 1) has been defined with Yu & Monkewitz (1990) paper to be able to make some comparisons. We know from Raynal & al. (1996) that relative positions of velocity and density profiles are of main importance. For numerical reasons, a small convection velocity outside the jet was adopted yielding a velocity ratio of $\Lambda = (U_j - U_\infty)/(U_j + U_\infty) = 0.93$ in all calculations. In that case, the critical density ratio predicted by the linear theory with zero Mach number is around $S_c = \rho_{jc}/\rho_\infty = 0.6$. According to Monkewitz & Sohn (1988), S_c decreases with Mach number in axisymmetric inhomogeneous jets. Hence, a density ratio of $S = \rho_j/\rho_\infty = 0.4$ was expected to give self-excited oscillations related to a sufficiently large pocket of absolute instability (Chomaz & Huerre 1988) for our 0.15 jet Mach number.

3.1. ABSOLUTE AND CONVECTIVE INSTABILITY

The hot (S=0.4) and cold (S=0.9) jet near-field velocity spectrum are shown in figure 2 (left). The same behaviour found in experimental work (Yu & Monkewitz 1993) is exhibited. A line dominated spectrum with harmonic contents is detected, characteristic of absolute instability, whereas for convective one, broadband spectrum with small peaks is observed. After a transient, a quasi-periodic longitudinal velocity on the jet centerline is seen for S=0.4, see figure 2 (right). The same qualitative structure is noticed everywhere in the potential cone.

Some comparisons were made with the experimental work of Yu & Monkewitz (1993). In figure 3, the weak dependence of the Strouhal number with jet velocity U_j is verified. So, the main frequency is well correlated to the jet

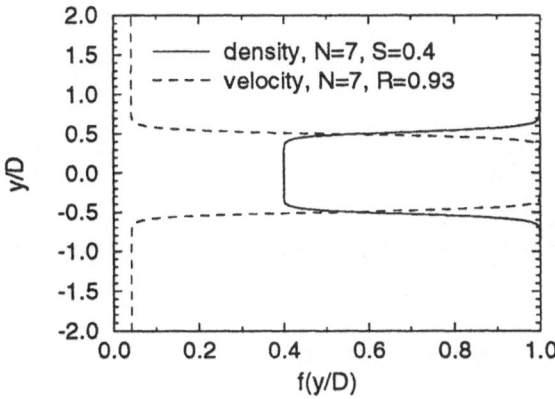

Figure 1. Velocity and density profiles at inlet boundary condition.

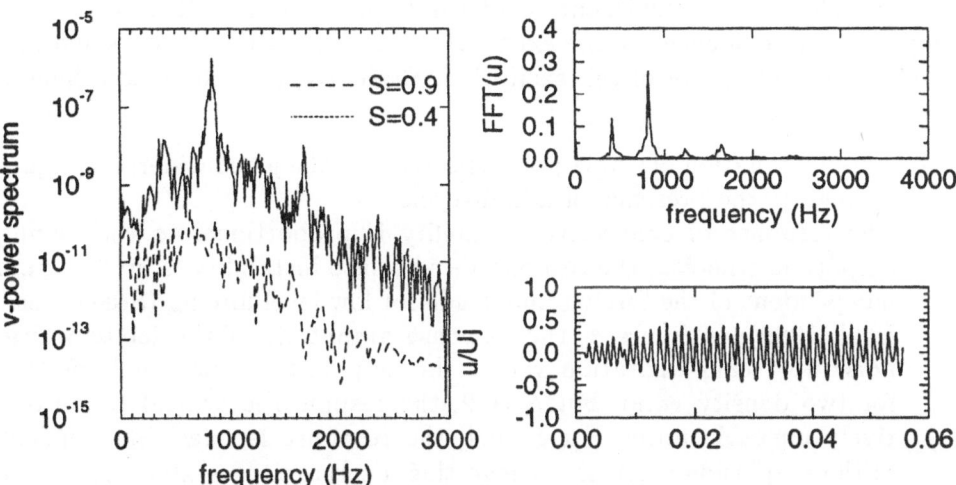

Figure 2. (left) Transversal velocity power spectrum at x/D=0.6 : (·····) S=0.9, (——) S=0.4, (right) Longitudinal velocity signal and spectrum at x/D=3.5, S=0.4

preferred mode. The value of the Strouhal number, $St=f_p.D/U_m=0.30$ is closed to the Yu and Monkewitz's (1993) value $St=0.32$. Moreover, the main frequency is also nearly independent of the density ratio S when $S < S_c$, see figure 3. This is a basic characteristic of a normal Hopf bifurcation (Bergé *et al.* 1984).

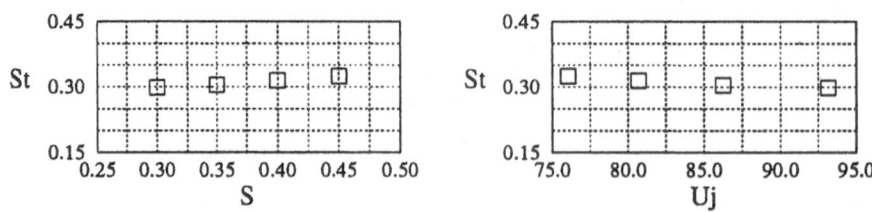

Figure 3. Strouhal number versus : (left) density ratio S, (right) centerline jet velocity U_j.

3.2. THE GLOBAL MODE OF THE JET

In this section, the jet asymptotic states for supercritical density ratio are investigated. For sufficiently low density ratio, numerical simulations have shown that, oscillations seem to saturate to a global self-excited mode (figure 2). Following Huerre and Monkewitz (1990), two classes of experiments are suitable for the identification of these global modes. The first class, only supports evidence for the existence of self-excitation. The second one yields conclusive proof of self-excitation. Presently, first class experiments have been used :

1. The examination of single-point spectrum in the jet considered in figure 2 suggests the presence of a limit-cycle.
2. The response of convective instability is proportional to the forcing amplitude whereas, the response of absolute instability is intrinsic *i.e.* independent of the forcing amplitude for low level forcing (Sreenivasan & al. 1989). In figure 4, the response amplitude of the jet to white-noise upstream excitation versus the amplitude excitation is plotted for two density ratio. For $S=0.9$, the response is typical of convectively unstable flows. For $S=0.4$, the response is quasi independent of the amplitude excitation suggesting a self-excited system. Response frequency f_r to periodic forcing at f_f, figure 4 (right), leads to the same conclusion.
3. Calculations were made with different density ratio (figure 5). Figure 6 shows a typical example of the jet global mode saturation amplitude versus density ratio S and identify a Hopf bifurcation. With a Mach number of 0.15, a critical density ratio $S_c = 0.49$ was found. This is consistent with a calculation for S=0.5 where self-excited oscillations were not observed (figure 5).

In a subsequent paper, second class experiments are planned to be investigated in order to positively caracterize the Hopf bifurcation, by deter-

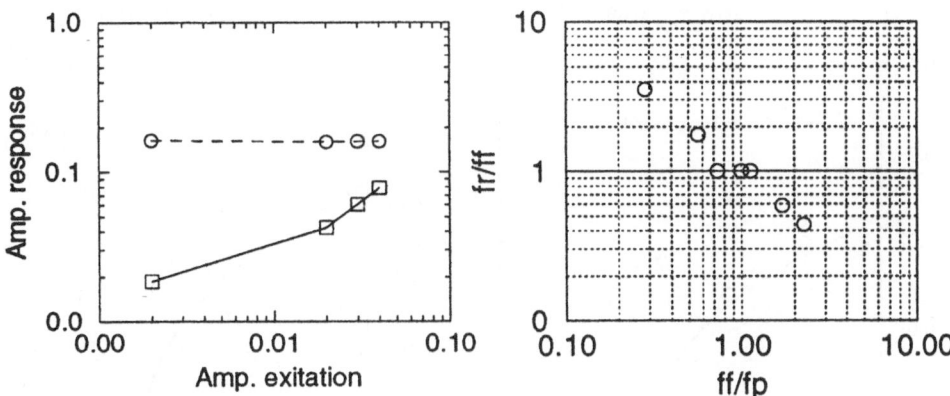

Figure 4.   Jet response to excitation : (left) White noise excitation with (——) $S=0.9$ and (– –) $S=0.4$, (right) Periodic forcing at f_f for $S=0.4$.

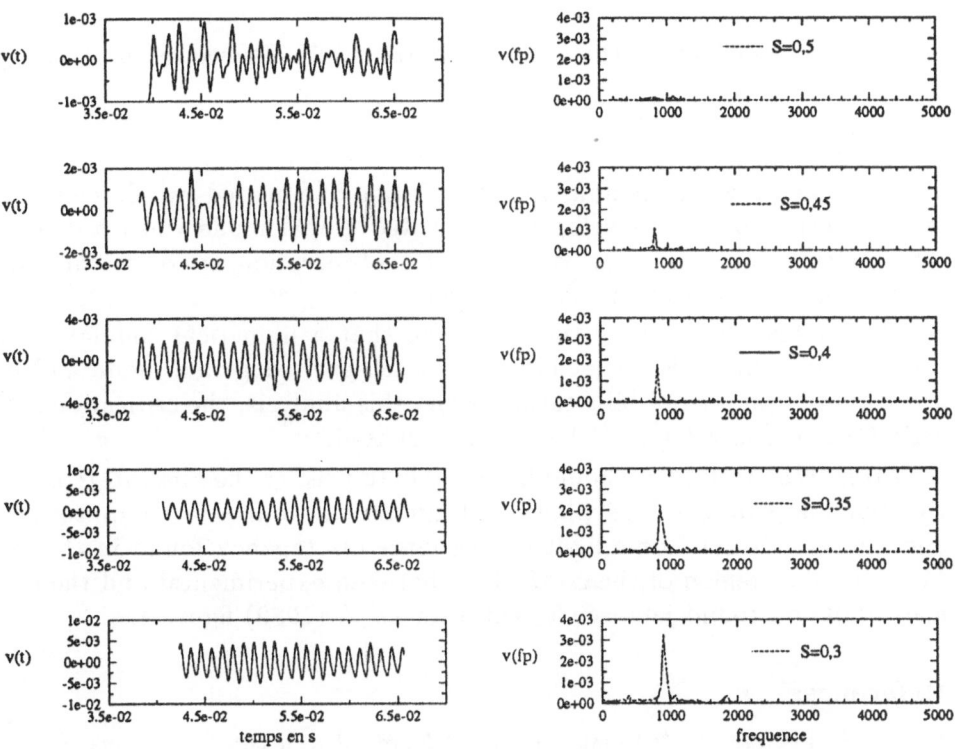

Figure 5.   Transverse velocity spectra versus density ratio S at x/D=1.2.

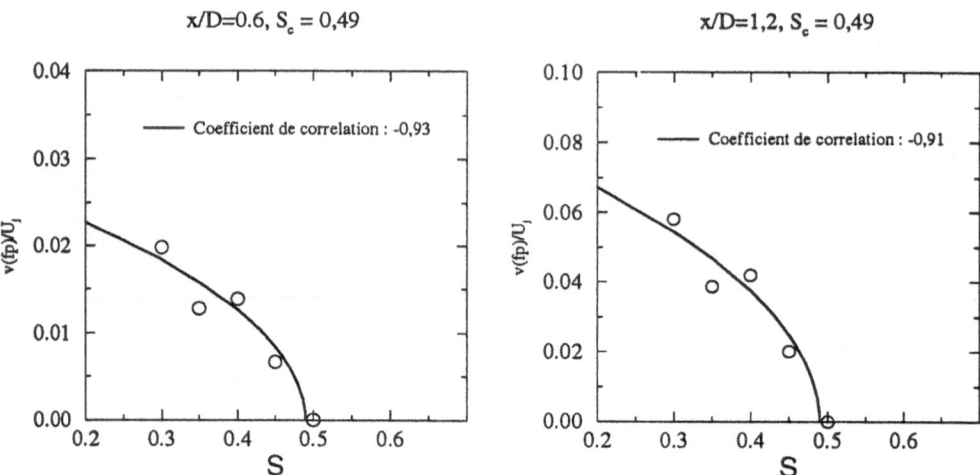

Figure 6. Transverse velocity amplitude of the spectral peak f_p versus density ratio S at x/D=0.6 and x/D=1.2.

mining the coefficients of the Landau-Stuart model (Provansal *et al.* 1987).

4. Conclusions

This first step toward the numerical simulation of a heated two–dimensional jet has been shown to develop a self–excited global mode when a density ratio below a critical value is used. This observation is consistent with experimental studies about plane hot jets.

To our knowledge, it is the first time that a numerical simulation of a heated two–dimensional jet identifies a self–excited global mode and its associated normal Hopf bifurcation. From this analysis, the critical density ratio for one doublet (Λ, M) has been evaluated.

After these results, it should be possible to predict the absolute/convective boundary in the S-M plane with numerical nonlinear computations, and perhaps, to put some light on the large discrepancy found in the low Mach number region of the S-M plane between experimental and theoretical results on round jets, see Sreenivasan *et al.* (1989) figure 4.

References

BERGÉ, P., POMEAU, Y. & VIDAL, CH. (1984) *L'ordre dans le chaos - Vers une approche déterministe de la turbulence.* Hermann, 1984.

CHOMAZ J.M. & HUERRE, P. & REDEKOPP, L.G. (1990) Bifurcations to local and global modes in spatially developing flows, *Phys. Rev. Lett. 60, 25-28.*

HUERRE, P. & MONKEWITZ, P.A. (1990) Local and global instabilities in spatially developing flows, *Ann. Rev. of Fluid Mech., vol 22, 473-537.*

JAMESON, A., SCHMIDT, W. & TURKEL, E. (1981) , *A.I.A.A. paper 81-1259.*

MONKEWITZ, P.A. & SOHN, K.D. (1988) Absolute instability in hot jets, *A.I.A.A. J. 26, 911-916.*

PROVANSAL, M., MATHIS, C. & BOYER, L. (1987) Bénard-von Karman instability : Transient and forced regimes, *J. Fluid Mech., vol. 182, 1-22, 1987.*

RAYNAL, L. & HARION, J.L. & FAVRE-MARINET, M. & BINDER, G. (1996) The oscillatory instability of plane variable-density jets, *Phys. Fluids 8(4), April 1996, 993-1006.*

SERS, F. (1995) Contribution à l'étude des instabilités dans les jets plans par simulation numérique, *Thèse de doctorat de l'Institut National Polytechnique de Toulouse, Décembre 1995.*

SREENIVASAN, K.R., RAGHU, S. & KYLE, D. (1989) Absolute instability in a variable density round jets, *Exps. Fluids 7, 309-317.*

YU, M.H. & MONKEWITZ, P.A. (1990) The effect of nonuniform density on the absolute instability of two-dimensional inertial jets and wakes, *Phys. Fluids A 2(7), July 1990, 1175-1181.*

YU, M.H. & MONKEWITZ, P.A. (1993) Oscillations in the near field of a heated two-dimensional jet, *J. Fluid Mech. vol. 255, 323-347.*

MIXING BEHAVIOR OF ABSOLUTELY UNSTABLE AXISYMMETRIC SHEAR LAYERS FORMING SIDE JETS

WILLIAM M. PITTS and ARTHUR W. JOHNSON*
National Institute of Standards and Technology
Gaithersburg, MD 20899

1. Introduction

Recently it has been shown that low-density axisymmetric jets can become absolutely unstable and develop highly coherent vortical structures in the near-field shear layer [1],[2],[3]. These jets have several unusual properties when compared to jets which are convectively unstable. The vortical structures grow quickly with downstream distance and rapidly develop three-dimensional structure. Intense pairing of alternate structures is observed. Perhaps the most interesting of all behaviors is the observation of strong ejections of jet fluid, termed "side jets", into the ambient surroundings [1],[4],[5]. Their formation is attributed to the generation of strongly coupled pairs of longitudinal vortices in the developing shear layer [5],[6].

To our knowledge, there has been only one detailed investigation of near-field real-time mixing in jets subject to an absolute instability and forming side jets. Richards et al. [7] have reported aspirated hot-film concentration measurements for downstream distances of two to five diameters for several helium jets. Here we summarize the findings of real-time concentration measurements in the developing shear layer and side jets. The results discussed are part of a broader effort to be described in detail elsewhere [8].

2. Experimental

Axisymmetric jets displaying absolute instability behavior were formed by flowing helium through a contoured 6.35 mm diameter nozzle [9]. Honeycomb, screens, and beads placed upstream of the nozzle exit smoothed the flow. The velocity profile had a uniform "top-hat" contour with fluctuations of less than 0.15%. A mass-flow controller was used to establish the helium flow. The nozzle was mounted on a computer-controlled positioning system which allowed it to be moved relative to the fixed Rayleigh light scattering system described below.

---

*Current address: GE Aircraft Engines, 1 Neuman Way, Cincinnati, OH 45215-6301

L. Fulachier et al. (eds.), IUTAM Symposium on Variable Density Low-Speed Turbulent Flows, 17–24.
© 1997 *Kluwer Academic Publishers.*

It has been shown previously [3] that the strength of the absolute instability in round jets depends on both the thickness of the nozzle boundary layer and the jet-to-ambient density ratio. In the current facility, absolute instability behavior was observed for jet velocities (U_o) of 15 m/s to 92 m/s (Reynolds numbers, Re = 800 to 4,900) with the strongest interaction observed for a velocity of 25 m/s and Re = 1,300 [8].

A spark schlieren system was used to visualize the near-field regions of the helium jets. The spark source was a xenon lamp generating an approximately 20 ns light pulse which effectively "froze" the flow. A standard schlieren system consisting of 7 cm diameter collimating and focusing lenses and a knife edge was used to create time-resolved images which were recorded by a cooled CCD camera.

A hot wire was placed near the jet boundary layer, and the signal from the anemometer electronics provided a distinct record of the passage of vortical structures shed by the nozzle. For the absolutely unstable flows, the resulting signal was highly coherent and could be used to trigger the flash lamp at particular phases of the flow development. In this way it was possible to track the growth of the vortical structures as a function of time.

Real-time point measurements of helium concentration in the jets were recorded using Rayleigh light scattering (RLS) [10] induced by a focused 20 W beam from an argon ion laser. Scattered laser light was collected at 90° by an f/2 optical system and focused 1:1 onto a 400 μm pinhole and then onto a photomultiplier tube (PMT). The output of the PMT was frequency filtered, digitized, and stored in a minicomputer. The optical components defined the sample volume to be a cylinder of 75 μm diameter and 400 μm length. Data were recorded at a frequency of 20 kHz. Calibration of the scattering signals for air and helium allowed an arbitrary helium concentration (mole fraction) to be determined using

$$X_{He}(t) = \frac{I(t) - I_{air}}{I_{He} - I_{air}} \quad , \tag{1}$$

where X_{He} is the helium mole fraction, $I(t)$ is the time varying RLS intensity from the observation volume, and I_{He} and I_{air} are the observed scattering intensities from air and helium, respectively.

The nozzle was positioned in the NIST Rayleigh Light Scattering Facility (RLSF) [11] which has a cylindrical working section with diameter and height of 2.4 m. Particles, which scatter light strongly and interfere with RLS measurements, are removed from the working section by air flows passing through high-efficiency particle filters, and the walls are painted black to limit glare. During an experiment the air flow was shut off, and the helium jet entered a quiescent air environment.

In the following discussions downstream distances (z) and radial positions (r) relative to the flow centerline are nondimensionalized by the nozzle radius (r_o = 3.175 mm).

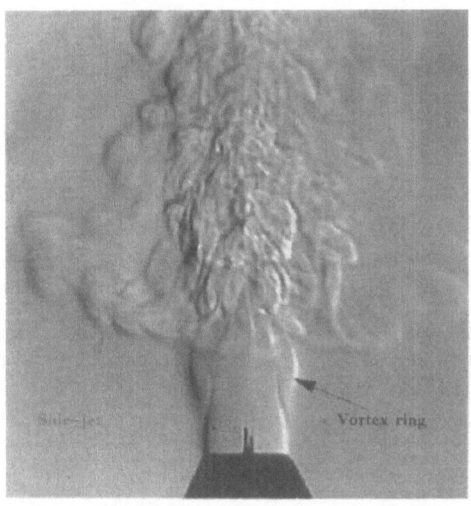

Figure 1. Schlieren photograph of an absolutely unstable helium jet recorded with a 20 ns light pulse showing shear layer structure and side jets. U_o = 24.6 m/s, Re = 1318, and f_o = 1275 Hz.

3. Results

Figure 1 shows a schlieren visualization of the near field of a helium jet with an initial jet flow velocity, U_o, of 24.6 m/s. A symmetrical vortical structure has formed near the nozzle. Phase-resolved images reveal that these structures develop very near the nozzle exit and grow rapidly with downstream distance. Alternate vortical structures are observed to pair further downstream .

Two side jets can be seen in Fig. 1 which have been ejected outward from the jet column to the left and right. There is a clear structure apparent in the side jets which suggests that they are formed by repeated pulses of helium from the jet core.

Time records of helium concentration were recorded using RLS at various locations in the near fields of the jets. Figure 2 shows an example of a short time record for z/r_o = 2 and r/r_o = 1.1, which lies within the shear layer close to the nozzle. U_o is 24.6 m/s. Helium concentration is plotted as a function of time nondimensionalized by multiplying by the primary shear-layer oscillation frequency, f_o = 1275 Hz, determined for this flow.

A highly repeatable oscillation of the helium concentration over a mole fraction range of 0.57 to 0.90 is observed. The full data record indicates that these fluctuations are extremely stable. The experimental data can be represented very well by a cosine function as seen in Fig. 2 by comparing a calculated curve having the appropriate amplitude, phase, and frequency with the experimental data.

The concentration time record shown in Fig. 3 was recorded at the same spatial location as Fig. 2, but the jet velocity was increased to U_o = 37.5 m/s. The shear-layer

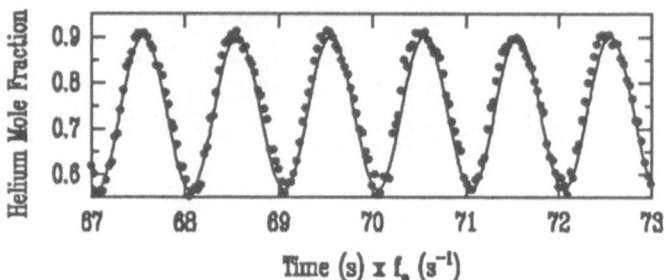

Figure 2. Helium mole fraction as a function of time nondimensionalized by the primary frequency, f_o, of the jet for $z/r_o = 2.0$ and $r/r_o = 1.1$. The solid line is a cosine function with phase and frequency adjusted to fit the data. $U_o = 24.6$ m/s, $Re = 1318$, and $f_o = 1275$ Hz.

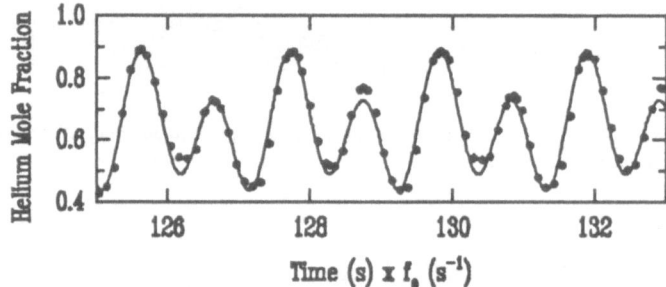

Figure 3. Helium mole fraction as a function of time nondimensionalized by the primary frequency, f_o, of the jet for $z/r_o = 2.0$ and $r/r_o = 1.1$. Solid line is Eq. (2) with $\phi = -18°$ and phase and frequency adjusted to fit the data. $U_o = 37.5$ m/s, $Re = 2010$, and $f_o = 2288$ Hz.

growth rate is known to be more rapid for this flow condition [8]. The observed signal for this case is also very stable, but the concentration time behavior is very different. The primary frequency of $f_o = 2288$ Hz is distinct, but there is clearly a variation at half this frequency as well. The data can be fit to a function having the following form:

$$X_{He}(t) = 0.64 + 0.17[\cos(f_o t) + 0.5\cos\left(\frac{f_o t}{2} + \phi\right)] \quad , \tag{2}$$

where ϕ is a phase-shift angle which is set to $-18°$ in order to reproduce the observed time dependence of the concentration fluctuations.

Figure 4 shows an example of concentration data recorded further downstream at $z/r_o = 3.0$ and $r/r_o = 1.4$ with $U_o = 37.5$ m/s. A short time record of a portion of the data is shown along with a much longer record in which the data have been averaged (smoothed) over fifty points to remove the high frequency contributions due to the primary frequency. The two time records show that while concentration fluctuations

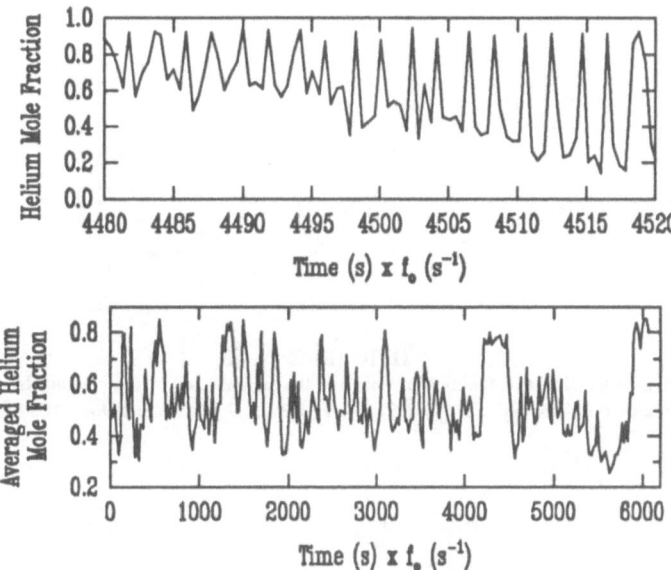

Figure 4. Helium mole fraction as function of time nondimensionalized by the primary frequency, f_o, of the jet for $z/r_o = 3.0$ and $r/r_o = 1.4$. A short real-time record and long-time smoothed (fifty-point average) are shown. $U_o = 37.5$ m/s, $Re = 2010$, and $f_o = 2288$ Hz.

are still present at the primary frequency, the overall mixing is no longer steady and has developed a great deal of both short-term and long-term randomness.

In regions well outside of the shear layer, any helium which is detected is due to the formation of side jets. The concentration time histories in these regions are very complex with a variety of different behaviors. Figure 5 shows an example plotted as a function of nondimensionalized time. A periodic variation is observed roughly every 40 primary vortical periods. There is also evidence for small concentration fluctuations at the primary frequency. Concentration maxima reach values greater than 60% helium.

Figure 6 shows time-averaged contours which have been calculated based on measurements over a range of near-field radial and downstream locations. The rapid spreading of the flow due to the occurrence of side jets is apparent.

4. Discussion

The concentration fluctuations in the boundary layer very close to the nozzle exit are consistent with the expected development of a shear layer subject to an absolute instability. Such layers develop initially highly repeatable azimuthal vortical structures at a single frequency which grow rapidly with downstream distance. The data in Fig. 2 demonstrate the high degree of coherence of these structures near the nozzle. The concentration signal, which closely obeys a cosine function, is the result of the radial

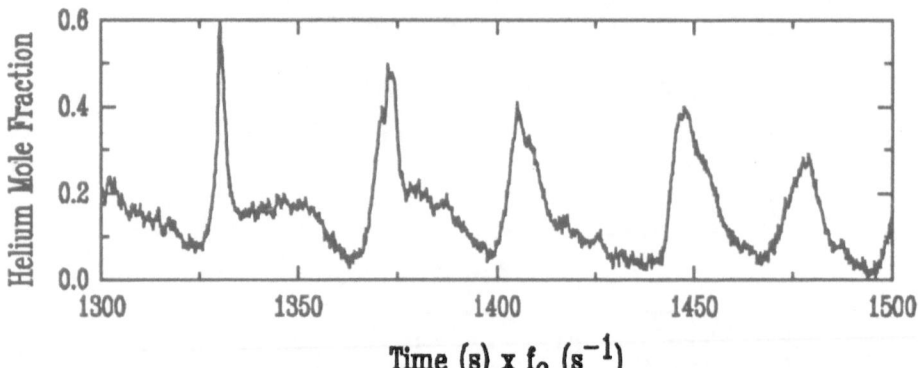

Figure 5. Helium mole fraction as a function of time nondimensionalized by the primary frequency, f_o, of the jet at a location outside the shear layer, $z/r_o = 3.5$ and $r/r_o = 3.0$. $U_o = 19.3$ m/s, $Re = 1034$, and $f_o = 958$ Hz.

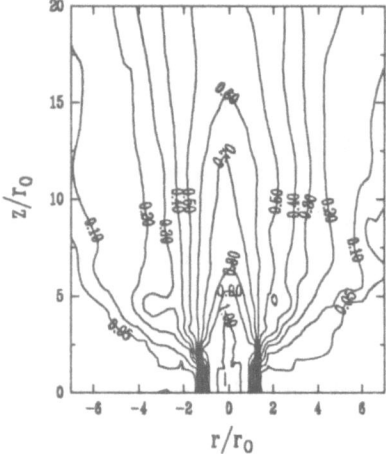

Figure 6. Time-averaged helium mole fraction contours measured as a function of radial and downstream position for a helium jet generating side jets. $U_o = 37.5$ m/s, $Re = 2010$, and $f_o = 2278$ Hz.

expansion and contraction of the concentration diffusion layer at the jet outer edge over the measurement point during the passage of the developing vortical structures. The absence of long-term variations shows that the structures are axisymmetric at this downstream distance. The large concentration fluctuations indicate that the structures have developed rapidly over the one jet diameter flow distance. Note that the smooth nature of the signals indicates that the growing structures have not yet "rolled up".

The data shown in Fig. 3 are for the same downstream distance as in Fig. 2, but the higher initial velocity of the flow results in faster development of the fluctuating shear layer. As a result, two adjacent structures have begun to interact, and a pairing process has been initiated. Observed concentration histories are consistent with a relative rotation of the vortices about a point located between them, with one structure moving toward the center, while the second rotates outward. This motion explains the alternating concentration pattern in the data. The diffusion layer adjacent to the structure which has moved closer to the center also moves in, and the concentration at the measurement point is reduced. The phase angle shift of $\phi = -18°$ necessary in Eq. 2 in order to fit the experimental data shows that the distance between the two interacting structures is decreasing as a result of the interaction. The smooth variations of the concentration indicate that pairing of the structures begins before significant roll up has occurred, while the absence of long-term variations demonstrates that the structures are still nearly axisymmetric about the centerline.

The concentration fluctuations in Fig. 4 show that rapid development of the shear layer continues as the flow moves further downstream. While the primary frequency is still present, the signals have become considerably more random. Concentration variations evident during the passage of individual structures show that vortex roll up has taken place with entrainment of large amounts of air. Large variations in concentration occur over periods long compared to the inverse of the primary frequency. These are attributed to development of azimuthal structure due to the Widnall instability [12], i.e, the structures are no longer axisymmetric. As these structures rotate about the jet center they result in the observed fluctuations.

These observations concerning shear-layer development are consistent with literature discussions of the growth of axisymmetic shear layers subject to an absolute instability [1]-[3],[6]. The availability of quantitative concentration measurements provides details concerning these processes which have been difficult to obtain previously.

Figure 5 is an example of the concentration fluctuations observed for positions near the jet exit, but well outside of the shear-layer flow. Note that only one of a wide variety of different behaviors which have been observed is shown. However, the data does serve to demonstrate several conclusions concerning side jet behavior.

Previous investigators [1],[4],[5] have shown that several side jets are formed simultaneously, which are highly localized in space and have a nonaxisymmetric structure which rotates about the centerline. The structure evident in Fig. 5 is consistent with such a rotation. The high concentration peaks are presumably due to direct side-jet impingement on the observation point, with lower concentrations observed between the side jets.

An unanswered question has been whether side jets are ejected continuously during multiple passages of primary vortices or intermittently during distinct phases of vortex passages. The continuous nature of the high concentrations observed during periods long compared to the inverse of the primary frequency strongly suggests that the ejections are continuous. Additionally, the high concentrations suggest the ejections occur from deep within the jet instead of in the shear layer, where lower helium concentrations are to be expected.

Close inspection of the time profile indicates that there are small variations in side-jet concentration at the frequency of the boundary-layer fluctuations observed in the primary jet. These fluctuations are thought to be associated with a pulsing of the side jet velocity which is responsible for the side-jet structure evident in Fig. 1.

Time-averaged concentration measurements such as shown in Fig. 6 demonstrate that the presence of an absolute instability and side jet formation leads to greatly enhanced spreading and mixing rates in the near fields of these jets.

5. Final Comments

The concentration measurements summarized here have provided new insights into the enhanced mixing as the result of the presence of an absolute instability in the near field of an axisymmetric jet shear layer as well as details concerning the effect of the absolute instability on shear-layer growth. Simultaneous velocity and concentration measurements in these flows have been used to characterize the dependence of side-jet strength on the intensity of the shear-layer oscillations due to the presence of the absolute instability. These measurements are discussed elsewhere [8].

6. References

1.   Sreenivasan, K.R., Raghu, S., and Kyle, D.: Absolute instability in variable density round jets, *Expts. Fluids* **7** (1989), 309-317.

2.   Monkewitz, P.A., Bechert, D.W., Barsikow, B., and Lehmann, B.: Self-excited oscillations and mixing in a heated round jet, *J. Fluid Mech.* **213** (1990), 611-639.

3.   Kyle, D.M., and Sreenivasan, K.R.: The instability and breakdown of a round variable-density jet, *J. Fluid Mech.* **249** (1993), 619-664.

4.   Barsikow, B., and Lehmann, B.: Orderly structures in a heated circular round jet, *Phys. Fluids* A1 (1989), 1445.

5.   Monkewitz, P.A., and Pfizenmaier, E.: Mixing by "side jets" in strongly forced and self-excited round jets, *Phys. Fluids* A3 (1991), 1356-1361.

6.   Brancher, P., Chomaz, J.M., and Huerre, P.: Direct numerical simulations of round jets: Vortex induction and side jets, *Phys. Fluids* **6** (1994), 1768-1774.

7.   Richards, C.D., Breuel, B.D., Clark, R.P., and Troutt, T.R.: Concentration measurements in a self-excited jet, to appear in *Expts. Fluids*.

8.   Johnson, A.W., and Pitts W.M., Manuscript to be submitted for publication.

9.   Richards C.D., and Pitts, W.M.: Global density effects on the self-preservation behavior of turbulent free jets, *J. Fluid Mech.* **254** (1993), 417-435.

10. Pitts, W.M., and Kashiwagi, T.: The application of laser-induced Rayleigh light scattering to the study of turbulent mixing, *J. Fluid Mech.* **141** (1984), 391-429.

11. Bryner, N., Richards, C.D., and Pitts, W.M.: A Rayleigh light scattering facility for the investigation of free jets and plumes, *Rev. Sci. Instrum.* **63** (1992), 3629-3635.

12. Widnall, C.D., and Sullivan, J.P.: The instability of short waves on a vortex ring, *J. Fluid Mech.* **66** (1974), 35-47.

INFLUENCE OF DENSITY CONTRAST ON INSTABILITY AND MIXING IN COAXIAL JETS

P. REYNIER AND H. HA MINH
Institut de Mécanique des Fluides de Toulouse
Av. du Pr. Camille Soula, 31400 Toulouse, France

1. Introduction

This paper focuses on numerical simulation of instability and mixing in heterogeneous coaxial jets. This flow is highly dependent on momentum, indeed the experimental study of Camano & Favre-Marinet (1994) and the numerical simulation of Reynier & Ha Minh (1994) show that a high momentum ratio between the two jets leads to the presence of a recirculation zone on the axis near the nozzle.

In rocket engine injectors, density and velocity contrasts have an important effect on instability and involve the atomization of the liquid oxygen round jet by the gaseous hydrogen of the annular jet. This phenomenon is at the origin of some droplets production mechanism. The experimental investigation of Shavit & Chigier (1996) shows the superiority (in the perspective of mixing and combustion) of coaxial jets against simple jet. Coaxial jets involve smaller droplets, a shorter liquid bulk and a larger spray angle. According to Gicquel *et al.* (1995) two-phase coaxial jets can be split in three regions. The first up to x/D=6 is the primary atomization zone where the breakup takes place. Then from x/D=6 to x/D=18 a strongly ligamental region is found. After this last location a second atomization zone is located, characterized by evaporation phenomenon. The understanding of these mechanisms is necessary for modelling the complex process of atomization-vaporization-combustion which takes place in a rocket engine.

In the present paper the two-phase character of rocket engine injectors is not taken into consideration. The study has for objective a first approach to this flow and it is restricted to numerical simulation of one-phase flows. Two levels of closure are used for the turbulence modelling : $k - \epsilon$ model and second order modelling taking account of the anisotropy of

L. Fulachier et al. (eds.), IUTAM Symposium on Variable Density Low-Speed Turbulent Flows, 25–32.
© 1997 *Kluwer Academic Publishers.*

this flow. Results are presented for two flows, an air-air configuration and hydrogen-oxygen coaxial jets. Firstly, coaxial jets investigated by Ribeiro (1972) have been computed to evaluate the reliability of the numerical results. Next hydrogen-oxygen coaxial jets are studied and the atomization of the oxygen located in the central jet by the hydrogen of the annular flow is investigated. Since for this flow pattern no quantitative data are available from experiment, only qualitative comparisons are allowed with results of experimental investigation carried out in presence of density effects.

2. Methodology

2.1. MODELLING ASPECTS

The equations of the physical problem are the unsteady Navier-Stokes equations with Favre's (1965) mass-weighted averaging, the equations of the turbulence model, the mass fraction rate of oxygen (for heterogeneous flows) and a state equation for a perfect gas. The problems to be treated are coaxial jet flows so it is natural to work in axisymmetric coordinates (e.g. Reynier 1995).

In coaxial jets, unsteadiness issuing from the separated shear layers is usually present. This existence of coherent structures, confirmed by the experiments of Gladnick *et al.* (1990), challenges the validity of the classical statistical turbulence models. Since DNS is not available for practical industrial flow patterns, we use an alternative approach, the semi-deterministic modelling, where the instantaneous motion is split in two parts, a naturally unsteady ensemble averaging motion (or deterministic part) and an incoherent (or random fluctuating part). Like the classical modelling, the use of a two-component splitting enables the time-dependent mean equations to be solved again with Favre's (1965) decomposition in the same form as those obtained in the classical way. However, since the new meaning of the turbulent fluctuations is not the same, the constants appearing in the model are not longer the same, and their calibration is again needed, using either theoretical considerations or experimental evidence (Ha Minh & Kourta 1993). In this contribution two closure levels are used : a two scale model $(k - \epsilon)$ with a new calibration of the constant C_μ proposed by Ha Minh & Kourta (1993) and the Reynolds stress model proposed by Launder *et al.* (1975) to include the anisotropy evolution aspects.

2.2. NUMERICAL ASPECTS

The finite volume method proposed by MacCormack in 1981 is used. This scheme uses the prediction-correction step technique, the equations are resolved in conservative form and the scheme is accurate to second order in

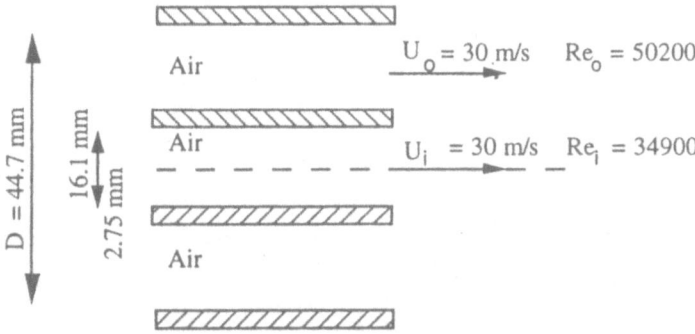

Figure 1. Configuration investigated by Ribeiro (1972)

time and space and does not require additional numerical dissipation for its stability. For the calculation of unsteady problems only the explicit part of the scheme is used.

At the nozzle, the inlet conditions are the experimental data of Ribeiro (1972) for air-air flow pattern. For heterogeneous jets laminar profiles are imposed. In this last configuration a wall is present outside the flow in the transverse direction (see figure 5). On this boundary Dirichlet conditions (zero) are used for velocity and turbulent quantities and zero gradients are applied for pressure and density. Ribeiro's (1972) configuration is a free flow, in consequence zero gradients are applied on this boundary. The lower surface (jet axis) is an axis of symmetry so zero gradients are assumed for all quantities. The upper surface is a free boundary so zero gradients are applied for all quantities. The outlet conditions are deduced from characteristic relationships (e.g. Thompson 1987).

3. Results

3.1. RIBEIRO CONFIGURATION

Firstly, the air-air configuration (figure 1) experimentally investigated by Ribeiro (1972) is numerically predicted. The injector diameter D is equal to 44.7 mm and the lip thickness between the two jets is 2.75 mm. The exit temperature is $300°K$, the pressure P_e at the exit and in the computational field is initially of 0.101 MPa and the initial density in all the field is $\rho_o = 1.28$ kg.m^{-3}. The computational grid used is 100×75 with grid stretching in the transverse direction outside the coaxial jets. The mesh is uniform in the x direction. It extends 16 diameters in the streamwise direction and 6 diameters in the transverse direction.

The calculations lead to the simulation of the unsteady flow, so we

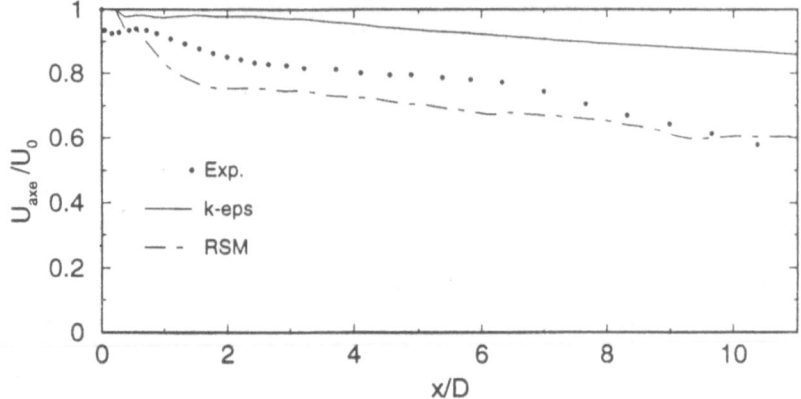

Figure 2.   Axis velocity decrease

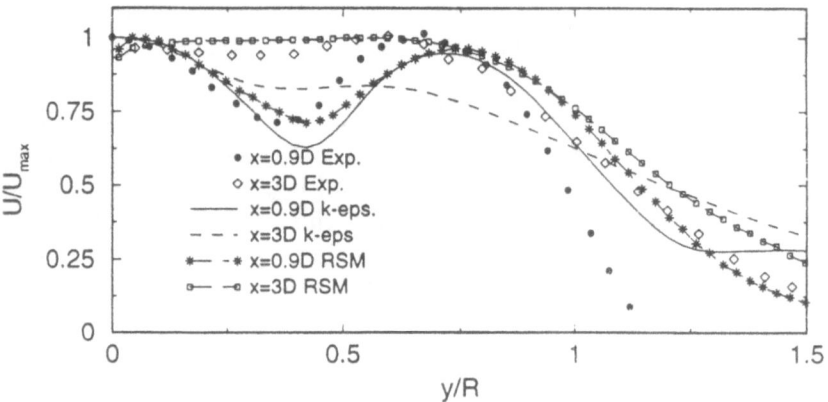

Figure 3.   Transverse velocity profiles at $x = 0.9D$ and $x = 3D$

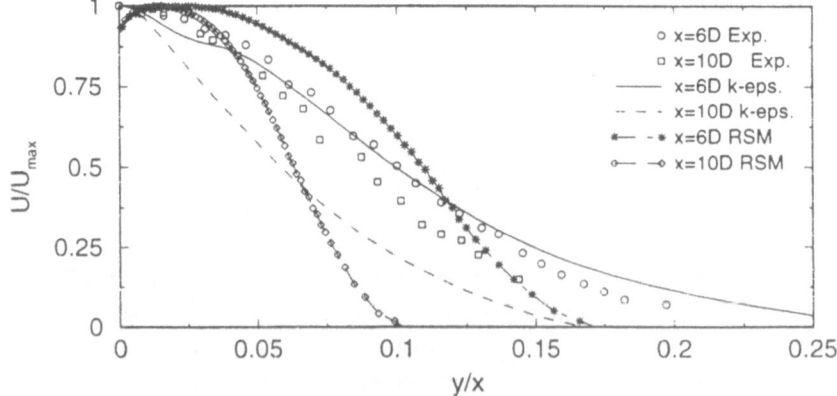

Figure 4.   Transverse velocity profiles at $x = 6D$ and $x = 10D$

take time-averaged quantitities for the comparisons. In figure 2 the axial velocity decrease and in figures 3 and 4 the transverse profiles of streamwise velocity for x=0.9D, x=3D, x=6D and x=10D are presented. The prediction of axis velocity decrease is better with the Reynolds stress model than with $k - \epsilon$ model. For the transverse velocity profiles the Reynolds stress model is not much better than the $k - \epsilon$ model. While the numerical results are in good agreement with experiment for x=0.9D and x=3D, this is not the case downstream at x=6D and x=10D. These discrepancies between numerical results and experimental data have two sources. The first is the boundary conditions used for this flow pattern. Indeed, the zero gradient conditions at the inflow don't work well : they involve an overestimate of velocities and turbulent quantities on this boundary (e.g. Reynier 1995). The characteristic conditions at the outflow are not suitable for low speed flows. The second source of the discrepancies is the defects of the models. Ha Minh & Kourta (1993) have recalibrated the constant C_μ on a backward-facing step and not on coaxial jets, its low value used here explains the unsatisfactory prediction of axis velocity decrease with this model. The results obtained from the model of Launder *et al.* (1975) are not in good agreement with the experiment of Ribeiro & Whitelaw (1975). According to these last authors the turbulent diffusion term is influenced by large scales and depends on the surrounding field where Reynolds stresses and their gradients can change, then this term is modelled with knowledge of the local values of Reynolds stresses and their spatial derivatives. These remarks explain the imperfect agreement between our results and those of Ribeiro (1972). However, the prediction of the near field is satisfactory and in the perspective of atomization in rocket engine injectors this zone of the flow has the greatest importance.

3.2. HETEROGENEOUS COAXIAL JETS

A hydrogen-oxygen coaxial jet injector (figure 5) is computed. The injector diameter D is equal to 8.76 mm and the lip thickness between the two jets is 0.5 mm. The Reynolds number of the round jet is 3220 and this of the annular jet is equal to 3460. The exit temperature is 300°K, the pressure P_e at the exit and in the computational field is initially equal to one atmosphere and the density in the annular jet and initially in all the field is the density of hydrogen, $\rho_{H2} = 0.08$ kg.m^{-3}. The density of the oxygen round jet is $\rho_{O2} = 1.3$ kg.m^{-3}. The computational grid is 100 × 97 with grid stretching in the transverse direction outside coaxial jets. The mesh is uniform in the x direction. It extends 17 diameters in the streamwise direction and 7 diameters in the transverse direction. The Reynolds stresses model of Launder *et al.* (1975) has been retained for the turbulence modelling.

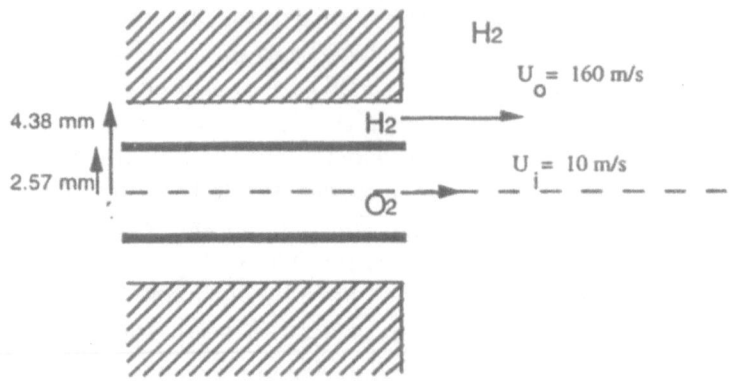

Figure 5. Hydrogen-Oxygen coaxial jets

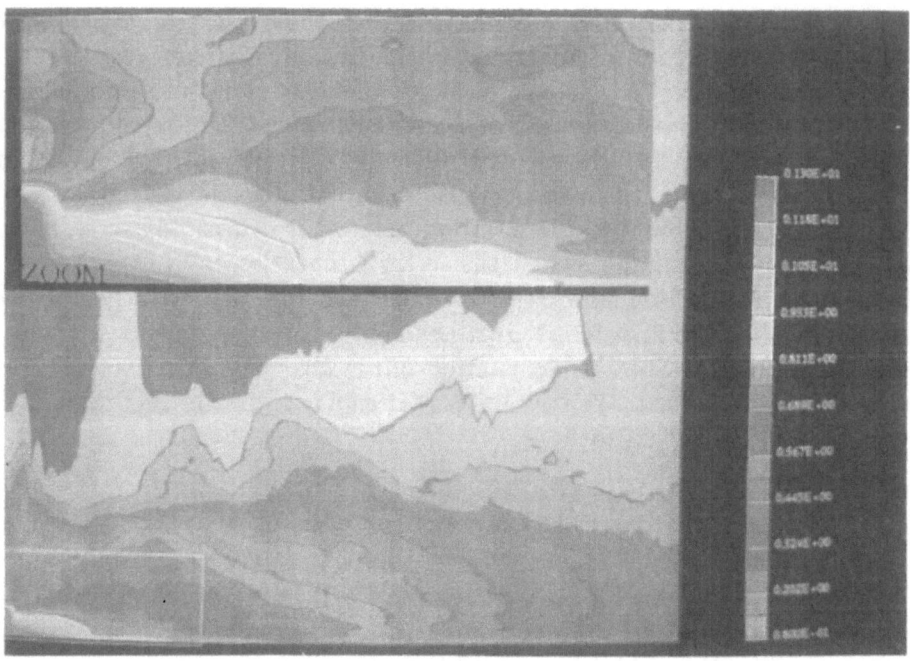

Figure 6. Time-averaged density field

In figure 6 the time-averaged density field is represented. The high density region where oxygen is dominant extends $7D$. The measurements of Gicquel *et al.* (1995) on two-phase coaxial jets show that the liquid core has a length of $6D$. In consequence the one-phase study on heterogeneous flow gives a good approximation of the liquid core length in cryotechnic injectors. In agreement with the experiment of Mansour & Chigier (1994)

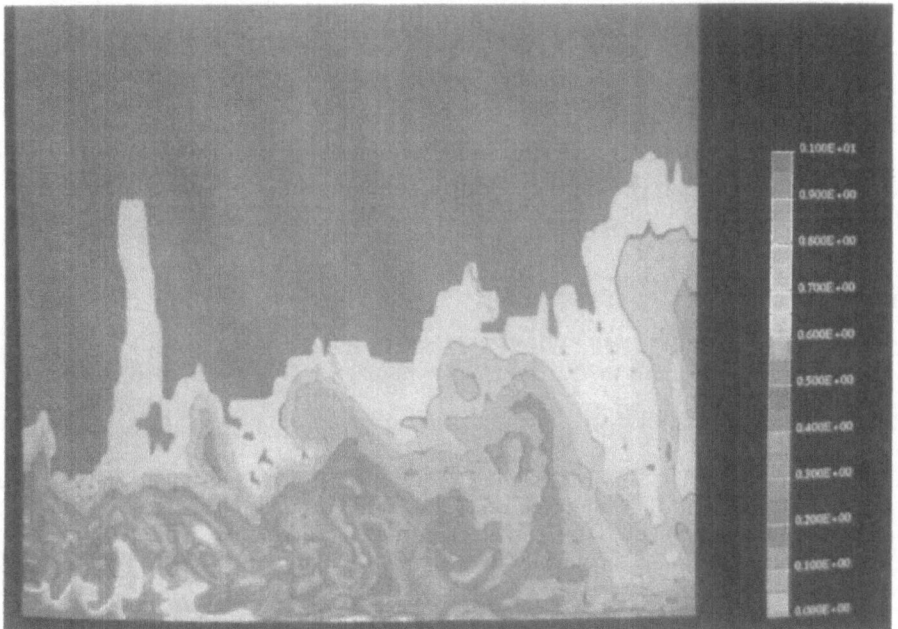

Figure 7. Unsteady oxygen concentration field

the random turbulence reaches its maximum near the nozzle then decays
very swiftly. The unsteady concentration field (figure 7) is strongly influenced by organized unsteadiness. Longmire & Eaton (1992) have observed
for a particle-laden round jet that the particle dispersion is dominated by
coherent structure convection rather than random turbulence. Downstream
$7D$, the concentration field shows a high degree of mixing which becomes
more and more intense, involving the pulling out of oxygen by hydrogen
then the tearing of oxygen packets in the shape of filaments. In this last
region the flow appears highly disorganized. The tearing phenomenon is violent. Indeed, Rupe (1962) already observed that laminar flows can be more
unstable and are atomized in a more violent fashion than fully developed
turbulent jets, due to higher spanwise velocity.

4. Conclusion

Two flows are numerically simulated : one homogeneous and one heterogeneous. The agreement between experience and numerical simulation is
not perfect, however the predictions are quite satisfying in the near field.
Natural unsteadiness has been simulated in heterogeneous coaxial jets. The
coherent structure growth involves the atomization of oxygen by hydrogen.

A strong interaction is obseved between organized unsteadiness and concentration field. Finally, the pulling out of oxygen by hydrogen is dominated by coherent structure convection rather than random turbulence.

Acknowledgement
We are grateful to the Société Européenne de Propulsion and to the CNES which support this work. Computer time on the Cray C94 was provided by IDRIS.

References
CAMANO E. B. and FAVRE-MARINET M., "On the initial region of inhomogeneous coaxial jets", Advances in Turbulence V, 58-62, R. Benzi Ed., Kluwer Academic Publishers, 1994.

FAVRE A., "Equations des gaz turbulents compressibles", J. de Mécanique Vol. 4 n° 3.3, pp. 361-390, Vol. 4 n°4 , pp. 391-421,1965.

GICQEL P., VINGERT L. and BRISSON E., "Caractérisation expérimentale d'un brouillard LOX-GH2 issu d'un injecteur coaxial en combustion", Note Onera, 1995.

GLADNICK P. G. , ENOTIADIS A. C., LARUE J. C. and SAMUELSEN G. S., "Near-field characteristics of a turbulent coflowing jet", AIAA Journal, Vol.28, n°8, pp1405-1414, Aug. 1990.

HA MINH H. and KOURTA A., "Semi-deterministic turbulence modelling for flows dominated by strong organized structures", 9th Symposium on Turbulent Shear Flows, Kyoto, 1993.

LAUNDER B. E., REECE G. J. and RODI W., "Progress in the development of a Reynolds stress turbulence closure", Journal of Fluid Mechanics, Vol. 68, Part 3, pp. 537-566, 1975.

LONGMIRE E. K. and EATON J. K., "Structure of a particle-laden round jet", Journal of Fluid Mechanics, Vol. 236, pp. 217-257, 1992.

MacCORMACK R. W., "A numerical method for solving the equations of compressible viscous flow", AIAA paper n° 81-0110, 1981.

MANSOUR A. and CHIGIER N., "Turbulence characteristics in cylindrical liquid jets", Phys. Fluids, Vol. 6, n°10, pp. 3380-3391, 1994.

REYNIER P. and HA MINH H., "Numerical simulation and physical analysis of coherent structures in compressible coaxial jets", Proceedings of the 1st International Conference on Flow Interaction, Sept. 5-9, Hong-Kong, 1994.

REYNIER P., "Analyse physique, modélisation et simulation numérique des jets simples et des jets coaxiaux, turbulents, compressibles et instationnaires", Ph. D. thesis, I.N.P.T., Toulouse, n° 1042, 1995.

RIBEIRO M. M., "Turbulent mixing of coaxial jets", M. Sc. thesis, University of London, 1972.

RIBEIRO M. M. and WhHITELAW J. H., "Coaxial jets with and without swirl", Journal of Fluid Mechanics, Vol. 96, Part 4, pp. 769-795, 1975.

RUPE J. H., "On the dynamics characteristics of free liquid jets and a partial correlation with orifice geometry", NASA Technical Report n°32, 1962.

SHAVIT U. and CHIGIER N., "Development and evaluation of a new turbulence generator for atomization research", Experiments in Fluids, Vol. 20, pp. 291-301, 1996.

THOMPSON K. W., "Time dependent boundary conditions for hyperbolic systems ", Journal of Computational Physics, Vol. 68 n°1, 1987.

A WAKE BEHIND A HEATED CYLINDER AT SMALL MACH NUMBERS: BIFURCATIONS OF THE VORTEX STREET AND POTENTIALITIES OF ACOUSTIC DIAGNOSTICS

A.B.EZERSKY, A.B.ZOBNIN, P.L.SOUSTOV AND V.V.CHERNOV
Institute of Applied Physics, Russian Academy of Science,
46 Uljanov Str., 603600 Nizhny Novgorod, Russia

Abstract. In this paper we investigate the influence of the temperature field on the vortex street behind a heated cylinder and present the results of remote acoustic sensing of the wake. It is demonstrated experimentally that the characteristics of the scattered sound can be used to determine the parameters of the vortex street, including the spatial period and repetition rate of vortices, the circulation of velocity, and the heat transferred by vortices.

1. Spatio-Temporal Structure of the Wake Behind a Heated Cylinder

A vortex street behind a cylinder is a traditional object investigated by classical hydrodynamics. Many novel experimental approaches and theoretical ideas were tested on this problem. The influence of heat fields on the vortex wake behind poorly streamlined bodies has attracted the attention of researchers recently. Can heat transfer at small Mach numbers modify the entire flow? What is the mechanism of this effect? These are fundamental questions in the problem of the wake behind a heated cylinder.

Much attention has recently been focused on the study of vortex shedding frequency. It was shown in Hamma, Lecordier & Paranthoen, 1991 and Ezersky, 1990 that heating of a streamlined body may lead to changes in vortex repetition rate or even to complete supresion of vortex formation. Such an effect of heating is due to the temperature dependence of density and dynamical viscosity of liquids. Combined action of these parameters may be reduced in some cases to variation of capillary viscosity.

L. Fulachier et al. (eds.), IUTAM Symposium on Variable Density Low-Speed Turbulent Flows, 33–41.
© 1997 *Kluwer Academic Publishers.*

It was found in particular that, as the temperature of a heated cylinder streamlined by an air flow is increased, the shedding frequency decreases. This effect may be explained as follows. As the temperature of the air grows, the kinematic viscosity of the latter is increased. Consequently, the effective Reynolds number is decreased. The shedding frequency being dependent on the Reynolds number, the frequency of unstable perturbations is decreased. As the temperature of the cylinder is increased in water, the kinematic viscosity decreases and the reverse effect is observed: the shedding frequency is increased by heating (Paranthoen & Lecordier, 1992).

It appears to be reasonable to explain the temperature dependence of shedding frequency in the context of effective Reynolds number (Re). Indeed, vortices may emerge either due to shear flow instability in a wake behind a cylinder or due to rolling up of the boundary layer separating from the cylinder (Chen, 1975). In the first case of temperature increase in the wake, viscosity and velocity profile change. Deformation of the velocity profile is analogous to that occurring when the Reynolds number is decreased. In the second case, vortices are formed immediately in the neighbourhood of the cylinder, where the temperature is much higher than in the wake and in the flow. In this case, we can regard the entire flow to be heated and this gives us grounds to introduce an effective number Re.

Besides the influence of cylinder heating on the repetition rate of vortices in the street that can be explained in terms of effective numbers Re, there exist the effects associated with rotation of a density stratified fluid. For a better understanding of how these effects may manifest themselves consider the analogy between the evolution of vorticity and temperature in the wake behind a heated cylinder. Strictly speaking, such an analogy may be drawn only for $Pr = \nu/\kappa = 1$, where ν is kinematic viscosity and κ is temperature conductivity. The Prandtl number for the air is $Pr = 0.71$ and it can be expected that the analogy holds. Measurements confirm this guess. The temperature pulsations, as well as the velocity pulsations, are increased in the wake, and a more heated (less dense) fluid is concentrated in vortex cores (Fig. 1). What are possible consequences of such a distribution in the core? The vortex street behind a heated cylinder was visualized in Ezersky, Gharib & Hammashi, 1994. The experiment was performed for Re = 61 and the temperature of the cylinder was varied up to 240°C. Oblique vortex shedding was obserbed without heating. For moderate heating (when the temperature of the cylinder was about $\Delta T_c = 90°C$ higher than that of the flow), flexural modes were formed at the vortices, and the vortex axes were, on the average, parallel to the cylinder. For higher temperatures $\Delta T_c \sim$ 240°C, the flexural waves disappeared and the vortex street became laminar (see Fig. 2).

The appearance of flexural modes at heated vortices in the street was explained on a simple model for a single vortex (Ezersky & Ermoshin, 1995). The instability of such a density stratified vortex was investigated on the following model: A higher-termperature gas was concentated in the vortex core that rotated as a solid and the periphery was a potential vortex. As was shown in Ezersky & Ermoshin, 1995, the instability in such a problem (i.e., generation of the flexural waves observed in experiments) arises as a result of the interaction of positive and negative energy waves. Both experiment and calculations verified that there is no instability and no flexural modes are observed at large density difference in the vortex core and at the periphery.

The characteristics of the vortices behind a heated cylinder were investigated by means of laser Doppler anemometer, hot-wire anemometer and cold-wire probe (see, e.g., Hamma, Lecordier & Paranthoen, 1991, Paranthoen & Lecordier, 1992). In this paper, we want to focus on the potentialities of acoustic diagnostics of the vortices transporting heat.

2. A Model of a Wake and Calculation of Scattering Coefficients

Is it possible to determine the parameters of the vortex street behind a heated cylinder by acoustic wave scattering? Acoustic diagnostics of density homogeneous flows was attempted in Gromov, Ezersky & Fabrikant, 1982. In the presence of heat transfer, the main problem is to determine the characteristics of velocity and temperature fields in the wake behind the cylinder by parameters of the scattered sound. In the general case, this problem, like many inverse problems, has no unique solution. However, the characteristics of the vortex street behind a heated cylinder may be determined from the scattered sound by parametrizing the street.

The simplest approximation of the wake behind a cylinder is a sequence of vortex filaments:

$$\Omega_z(\mathbf{r}) = 2\pi\Gamma \sum_{m=-\infty}^{\infty} [\delta(x-l/4-nl-u_0t)\delta(y-h/2)-\delta(x+l/4-nl-u_0t)\delta(y+h/2)]$$

(1)

Here $\delta(...)$ is the delta function, Γ is vortex circulation in the vortex sheet, l is the spatial period, h is the distance between the rows of vortices, u_0 stands for the velocity of the vortex street, x is the coordinate along the vortex wake, and y is the transverse coordinate. The temperature field of the vortex street can be written in the form

$$T(\mathbf{r}) = 2\pi\frac{Q'}{\rho_0 c_p}l_z \sum_{m=-\infty}^{\infty} [(x-l/4-nl-u_0t)\delta(y-h/2)+\delta(x+l/4-nl-u_0t)\delta(y+h/2)],$$

(2)

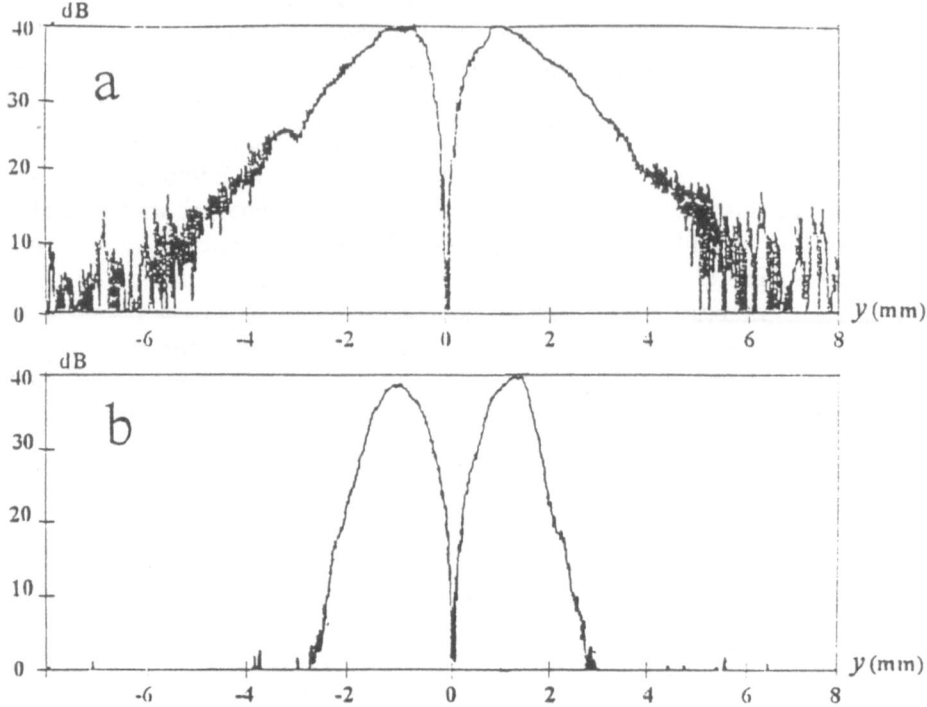

Figure 1.   The amplitudes of velocity pulsations at $\Delta T = 0$ (a) and of temperature pulsations at $\Delta T = 27°C$ (b) along the transversal coordinate y. The distance from the cylinder is 1 cm.

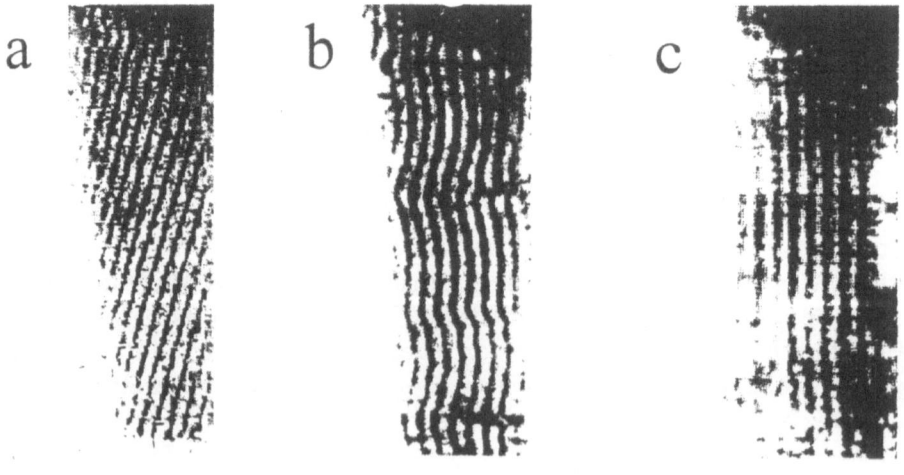

Figure 2.   Visualization of a vortex street behind the cylinder at Re = 61: (a) $\Delta T_c = 0°C$, (b) $\Delta T_c = 90°C$, (c) $\Delta T_c = 240°C$ [5].

where Q' is the amount of the heat transferred by a single vortex, ρ_0 is the density of the environment, $c_p = (\partial Q/\partial t)_p$ is the heat capacity of the medium at constant pressure, and l_z is the length of the cylinder. The scattering of a plane sound wave at the vortices (1)–(2) may be calculated in the Born approximation. Then, the scattered wave may be represented as a set of harmonics propagating at different angles that are determined by the vortex street period. The amplitude ρ_1 of the scattered wave $\tilde{\rho} = \rho_1 \exp(i(\omega_0 t - \mathbf{kr}))$ in the far field is written as $\rho_1/\rho_{wo} = \sum_m T_m e^{i((\omega_0 + \Delta\omega_m)t - |\vec{k}_x|x - k_y y + \pi)}$, where $\Delta\omega_m = (2\pi m)/l \cdot u_0 = 2\pi f_0 \cdot m$ (here f_0 is the frequency of vortex shedding and φ is the angle of scattering). If the sound is incident normal to the vortex street velocity, the scattered field is a sum of the harmonics propagating at angles symmetric to the direction of the incident sound. The frequency of each harmonic is shifted relative to the frequency of the incident sound due to the Doppler effect because the vortex street for the sound is a moving grating. The direction of propagation of each harmonic is determined from the condition $\omega_0/c|\sin\varphi_m| - (2\pi m)/l = 0$, i.e. the propagation angle of each harmonic is determined unambiguously by the spatial period of the vortex street. The amplitude of each harmonic may be expressed through the coefficients T_m which were calculated explicitly in Ezersky, Zoustov & Zobnin, 1995:

$$T_m = i\frac{\omega_0}{8c^2}\frac{L_x L_z}{8R}\frac{\Gamma \sin\varphi_m}{1 - \cos\varphi_m}\sin\left(\frac{\pi m}{2} + \frac{h\omega_0}{2c}(\cos\varphi_m - 1)\right) +$$

$$\frac{\omega_0}{T_0}\frac{Q'}{\rho_0 c_p}l_z \cos\left(\frac{\pi m}{2} + \frac{h\omega_0}{2c}(\cos\varphi_m - 2)\right)\cos\varphi_m,$$

where L_x and L_z are the horizontal and vertical widths of the sound beam the amplitude of which $S(x, z)$ is approximated by the formula

$$S(x, z) = \exp[-(x/L_x)^2 - (z/L_z)^2]$$

Analysis of the dependences of the coefficients T_m at the values of l and h close to those observed in experiments shows that $|T_1| \gg |T_{2,3}|$ without heating. In the case of temperature pulsations, the value of the coefficient T_2 may be comparable to T_1. This has a simple qualitative justification. The vortices are staggered in the street. The distance between the rows of vortices is much smaller than the period $h/l \simeq 0.28$. The neighbouring vortices have opposite circulations but equal temperatures in the cores. Therefore, temperature pulsation scattering occurs, actually, at the grating whose spatial period is half the period of the vortex street. Consequently, T_2 is determined, primarily, by the temperature inhomogeneity scattering.

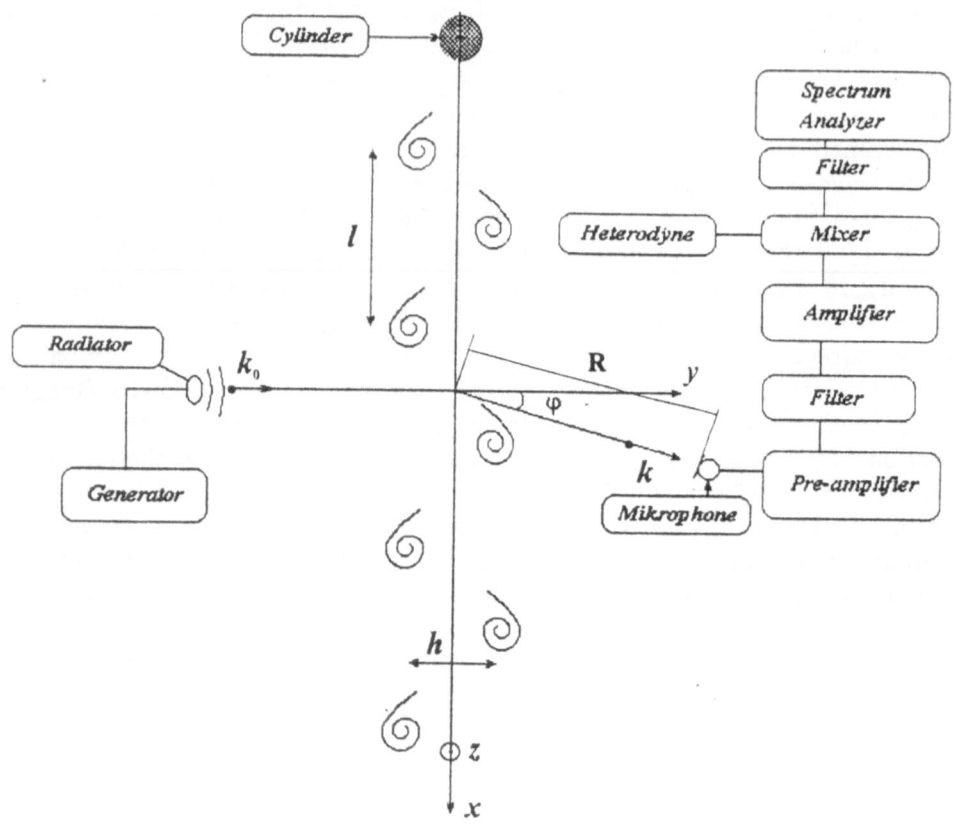

Figure 3.   Scheme of the experiment.

3. Experiment on Remote Acoustic Diagnostics

Experiments were made in a low-turbulence tunnel in the Institute of Applied Physics of the Russian Academy of Science. The scattering of ultrasound with frequency $F_0 = \omega_0/2\pi \approx 120$ kHz in a vortex street behind a heated cylinder 2 mm in diameter was investigated. The flow velocity amounted to $U = 90$ cm/s, the Reynolds number Re = 120, and the vortex shedding frequency $f_0 = 78$ Hz. The cylinder was heated by electric current. A piezoceramic emitter 4 cm in diameter was used as an ultrasonic source. The emitter was placed at a distance L_y from the vortex street. The parameters of the ultrasound were measured by a Bruel & Kijer mi-

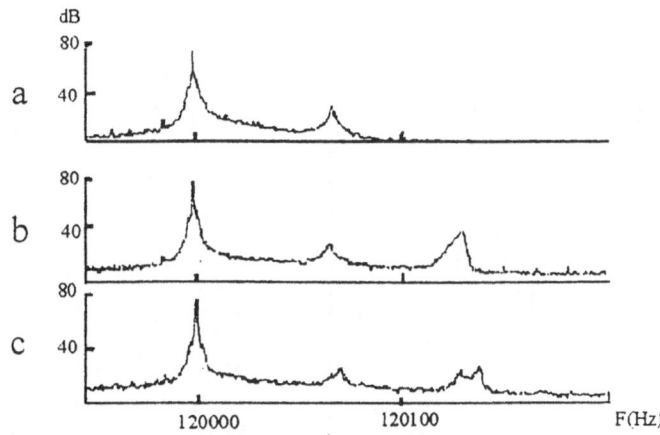

Figure 4. Spectra of the scattered sound: (a) unheated cylinder, (b,c) heated cylinder (Re = 120).

crophone placed at the distance $R \approx 120$ cm from the vortex street. The signal received by the microphone was transformed and analyzed on the Bruel & Kijer spectrum analyzer 2034. The scheme of the experiment is presented in Fig. 3.

The spectra of the ultrasound scattered at the vortex street behind the heated and unheated cylinders were compared. In the absence of heating, only the harmonics of frequency $F_0 \pm f_0$ (see Fig. 4a, where $F_0 = 120$ kHz and $f_0 = 78$ Hz) could be detected. The amplitudes of all the other harmonics were detected only as noise.

If the vortices transport heated gas, then the spectrum of the transmitted sound contains harmonics with frequencies $F_0 \pm 2f_0$ (Fig. 4b). The coefficients $T_{\pm 1}$ and $T_{\pm 2}$ were determined in experiments by the amplitudes of the harmonics with frequencies $F_0 \pm f_0$ and $F_0 \pm 2f_0$ for the cases of scattering in the wake behind heated and unheated cylinders (see Fig. 5). Measurements of the amplitudes of harmonics with frequencies $F_0 + f_0$ and $F_0 + 2f_0$ as a function of angle of scattering, verify that heating does not produce any significant effect on the spatial period of the vortex street. The spatial period calculated by φ_m is $l = 1.1$ cm, which is in a good agreement with empirical data. The frequency of vortex shedding determined from the scattered sound spectrum almost coincides with that obtained in direct measurements. The quantity Γ was calculated by the coefficients $T_{\pm 1}$ in the absence of heating and amounted to 76 cm^2/s. While the empirical data give $\Gamma = 80 - 90$ cm^2/s the quantity $T_{\pm 2}$ was used to estimate the amount of heat transferred by an individual vortex, this is possible because the vorticity field makes an infinitesimal contribution to the coefficients $T_{\pm 2}$.

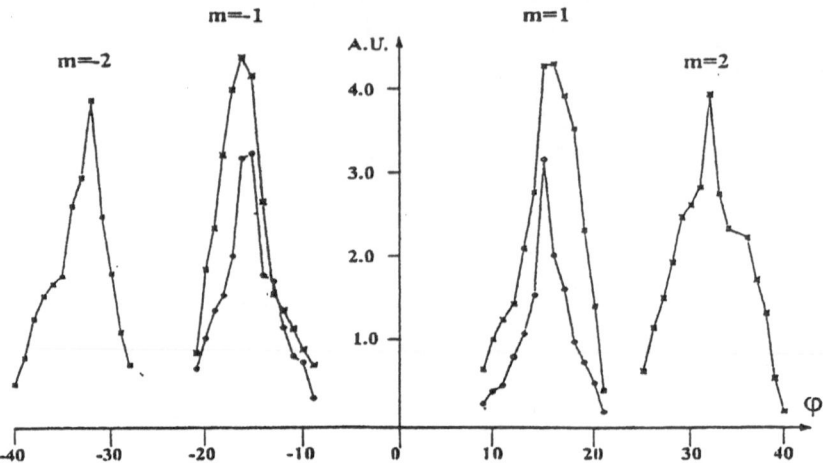

Figure 5.   Amplitude of the harmonics versus angle: ⋆ – heated cylinder, ● – unheated ctylinder.

Note that only integral characteristics, such as vortex circulation and amount of heat, may be estimated to a sufficiently high accuracy by means of acoustic diagnostics. For instance, if one attempts to estimate the temperature in the vortex cores, the data are likely to have a pronounced scatter. The point is that the vortex core radius changes with distance, whereas it is necessary to know its magnitude exactly for calculation of temperature in the core. On the other hand, the fact that acoustic diagnostics gives integral characteristics is one of its primary advantages. In order to obtain such characteristics by means of a temperature probe, for example, one needs to make measurements at very many points.

The spectral peaks of the scattered harmonics were observed to split in a definite range of heating temperatures and Reynolds numbers (see Fig. 4c). Calculations of the spectrum of the sound scattered at the street show that such a splitting may be caused by flexural oscillations of the vortices depicted in Fig. 2b. There is another explanation of spectrum splitting. As the temperature of the cylinder is increased,intermittency may occur (i.e., the vortex street consists of pieces having different frequencies). In this case the averaged spectra of the scattered sound have the form shown in Fig. 4c. This problem, however, needs further investigation.

To conclude, the experiments on acoustic scattering in a wake behind a heated cylinder demonstrate the possibility of remote determinition of parameters of the vortices whose cores have a temperature different from the temperature of the flow.

The authors are grateful to Prof. A.P.Kozlov (the Kazan Technical University) and Prof. P.Paranthoen (the Rouen University) for discussions.

This research was supported by the Russian Foundation for Basic Research (grant N 960216834).

References

Chen, C.N.: 60 Jahre Forschung Über die Karmanschen Wirbelstrassen – Ein Rückblick, *Schweizerische Bauzeitung* 44 (1975) 1079–1096.

Ezersky, A.B.: Flow around a heated cylinder at small Mach numbers, *Zh. Prikl. Matem. i Tekhn. Fiz.*, 5 (1990) 56–62.

Ezersky, A.B. and Ermoshin, D.A.: The instability of density stratified vortices, *Europ. J. Mech. /B Fluids* 14, 5 (1995) 667–628.

Ezersky, A.B., Gharib, M., and Hammashi, M.: Space–time structure of the wake behind a heated cylinder, *Zh. Prikl. Matem. i Tekhn. Fiz.*, 1 (1994) 74–84.

Ezersky, A.B., Soustov, P.L., and Zobnin, A.B.: Remote sensing acoustic diagnostics of a wake behind a heated cylinder *Izv. VUZov, Radiofizika* 28, 8 (1995) 832-840.

Gromov, P.R., Ezersky, A.B., and Fabrikant, A.L.: Sound scattering by the vortex wake behind a cylinder, *Akust. Zh.* 28, 6 (1982) 763–769.

Hamma, L., Lecordier, J.C., and Paranthoen, P.: The control of vortex shedding behind heated cylinder at low Reynolds numbers, *Experiments in Fluids*, 10 (1991) 224–229.

Paranthoen, P. and Lecordier, J.C.: Bluff Body Wakes Dynamics and Instabilities, IU-TAM Symposium, Göttingen 1992, Springer – Verlag, 1992.

EXPERIMENTAL STUDY OF THE BENARD-KARMAN INSTABILITY DOWNWIND OF A HEATED CYLINDER

J.C. LECORDIER, F. DUMOUCHEL, F. WEISS, P. PARANTHOËN,
U.R.A. 230 C.N.R.S. Université de Rouen
76821 Mont Saint Aignan Cedex (France)

1 INTRODUCTION

The periodic pattern of the vortex street downstream of a circular cylinder at low Reynolds numbers (30<Re<80) has always received significant attention of researchers because of its practical importance as well as theoretical interest. Here the Reynolds number is defined by: Re= $U_\infty d/\nu_g$ where U_∞ is the free stream velocity, d is the cylinder diameter and ν_g is the kinematic viscosity calculated at the free stream temperature. This interest has recently increased when this configuration has been widely used to test new development in the theory of hydrodynamic instability, Mathis et al. [1], Huerre and Monkewitz [2], Oertel [3]. When the cylinder is heated, many aspects of the velocity and temperature fields have been studied in detail only at high Reynolds numbers (Re>1000), Freymuth and Uberoi [4], LaRue and Libby [5], Matsamura and Antonia [6], Xenopoulos et al. [7]. Over the low Reynolds numbers range mentioned above, corresponding investigations have not yet widely been achieved and experimental results are sparse, Eckert and Soehngen [8], Vilimpoc et al. [9].

One of the main reason of this situation could be linked to the experimental difficulties to carry out such experiments at low velocities in a wake of small dimensions. Furthermore, preliminary works carried out by the authors have shown that, over this Reynolds numbers range, the structure of the wake behind an heated cylinder was very sensitive to the heat input and that heat was never a passive contaminant. Due to heat input to the cylinder the characteristics of vortex shedding can be significantly altered and, in air, total suppression can be achieved by increasing the heat input sufficiently, Lecordier et al. [10]. Up to now this effect has been mainly related to the increase of the kinematic viscosity of the gas and to the corresponding decrease of the effective Reynolds number, Sreenivasan et al. [11], Lecordier et al. [10], Ezerski [12]. Other authors reported similar results but related this effect only to the density change in the near wake, Yu and Monkewitz [13], Schumm et al. [14]. However the opposite behaviour found in water experimentally by Paranthoën et al. [15] and numerically by Socolescu et al. [16] suggests that this control is due to both changes of the dynamic viscosity and density with temperature.

In parallel a large number of numerical calculations have reported the influence of buoyancy effect on the development of the vortex shedding phenomenon, Chang and Sa [17], Vilimpoc et al. [9]. The upper limit of the forced convection regime is characterized by a value of the Richardson number Ri of about 5.10^{-2}, Merkin [18] .

L. Fulachier et al. (eds.), IUTAM Symposium on Variable Density Low-Speed Turbulent Flows, 43–50.
© 1997 *Kluwer Academic Publishers.*

Here $Ri=g\beta(T_W-T_\infty)d^3/Re^2v_g^2$ where g is the acceleration due to the gravity, β is the temperature coefficient for volume expansion and (T_W-T_∞) is the temperature difference between the cylinder and the upstream flow. The above limit leads for the laminar vortex street range (Re~50) to : $(T_W-T_\infty)< 12.5\ v_g^2 / \beta d^3$. It results that conditions where buoyancy contamination can be neglected strongly depend on the nature of the fluid. For an example, with $d=4.10^{-3}$m the values of $(T_W-T_\infty)_{limit}$ are 12K and 1K in air and in water respectively. In order to have significant change of thermal properties of the fluid in the wake, without buoyancy effects, higher cylinder temperature can be chosen by selecting smaller diameter cylinders. When free convection effects can be ignored this thermal control may reveal one of the keys to the onset of vortex shedding. From a practical point of view, it is also essential to relate the influence of the cylinder heating to the existence of an effective temperature used to calculate an effective Reynolds number. This would then allow to know the effective Reynolds number for a heated bluff body and determine the flow regime downwind of this obstacle. Up to now the notion of effective temperature has especially been used in hot-wire anemometry. Depending on authors, it is customary to use in the Nusselt-Reynolds relationships either fluid properties at the mean film temperature, Collis and Williams [18] or the dynamic viscosity at the mean film temperature and the density at the free stream temperature, Mac Adams [19]. In the present communication the structure of the near wake behind a heated or unheated cylinder is investigated experimentally in the regime corresponding to the transition from a 2-D steady to a 2-D periodic wake. We present some results obtained in air concerning some aspects of the velocity field over the Reynolds numbers range (45<Re<75). Particular attention is paid to the influence of the heating on the structure of the wake in relation with the change of physical properties of the fluid with temperature. The notion of effective Reynolds number is discussed.

2 EXPERIMENTAL SET-UP

The experiments were carried out in air in the potential core of a laminar plane jet. As shown in figure 1 the jet exits normally to an end plate (17 cm x 35 cm) from a slit of width 1.5 cm and span 15 cm centrally located in this plate. The vortex shedding bluff body was a smooth stainless steel 1mm diameter tube (l/d=150) mounted horizontally in the middle of the jet close to the exit plane. This obstacle could be heated by Joule effect by means of direct current. The cylinder temperature was obtained by inserting a chromel-constantan thermocouple within the tube at the center position. In order to avoid vibrations, the circular cylinder was damped with pieces of foam located at its ends. With the selected diameter, d=1mm, Reynolds numbers from 40 up to 80 could be obtained by varying the upstream velocity between 0.60 m/s and 1.20 m/s. The critical Reynolds number was about 46.3. The 1 mm cylinder temperature excess could reach 150K at the maximum leading to a maximum value of Ri of about 0.012 corresponding to the forced convection regime. Velocity measurements were made using an LDA TSI system incorporating a 1.5 W Spectra-Physics laser system, an integrated optical transmission unit and a light collecting system for forward scattering mode. The optical measuring volume was $0.08x0.08x1mm^3$. As shown in figure 1 the origin of the coordinate system was taken at the center of the cylinder. The x axis was measured in the direction of the flow, the y axis was perpendicular to the flow. In our study all the lengths and the velocities are non-dimensionalized by the diameter of the cylinder d and the the upstream velocity U_∞ respectively. Re is the Reynolds number in isothermal conditions, i.e. calculated with the viscosity of the upstream flow. Re_c is the critical Reynolds number corresponding to the transition from a 2-D steady to a 2-D periodic wake. Re_{eff} is the effective Reynolds numbers calculated at the effective temperature T_{eff} defined in section 5.

3. EXPERIMENTAL RESULTS

3.1 Results obtained in the Isothermal Situation

In this section we present some results concerning the velocity field downstream of the unheated circular cylinder at various x/d positions (1<x/d<20)in the wake. The profiles of the longitudinal mean velocity , rms values of the transverse velocities and Reynolds stresses, measured for Re=Rec + 6 and Re=Rec +15, are presented in figures 3-8. Figures 3 and 4 show that the profiles of the longitudinal mean velocity are always symmetrical about the center line. They put in evidence the existence of the recirculation zone close to the cylinder and then the evolution of the velocity defect in the wake. The distribution of v' is quite similar in shape at all the sections (figures 5, 6) with v' is always maximum on the center line. Figures 7 and 8 show the profiles of the Reynolds stresses. The distribution of u'v' has the same sign as that of $\partial U/\partial y$ except in the mean recirculation zone (x/d<3, |y/d|<0.8). Longitudinal evolution of the centerline rms tranverse velocities measured for various Reynolds numbers is presented in figure 9. Maximun values and streamwise location of the maximum of rms transverse velocities are shown in figure 10 and are function of the Reynolds number as found by Goujon-Durand et al.[20]. Strouhal number, wave length and convection velocity measured in the wake are presented in figures 11 and 12 agree with the results of Williamson [21].

3.2 Results obtained in the Nonisothermal Situation

As shown in previous studies, Lecordier et al [10], Ezerski [13] in presence of a 2-D periodic wake, heating the cylinder in air is found to stabilize the flow. Suppression of the instability is characterized by a continuous decrease of the amplitude and of the frequency when the heating is increased. As for an example, the profiles of the rms values of the transverse velocities v', measured at Re=Rec +15 for various levels of heating (P/l=0 and 44W/m), show in figure 13 this stabilization. As already mentioned by the authors the change of physical properties of the fluid in the near wake, due to the increase of temperature, could change the velocity field and reduce the interaction between the two shear layers. An example of this modification can be observed in figure 14 on the profiles of the longitudinal velocity measured at the end of the recirculation zone at Re=Rec+15 with ((P/l=0 and 44W/m). When the cylinder is heated the distance between the shear layers Δ, defined in figure 2, is found to increase.

3.3 Effective Reynolds Number

In order to characterize the wake when the cylinder is heated it is possible to define simply an effective temperature and an effective Reynolds number from the following manner. When the heating is just sufficient to stabilize the flow, the effective Reynolds number $Re_{eff}=U_\infty d/\nu_g(T_{eff})$ is equal to the critical Reynolds number Re_c. By knowing the temperature dependance of ν_g it is then possible to determine T_{eff}. Experimental results show that $\Delta T_{eff}/\Delta T_w = 0.24 \pm 0.02$ where $\Delta T_{eff}=T_{eff}-T_\infty$ and $\Delta T_w=T_w-T_\infty$. This result confirms the qualitative observation of Sreenivasan et al. [12] and the previous quantitative estimates of Ezerski [13], Lecordier et al. [10] and Paranthoën et al. [11] where $\Delta T_{eff}/\Delta T_w$ was respectively 0.23, 0.3 and 0.27. It is worth to note that this effective temperature T_{eff} is much lower than the film temperature generally used to take into account the influence of temperature on fluid properties in usual Nu-Re relationships, Collis and Williams [13].

3.4 Use of an Effective Reynolds Number to Characterize the Flow Regime

In order to test this effective Reynolds number, experiments have been performed successively in isothermal conditions and with the heated cylinder. In these experiments the Reynolds numbers Re were different but the same effective Reynolds numbers were obtained by selecting the temperature of the cylinder in order to get the needed effective temperature. The profiles of the longitudinal mean velocity and the rms values of the transverse velocities v' measured for two values of the effective Reynolds number : $Re_{eff}=Re_c + 6$ and $Re_{eff}=Re_c +15$, are shown in figures 15-18. The cases $Re_{eff}=Re_c+6$ and $Re_{eff}=Re_c+15$ have been obtained for the following conditions (Re=Re_c+10, P/l=20W/m; Re=Re_c+15, P/l=44W/m) and (Re=Re_c+15, P/l=0W/m; Re=Re_c+20, P/L=22W/m) respectively. For $Re_{eff}=Re_c$+6 results gather well while for $Re_{eff}=Re_c$+15 some slight discrepencies appear. This tendancy exists also in figure 19 where the maximun value of rms transverse velocities has been plotted versus Re_{eff}.

An other method can be used to define an effective Reynolds number based on the interaction term defined by Paranthoën et al. [15] as the ratio between the rate of circulation in the shear layer $d\Gamma/dt$ and the shear layer spacing Δ. For the "sake" of simplicity this interaction term I was estimated at the end of the recirculation zone as : $I=U(\Delta/2)(dU/dy)_{max}\delta/\Delta$. A second effective Reynolds number Re_{inter} can be deduced from the value of I assuming that an universal dependance between Re_{inter} and I exists whatever the level of heating. Results presented in figure 20 show that this effective Reynolds number does not allow to gather in a best manner the results. This could be due to some inaccuracies in the determination of I in our narrow wake.

Nevertheless, from the knowledge of the characteristics Δ and U_{max} of the velocity profile at the end of the recirculation zone, an effective Strouhal number $St^*=f\Delta/U_{max}$ can be calculated which is almost constant $St^*=0.14\pm0.005$ for the heated or unheated cases.

4. CONCLUSION

Experimental study in air of the Benard-Karman vortex street downwind an heated or unheated cylinder has shown the following facts :
- The flow regime downstream of the heated circular cylinder is strongly dependent on the level of heating and, even in absence of buoyancy effects, heat is never in this situation a passive contaminant.
- When the cylinder is heated the flow regime in the wake can be characterized by an effective Reynolds number calculated with fluid properties at an effective temperature

$T_{eff} = T_{\infty} + 0.24\ (T_W - T_{\infty})$. Here T_{∞} is the temperature of the upstream flow and T_W is the cylinder temperature.

5. REFERENCES

1. Provansal M., Mathis C. and Boyer L.: Bénard-von Karman instability : transient and forced regimes, *J.Fluid.Mech.* **31** (1987), 1-22.
2. Huerre P. and Monkewitz P.:Local and global instabilities, *Ann.Rev.Fluid Mechanics* **22** (1990), 473-537.
3. Oertel H.: Wakes behind blunt bodies, *Ann.Rev.Fluid Mechanics* **22** (1990), 539-564.
4. Freymuth P.and Uberoi M.S.: Structure of temperature fluctuations in the turbulent wake behind a heated cylinder, *Phys. of Fluids* **14** (1971), 2574-2580.
5. LaRue J.Cand Libby P.A.: Temperature fluctuations in a plane wake, *Phys. of Fluids* **17** (1974), 1956-1967.
6. Matsumura M.and Antonia R.A.: Momentum and heat transport in the turbulent intermediate wake of a circular cylinder, *J. Fluid. Mech*, **250** (1993), 651-668.
7. Xenopoulos G., Stapounzis H., Salpistis C.and Goulas A.:: Flow in the wake of a heated circular cylinder in rotary oscillation, *International Symposium on Turbulence, Heat and Mass* , *Lisbonne* (1995), 1.3.11.3.6,
8. Eckert E.R.G. and Soehngen E.: Distribution of heat transfer coefficients around a circular cylinders in crossflow at Reynolds numbers from 20 to 500, *Trans. ASME* **74**, (1952), 343-347,
9. Vilimpoc V. Cole R.and.Sukanek P.C.: Heat transfer in newtonian liquids around a circular cylinder, *Int. J .Heat Mass Transfer* **33** (1990), 447-456, .
10. Lecordier J.C., Hamma L. and Paranthoën P.: The control of vortex shedding behind heated circular cylinders at low Reynolds numbers, *Exp. in Fluids* **10** (1991), 224-229.
11. Sreenivasan K., Tavoularis S., Henry R.and Corrsin S.:, Temperature fluctuations and scales in grid-generated turbulence, *J. Fluid. Mech*. **100**, (1980), 597-621.
12. Ezerski A. B.: Detached flow around a heated cylinder at small Mach number, *Prikladnaya Mekhanika i Tekhn.Fizika*, 5, (1990.), 56-62,
13. Yu M. H., Monkewitz P. A.: The effect of non-uniform density on the absolute instability of two-dimensional inertial jets and wakes, *Phys. of. Fluids A* **2** (1990), 1175-1181.
14. Schumm M., Berger E.,. Monkewitz P.: Self excited oscillations in the wake of two-dimensional bluff bodies and their control, *J.Fluid Mech* **271** (1994.), 17-53.
15. Paranthoën P., Browne L., Le Masson S.and Lecordier J. C.: Control of vortex shedding by thermal effect at low Reynolds number, *Int Rept. MT1*, Rouen University (1995)
16. Socolescu L., Mutabazi I., Daube O. and Huberson S.: : Etude de l'instabilité du sillage 2D derrière un cylindre faiblement chauffé, *C.R.Acad.Sciences Paris* t**322** Serie IIb, (1996), 203-208.
17. Merkin J.H.: Mixed convection from an horizontal cylinder, *Int. J. Heat Mass Transfer* **20** (1977), 73-77.
18. Collis D.C., Williams M. J.: Two dimensional convection from heated wires at low Reynolds numbers, *J. Fluid. Mech* **6** (1959.), 357-384.
19. Mc Adams W.H. : Heat Transmission, third edition, McGraw-Hill, (1954)
20. Goujon-Durand S., Jennfer P. and Weisfreid E. :
. Downstream evolution of the Benard-Karman instability, *Phys.Rev.E* . **7** (1994), 308-313.
21. Williamson C.H.K. : Oblique and parallel modes of vortex shedding in the wake of a circular cylinder at low Reynolds numbers, *J.Fluid Mech*. **206** (1989), 579-627

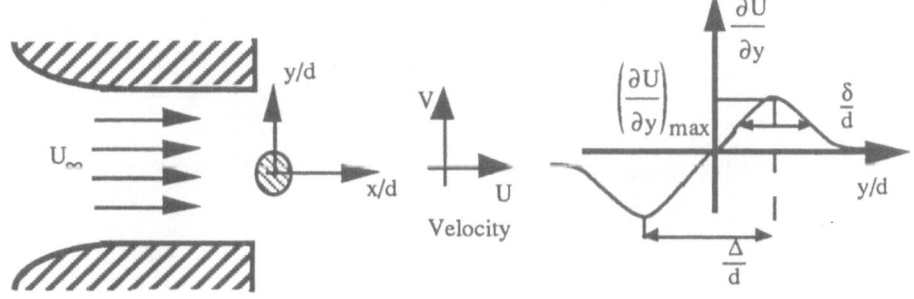

Figure 1. Experimental set-up   Figure 2. wake characteristics

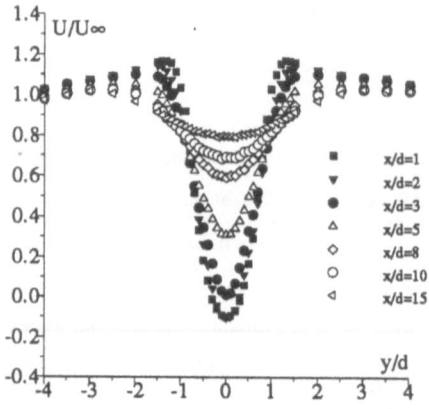

Figure 3. Mean velocity profiles Re=Rec+6

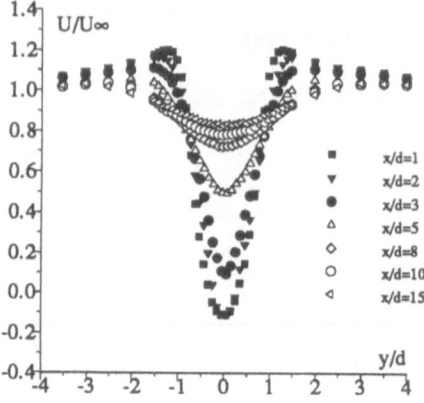

Figure 4. Mean velocity profiles Re=Rec+15

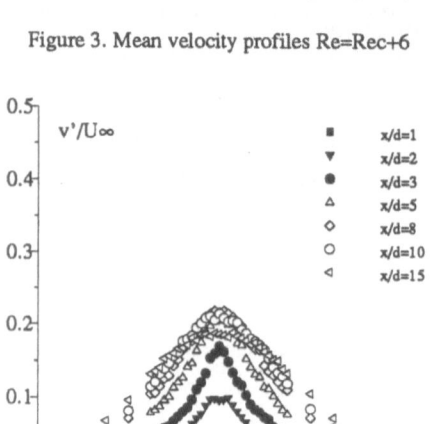

Figure 5. rms transverse velocity Re=Rec+6

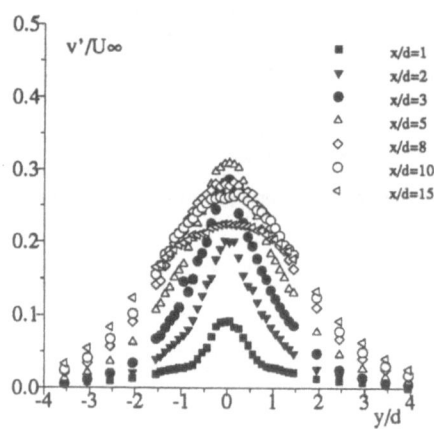

Figure 6. rms transverse velocity Re=Rec+15

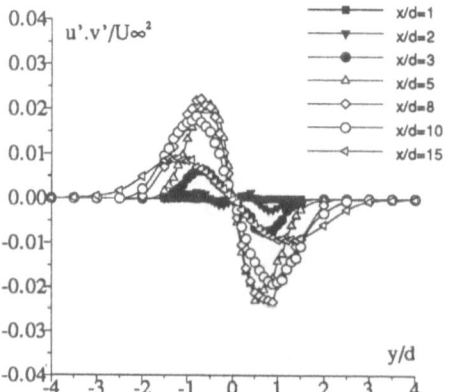

Figure 7. Reynolds stresses Re=Rec+6

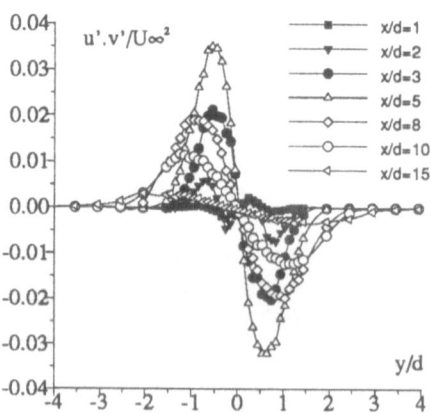

Figure 8. Reynolds stresses Re=Rec+15

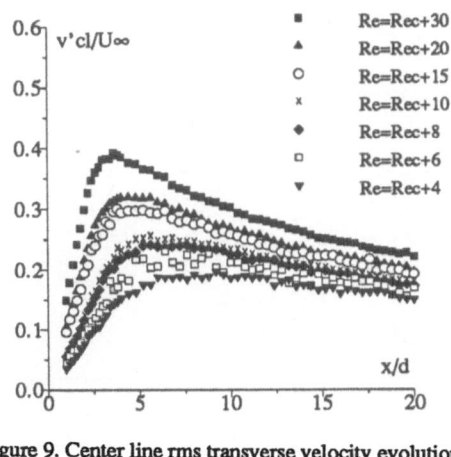

Figure 9. Center line rms transverse velocity evolution

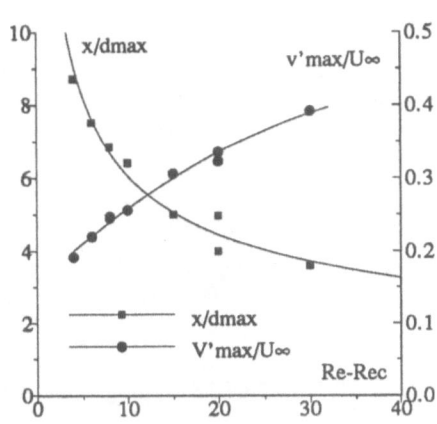

Figure 10. x/dmax v'max/u∞ evolution

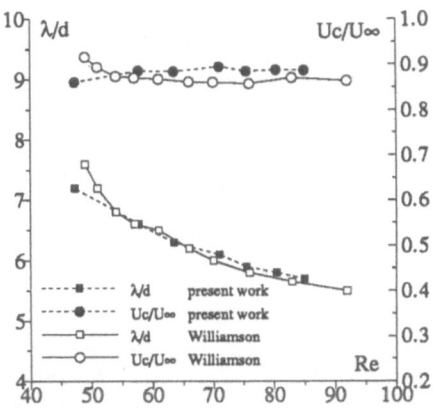

Figure 11. Wavelength and convection velocity

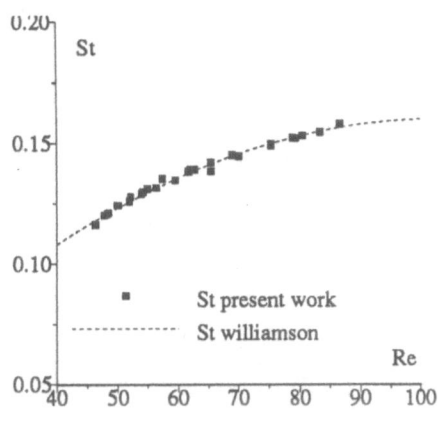

Figure 12. Strouhal number evolution

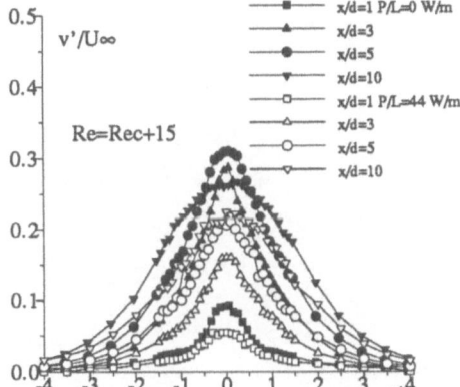

Figure 13. rms transverse velocity with heat input

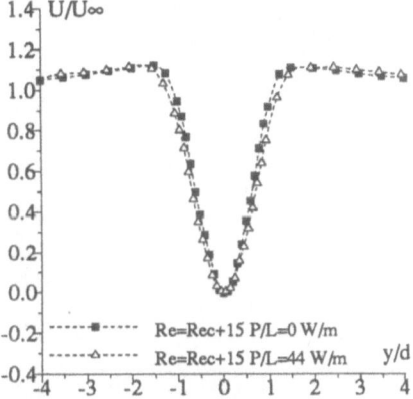

Figure 14. Mean velocity at the end of the wake bubble

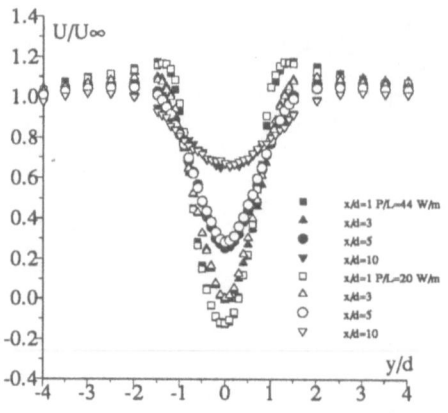

Figure 15. Mean velocity profiles Re=Reff+6

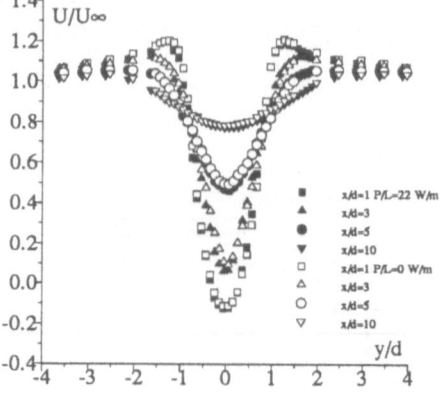

Figure 16. Mean velocity profiles Re=Reff+15

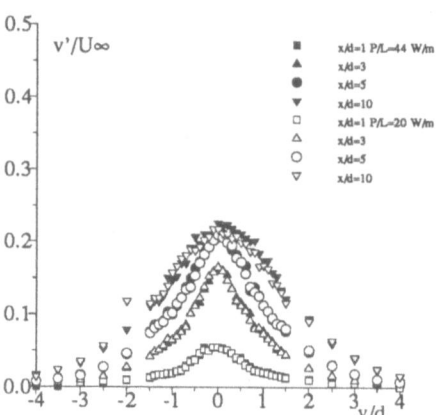

Figure 17. rms transverse velocity Reff=Rec+6

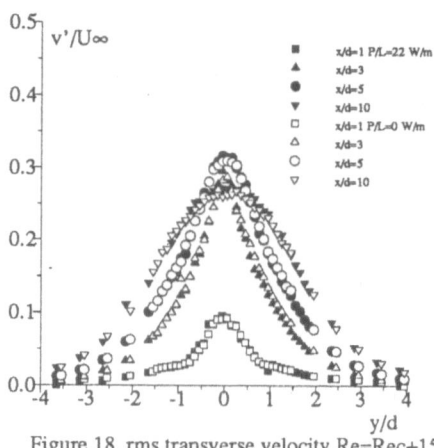

Figure 18. rms transverse velocity Re=Rec+15

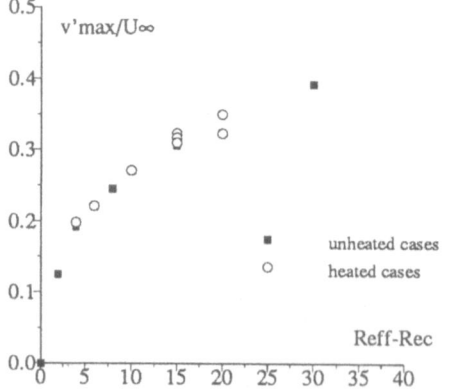

Figure 19. x/dmax v'max/U∞ evolution with Reff

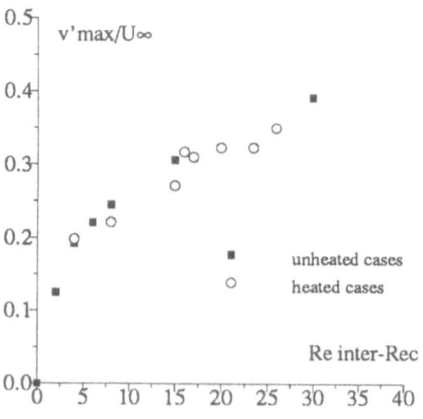

Figure 20. x/dmax v'max/U∞ evolution with Re inter

DIRECT SIMULATION OF MULTIPHASE FLOWS WITH DENSITY VARIATIONS

S. ZALESKI, J. LI, R. SCARDOVELLI AND G. ZANETTI
Laboratoire de Modélisation en Mécanique, CNRS URA 229,
Université Pierre et Marie Curie, 4 place Jussieu,
75252 Paris Cedex 05, France.(zaleski@lmm.jussieu.fr)

Abstract. Interfacial instability plays an important rôle in the primary atomization of high-speed liquid jets. We present simulations of the Kelvin-Helmholtz instability of a sheared liquid-gas interface. The two-dimensional simulations of the Navier-Stokes equation with surface tension show the formation of liquid sheets at sufficiently high Weber and Reynolds numbers. The three-dimensional simulations show two scenarios, depending on the degree of three dimensionality in the initial conditions. The first scenario involves the Rayleigh instability of cylinders that are formed when the rim detaches from the sheets. The second involves a faster distortion of the rim; the precise mechanism that yields this behavior is still unknown.

1. Introduction

Spray formation processes are a particularly interesting case of multiphase variable-density turbulent flow. In air-assisted nozzles, droplets are formed in the manner shown on Fig. 1. Several methods have been proposed for the study of such flows.

- The coarse-grained or averaged equations such as k-ϵ, Reynolds stress and PDF models are the only recourse in engineering studies. However they mostly apply to regions where droplets are sufficiently far apart.
- The particle methods, in which individual particles are followed. These methods allow to obtain quantitatively significant results in complex situations. However they cannot treat the situation near the liquid core, where droplets are formed.

L. Fulachier et al. (eds.), IUTAM Symposium on Variable Density Low-Speed Turbulent Flows, 51–58.
© 1997 *Kluwer Academic Publishers.*

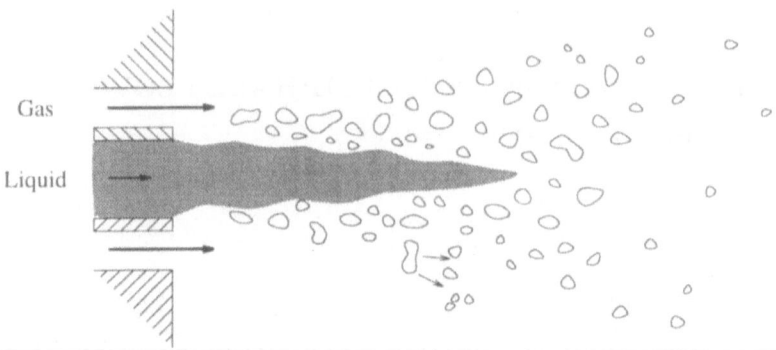

Figure 1.   Schematic representation of atomization in a coflowing jet atomizer.

— Direct Numerical Simulation, or DNS, which involves the unaveraged Navier-Stokes equations. This approach can only treat a small portion of the jet structure, but it yields a detailed view of the droplet-forming process. It thus complements the previous approaches.

In this communication, we present results obtained by DNS. We solve the Navier-Stokes equation with a surface tension term $\mathbf{F}_c$. They read

$$\rho(\partial_t \mathbf{u} + \mathbf{u} \cdot \nabla \mathbf{u}) = -\nabla p + \nabla \cdot (2\mu S) + \mathbf{F}_c, \tag{1}$$

where ρ is the local fluid density, μ the viscosity, p the pressure, and $S_{ij} = (\partial_i u_j + \partial_j u_i)/2$ the rate of strain tensor. The capillary force term $\mathbf{F}_c$ may be written

$$\mathbf{F}_c \;=\; \sigma \kappa \delta_S \mathbf{n} \tag{2}$$
$$\;=\; \nabla \cdot (\mathbf{I} - \mathbf{n} \otimes \mathbf{n}) \sigma \kappa \delta_S, \tag{3}$$

where σ is the surface tension, κ the mean curvature of the interface, δ_S a Dirac distribution concentrated on the interface, $\mathbf{I}$ the unit tensor and $\mathbf{n}$ the unit vector normal to the interface. Expressions (2) and (3) are both valid when σ is a constant. Expression (2) is the most familiar representation. When the δ_S term just balances a jump in pressure, Laplace's law is recovered. On the other hand expression (3) expresses the capillary effects in tensorial form and is used in our numerical discretization (Lafaurie et al., 1994).

Despite the sometimes high gas velocities (typically 100 m/s), we choose for reasons of simplicity to use the incompressible flow limit

$$\nabla \cdot \mathbf{u} = 0. \tag{4}$$

This condition may be relaxed on the interface if evaporation occurs. However evaporation concerns time scales much larger than the time scale for droplet formation near the nozzle, which mostly concerns us.

Our numerical method for the solution of these equations has been tailored for the purpose of studying qualitatively the droplet formation process. It is briefly detailed in the first section. We then present our results and a brief theoretical description of the scenarios observed.

2. Numerical Method

We use a second-order version of the Volume of Fluid (VOF) method for the tracking of interfaces. The second order version is sometimes called Piecewise Linear Interface Calculation (PLIC)(Li, 1995; Li, 1996). This allows substantial accuracy for the tracking of interfaces. It allows to deal with reconnection in an arbitrary, but simple way. It is also possible to achieve exact microscopic mass conservation with this method.

For the Navier-Stokes equation, we use explicit finite differences on a square staggered grid of "Marker and Cell" (MAC) type as described in (Lafaurie et al., 1994). A projection method ensures the incompressibility. The density variations are a source of some difficulty in this method, since they result in a Poisson-type equation of the form

$$\nabla \cdot \frac{1}{\rho} \nabla p = \tau \nabla \cdot \mathbf{u}^* \tag{5}$$

where $\mathbf{u}^*$ is a provisional field computed by an explicit finite diffence discretization where ρ may vary by several orders of magnitudes in the interface region. The linear problem (5) is solved by a multigrid method(Wesseling, 1992).

3. Results

We consider the temporal development of the Kelvin-Helmholtz instability of a two-phase parallel flow. The base flow consists of two layers of relative velocity U and densities $\rho_{L,G}$. The interface deviates by a small displacement $\xi(x, y)$ from the horizontal. Initial conditions are defined in the box $(-L/2, L/2)^2 \times (-H/2, H/2)$ by

$$\mathbf{u}_0 = U[\Theta(z)/2 - 1]\mathbf{e}_x, \tag{6}$$
$$\rho = \rho_L + (\rho_G - \rho_L)\Theta[z - \xi(x, y)], \tag{7}$$

where Θ is the step function. We use two methods to initiate the instability:

- In some cases we displace the interface, but do not perturb the velocity field, which is left in the form (6). The perturbation of the interface is

$$\xi(x, y) = A_1 \cos[2\pi x/L + A_2 \cos(2\pi y/L)]. \tag{8}$$

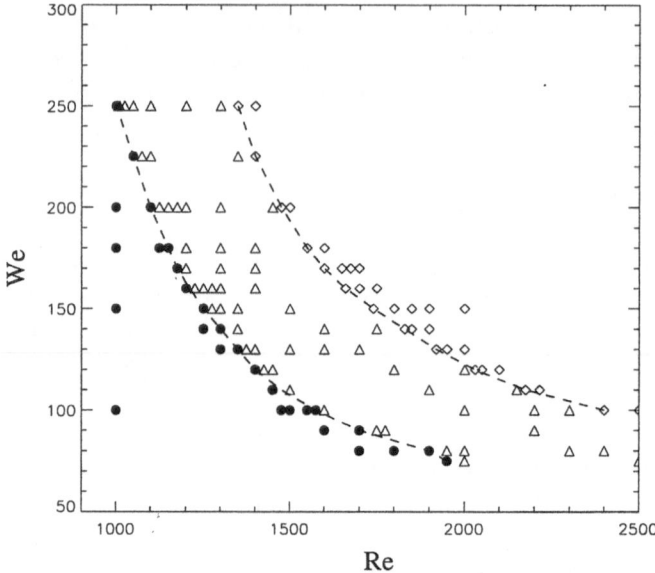

Figure 2.   The various liquid sheet and droplet formation regimes observed in two dimensions. Full circles indicate the stable regime. Triangles indicate the laminar-ejection regime and diamonds the turbulent-ejection regime. The dotted lines are drawn to guide the eye.

The parameter A_2 controls the amount of three-dimensional structure in the initial conditions. We call this initial condition an "interface only" perturbation.

- In other types of perturbation, we also perturb the velocity field. The velocity field is given by the solution of linear theory for the *two-dimensional* instability problem in each $y =$Cst plane. In other words, y is fixed, the linearized two-dimensional problem is solved analytically and the two-dimensional flow in that y-plane is added to $\mathbf{u}_0$. We call this initial condition a "full" perturbation.

The motivation for these various types of initial condition is that we do not know, *a priori*, which simple initial condition represents best the conditions at nozzle exit. We found great variability of the results depending on initial conditions and accordingly explored a large number of initial configurations. With these conditions, the problem has four basic dimensionless numbers, the Weber numbers $We_i = \rho_i U^2 L/\sigma$ and the Reynolds numbers $Re_i = \rho_i U L/\mu_i$ being defined for each phase i. Furthermore, the aspect ration H/L and the initial amplitudes A_1, A_2 may play a rôle.

It is interesting to make a preliminary study of the corresponding 2D problem (some results have already been reported by Keller et al. (1994)). For an inviscid, linear theory (Chandrasekhar, 1961) predicts an instability for We $> 2\pi$. However numerous simulations nonlinear and viscous effects very quickly set in. Above $We_G \simeq 50$, two-dimensional simulations show the appearance of liquid sheets erupting from the liquid layer. These sheets end in a thick rim-like edge (Rayleigh, 1891; Taylor, 1959; Culick, 1960; Ting and Keller, 1990). We distinguish several regimes for the subsequent evolution of these sheets, with increasing Weber number

1. *Stable regime.* At low We_G, the sheets reconnect with the liquid surface and droplets are not ejected. The flow appears laminar.
2. *Laminar-ejection regime.* At moderate We_G, the sheets tend to create circular patches (due to the two-dimensional character of the computation these are the section of cylinders). The droplets are systematically ejected towards the bulk of the gas phase.
3. *Turbulent-ejection regime.* At higher We_G, the production and the fate of droplets is irregular. The turbulence in the flow may let some droplets fall back on the bulk of the liquid phase.

A series of computations were performed for $Re_L = 1000$ and $\rho_L/\rho_G = 10$ in a 64^2 box. The various regimes in We_G, Re_G space are indicated on Figure 2, which complements a similar diagram in (Keller et al., 1994).

However, the pinching mechanism by which liquid droplets detach from the edges of sheets in two-dimensional flow is not necessarily the realistic three-dimensional regime. We selected parameters in the stable and laminar-ejection regime to perform three-dimensional computations. We were able so far to distinguish two scenarios by which three-dimensional flow sets in. We report the results of two simulations in a 64^3 box. All parameter are listed in Table 1. Figure 3 shows a case which A_2 is small. The simulation remains two-dimensional until the sheet breaks and a cylinder pinches of. Subsequently, the capillary instability of jets gives three-dimensional structure to the flow. In contrast, for higher A_2 and an interface-only perturbation, a different scenario is observed, as shown on Figure 4. The rim-like edge concentrates in a protruding finger that will be subject to the capillary instability. However, the unstable cylinder is now streamwise. Comparing this problem to studies of the disintegration of liquid sheets and planar liquid jets one seems to observe a universal scenario by which these sheets disintegrate into droplets (the clearest picture we could find so far is still Figure 7 of Rayleigh (1891). This scenario corresponds to what is observed in real experiments on shear layers (Raynal et al., 1996) but it is also reminiscent of the the crown formation in droplet impact experiments.

A theoretical analysis has been proposed by Yarin and Weiss (1995). They remark that the rim moves at a constant velocity $V \simeq (2\sigma/\rho_L e)^{1/2}$ where e is the film thickness. Thus any rim shape perturbed from a straight line should lead to cusp formation as in the usual front propagation problem. Our simulations and the current status of their interpretation do not

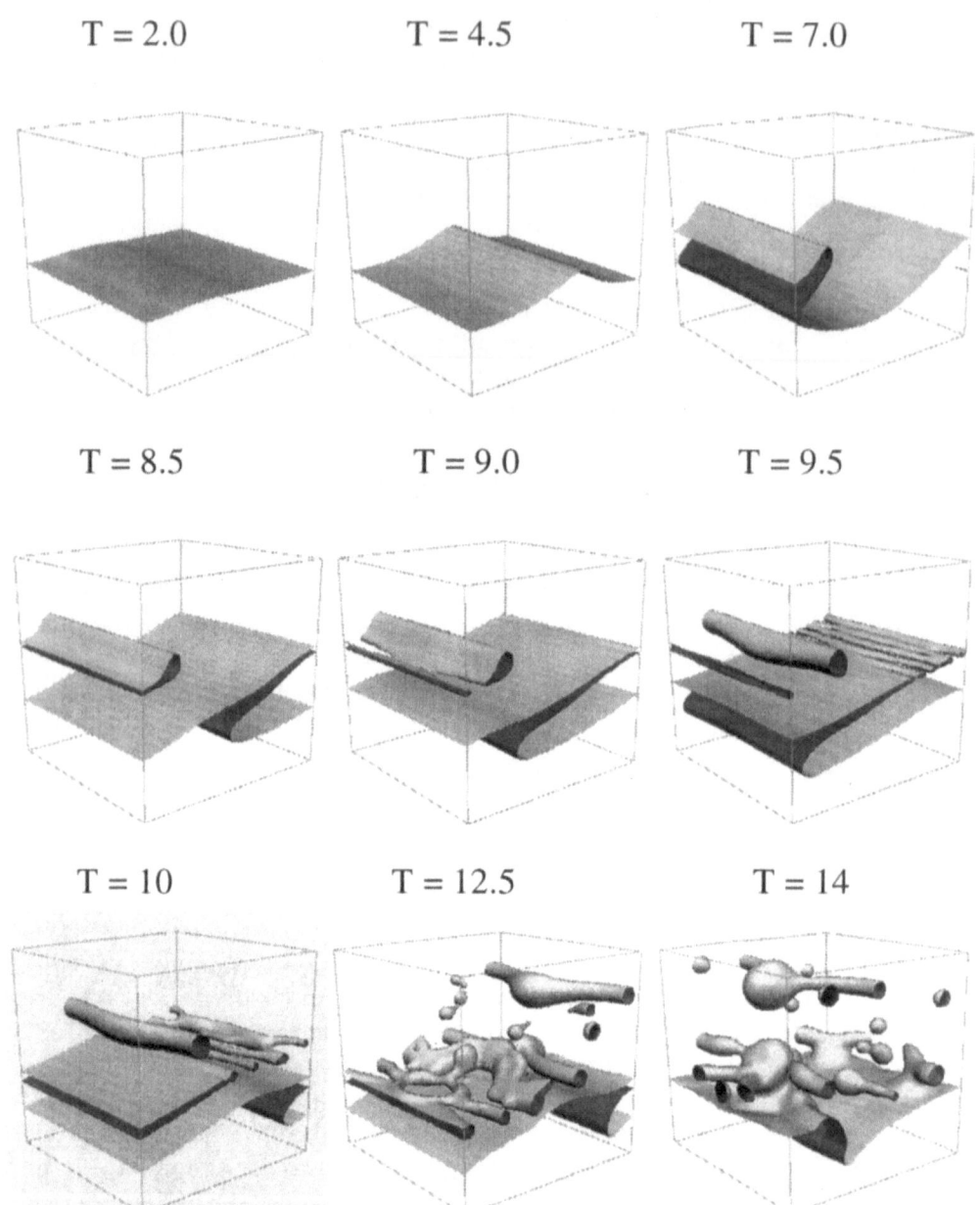

Figure 3. Run A shows a quasi-two-dimensional breakup of the interface. See parameters in table.

allow us to either verify or falsify this, but they do suggest a more complex dynamics. In particular the rim motion may be more complex than

T = 2.0   T = 4.5   T = 6.0

T = 7.0   T = 8.5   T = 9.0

T = 9.5   T = 10.0   T = 12

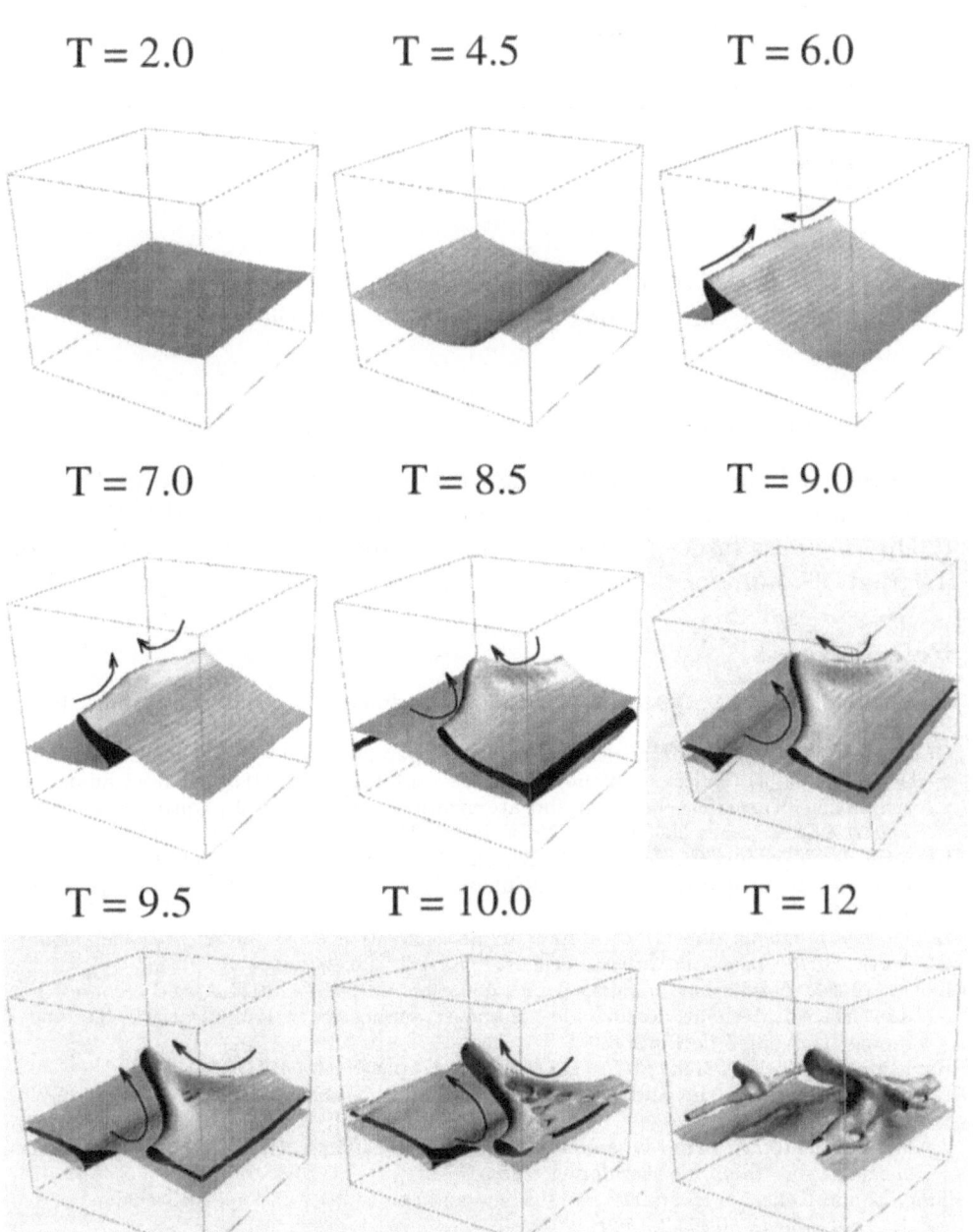

Figure 4. Run B shows breakup of the interface through a secondary instability of the sheet. See parameters in table.

TABLE 1. Parameter values for all runs.

Initial perturbation	Run	We_G	Re_G	A_1	A_2	Re_L	We_L
full	A	150	1500	0.001	0.1	1000	1500
interface only	B	200	1000	0.001	0.4	1000	2000

the simple constant velocity kinematics, since the shearing flow diminishes e. More detailed comparisons of simulations with theory should soon be possible.

Acknowledgements

S. Zaleski and Jie Li gratefully acknowledge Société Européenne de Propulsion, Centre National d'Études Spatiales and the Groupe de Recherche (GDR) moteurs fusées for their combined support and IDRIS (Orsay) for its grant of computer time.

References

Chandrasekhar, S. (1961). *Hydrodynamic and hydromagnetic stability*. Oxford Univ. Press, Oxford.

Culick, F. E. C. (1960). Comments on a ruptured soap film. *J. Appl. Phys.*, 31:1128–1129.

Keller, F. X., Li, J., Vallet, A., Vandromme, D., and Zaleski, S. (1994). Direct numerical simulation of interface breakup and atomization. In Yule, A. J., editor, *Proceedings of ICLASS94*, pages 56–62, New York. Begell House.

Lafaurie, B., Nardone, C., Scardovelli, R., Zaleski, S., and Zanetti, G. (1994). Modelling merging and fragmentation in multiphase flows with SURFER. *J. Comp. Phys.*, 113:134–147.

Li, J. (1995). Calcul d'interface affine par morceaux (piecewise linear interface calculation). *C. R. Acad. Sci. Paris, série IIb, (Paris)*, 320:391–396.

Li, J. (1996). Résolution numérique de l'équation de Navier-Stokes avec reconnection d'interfaces. Méthode de suivi de volume et application à l'atomisation. Technical report, Université de Paris 6.

Rayleigh, L. (1891). *Scientific Papers*, volume III, pages 441–451.

Raynal, L., Hopfinger, E., and Villermaux, E. (1996). Technical report, IMG, Grenoble, France.

Taylor, G. I. (1959). The dynamics of thin sheets of fluid III. Disintegration of fluid sheets. *Proc. Roy. Soc. London A*, 253:313–321.

Ting, L. and Keller, J. B. (1990). Slender jets and thin sheets with surface tension. *SIAM J. Appl. Math.*, 50:1533–1546.

Wesseling, P. (1992). *An introduction to multigrid methods*. Wiley.

Yarin, A. L. and Weiss, D. A. (1995). Impact of drops on solid surfaces. *J. Fluid. Mech.*, 283:141–173.

SELF-TURBULENT FLAME SIMULATION BY A CELLULAR AUTOMATON

A. LÓPEZ-MARTÍN, P. L. GARCÍA-YBARRA, J. L. CASTILLO and
J. C. ANTORANZ
*L.C.D.I., Dept. Física Fundamental, U.N.E.D., Apdo. 60141,
28080 Madrid, Spain*

1. Introduction

The analogy between a premixed flame front and a propagating line of sources of fluid volume [1] has been exploited to simulate premixed flames by cellular automata [2,3]. The automaton consists of individual particles moving in a two-dimensional square lattice. Two binary variables (accounting for temperature and fuel concentration) are associated to each individual particle. Thus, a normalized temperature equal to one is assigned to every hot particle and equal to zero to every cold particle. Mean local temperature values, corresponding to the average mesoscopic diffusive-convective thermal field, are obtained by averaging over the lattice on 7×7 boxes around each particle. Analogously, the mesoscopic fuel concentration is obtained by averaging the two-valued scalar variable which is equal to one for unburned particles and equal to zero for burned particles. Every time step, at the border between the unburned and burned regions, some unburned particles transform to burned particles. At these locations a source of fluid is inserted to account for the volume increase of the gas. The transformation from unburned to burned is simulated either by an inflammation temperature criterion or by a probability law which depends on the local mean temperature and concentration values, such as to mimic an Arrhenius law. The local particle sources generate extra (burned and hot) particles which push the neighbor particles along random and radial expansion directions, thus accounting for the potential part of the self-generated convective flow of flame fronts. The simulations have been performed for 800×1600 lattices with periodic lateral boundaries and open front and rear boundaries. The influence of the Lewis number *Le* (ratio of thermal diffusivity to reactant diffusivity) on the morphological instabilities of planar fronts is studied.

2. Dispersion Relation and Long Time Evolution

When the Lewis number is equal to unity only one binary variable is required (the temperature, say). The particle transformation from unburned to burned takes place when the local mean temperature reaches the prescribed inflammation temperature. In the simulation of these fronts, any planar front is shown to be unstable.

L. Fulachier et al. (eds.), IUTAM Symposium on Variable Density Low-Speed Turbulent Flows, 59–62.
© 1997 *Kluwer Academic Publishers.*

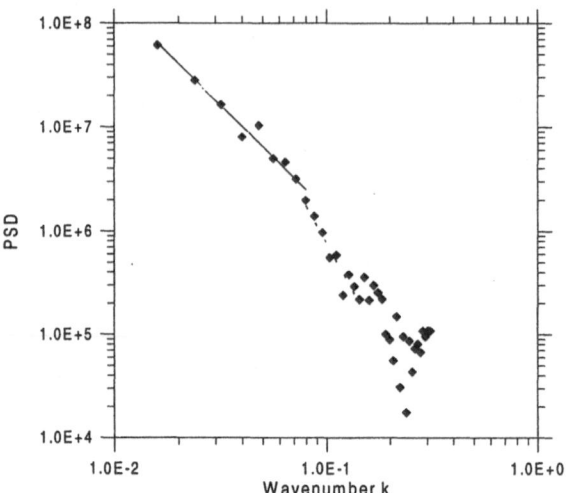

Figure 1. Power spectral density of the front distortion, after 209 time units and with a normalized ignition temperature equal to 0.75, compared with the power law distributions k^{-2} (solid line) and $k^{-11/3}$ (broken line).

Initially, the amplitude of the noise-induced disturbances of the planar front evolves exponentially in time. The dependence of the growth rate on the disturbance wavenumber provides the dispersion relation which shows the same wavenumber dependence as the Darrieus-Landau instability of actual flames [3]. To characterize the non-linear stage, we have simulated the long-time evolution of fronts through the Fourier analysis of the front shape. The noise-induced distortions associated to the linearly unstable wavenumber band are amplified by the front instability whereas the non-linearity redistributes the front deformation energy leading to a steady power spectrum. The long-time front morphology takes the form of a distorted parabolic front, filling the whole cross section of the duct. In the Figure 1, the long time power spectral density of a distorted front is compared with power law distributions. A double cascading spectrum is observed similar to the passive scalar spectra observed in forced turbulence (see, for instance, [4]). Deformation energy is injected by the instability around the wavenumber where the linear dispersion relation peaks up ($k_m \approx 0.016$). Then, the non-linear mode interaction generates two energy cascades. A k^{-2} inertial-convective cascade transports energy towards the smaller scales up to the diffusive flame thickness scale where a $k^{-11/3}$ inertial-diffusive cascade establishes.

3. The Diffusional-Thermal Instability

For the case of Lewis number different from unity, two binary variables are needed. One of them accounts for the temperature distribution whereas the second one represents the fuel concentration. Both fields are used to perform, each time step, the computation of the local values of the chemical reaction rate term.

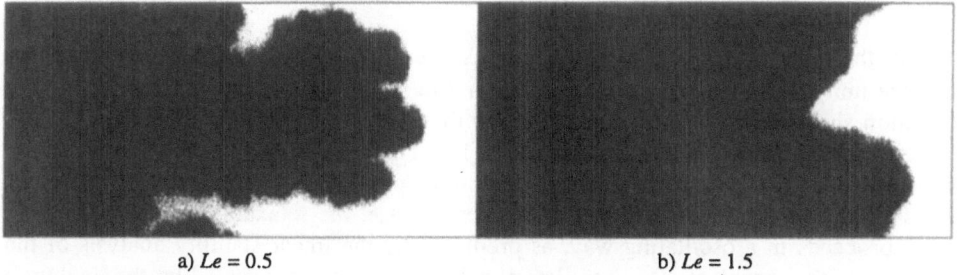

a) $Le = 0.5$ b) $Le = 1.5$

Figure 2. Front shapes, after 500 time units, of an initially planar front exhibiting diffusional-thermal instability a) Cellular front ($Le = 0.5$) and b) Overstability ($Le = 1.5$).

The differential diffusion between heat and mass adds a new front phenomenology associated to the so called diffusive-thermal instability, that operates even in the absence of gas thermal expansion. For Lewis numbers sufficiently smaller than unity, this instability leads to cellular flame fronts through a steady bifurcation [5]. In slightly supercritical conditions, asymptotically valid differential equations are known to hold (see the summary in [6]) but computational studies under strongly non-linear conditions do not exist. The automaton provides a tool to simulate arbitrarily deformed fronts for any value of the Lewis number. Figure 2a shows the long-time front evolution when $Le = 0.5$, which corresponds to a very unstable front. Any initially planar front leads to the formation of dendritic structures. The area of the reactive surface is continuously increasing, in such a way that, the overall burning rate does not show signs of saturation, as can be seen in the Figure 3. With regard to the properties of the usual gaseous combustible mixtures, only mixtures containing hydrogen as fuel can have such small values of the Lewis number.

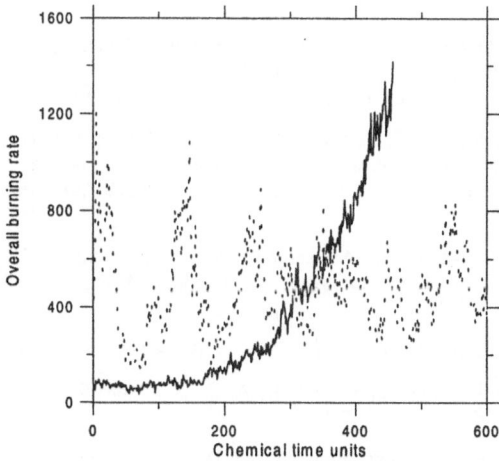

Figure 3. Time evolution of the burning rate. Solid line: $Le = 0.5$. Broken line: $Le = 1.5$.

In the opposite case of gaseous combustible mixtures with Lewis numbers larger than the unity, overstability may occur leading to oscillatory phenomena [5]. The time evolution shows a dense non planar growing that leads to a distorted unique large cell, depicted in the Figure 2b for the case $Le = 1.5$. This final shape is very similar to the shape observed in the case of $Le = 1$ studied in the previous section. However, there is a non surprising difference because, in the present case, the relaxation towards the final shape proceeds in a oscillating way, as predicted by the linear stability analysis of the analogous flame problem. This oscillatory behavior is clearly revealed by the evolution of the overall burning rate, shown in Figure 3, which can be compared with the case of the stationary instability.

4. Conclusions

A cellular automaton has been developed to mimic unstable flame fronts. The automaton simulates the combined effect of gas thermal expansion and diffusive-thermal instabilities. Well differentiated front morphologies appear depending on the sign, positive or negative, of the departure of the Lewis number from the unity. However, the automaton does not account for the vorticity generation through the flame front. The automaton is actually being improved in this direction.

5. Acknowledgments

This work has been sponsored by the Spanish DGICYT under project No. PB94-0385.

6. References

1. García-Ybarra, P. L., Antoranz, J. C. and Castillo, J. L. (1991) Simulation of Flame Fronts by Sources of Fluid Volume, in M. Ben Amar, P. Pelcé and P. Tabeling (eds.), *Growth and Form: Nonlinear Aspects*, NATO ASI Series B vol. 276, Plenum Press, New York, pp. 253-259.

2. Antoranz, J. C., López-Martín, A., Castillo, J. L. and García-Ybarra, P. L. (1993) Discrete Potential Flow Simulation of a Premixed Flame Front, in J. M. Garcia-Ruiz, E. Louis, P. Meakin and L. M. Sanders (eds.), *Growth Patterns in Physical Sciences and Biology*, NATO ASI Series B vol. 304, Plenum Press, New York, pp. 119-126.

3. García-Ybarra, P. L., López-Martín, A., Antoranz, J. A. and Castillo, J. L. (1994) Unsteady Potential Flows Computation by Cellular Automata: the Premixed Flame Instability, *Transport Theory & Statistical Physics* **23**, 173-193.

4. Lesieur, M. (1990) *Turbulence in Fluids*, pp. 276-279, Kluwer Academic Publishers, Dordrecht.

5. Joulin, G. and Clavin, P. (1979) Linear Stability Analysis of Nonadiabatic Flames: Diffusional-Thermal Model, *Combustion and Flame* **35**, 139-153.

6. Frankel, M. L. (1991) Free Boundary Problems and Dynamical Geometry Associated with Flames, in P. C. Fife, A. Liñán and F. Williams (eds.), *Dynamical Issues in Combustion Theory*, The IMA Volumes in Mathematics and its Applications vol. 35, Springer-Verlag, New York, pp. 107-126.

II. Modelling and Experiments

SOME PROBLEMS ON SINGLE POINT MODELING OF TURBULENT, LOW-SPEED, VARIABLE DENSITY FLUID MOTIONS

P. CHASSAING
Professeur à l'INPT-ENSEEIHT, Chef du département Mécanique des fluides de l'ENSICA
Institut de Mécanique des Fluides de Toulouse UMR CNRS/INPT-UPS 5502, Allée du Professeur Camille SOULA, 31400 TOULOUSE ;
École Nationale Supérieure d'Ingénieurs de Constructions Aéronautiques, 1 Place Émile BLOUIN, 31056 TOULOUSE CEDEX

1 Introduction

Turbulent flows of variable density fluids are widely present in various domains of human activity. From an engineer point of view, one of the most commonly quoted is probably aeronautics, where *compressible* flows are present in various applications, such as high speed aircraft and supersonic combustion ramjet engines for instance. However, as far as general industrial fluid applications are concerned, *variable density fluid motions* are part of chemical engineering, energetics, science of heat . . .
Fluid density variations also occur in geofluids motions, environment, and MHD turbulence (plasma fusion studies, solar wind . . .).

Face to such a variety of natural or practical situations of variable density fluid motions, it is not easy to define and analyze the whole scope of density effects on turbulence. Moreover, and if this should have been done, it is not clear whether the modeling of such density effects should be addressed independently of their origins, must be treated separately, or could be taken hierarchically according to some analogies and differences between various types of variable density flows.

This explains why, for both fundamental and modeling reasons, it is necessary to bound the domain of variable density fluid turbulence to some tractable situation, without loosing to much of the specificity of the problem. This question will be discussed in the first section of the paper. The second section is devoted to the correlations with the density fluctuations (d.f.c.). The next one (4) deals the deriving and the physical interpretation of the open set of equations. The last section is devoted to the enlightening of some consequences of the statistical formulation of the equations on second order modeling.

2 A specific class of low speed turbulent motion of variable density fluid.

2.1 THE SCOPE OF DENSITY VARIATION IN FLUID MOTIONS

Starting simply from the definition of the density $\rho = M/V$, two specific situations can be introduced according to whether (i) the density of a constant mass fluid element

L. Fulachier et al. (eds.), IUTAM Symposium on Variable Density Low-Speed Turbulent Flows, 65–84.

changes in response to variations of its volume, or (ii) the mass of the fluid element is changed within a given volume.

The latter situation is observed when two (or more) non reactive gases mix with constant temperature and pressure. Analogous to this case is the mixing of different temperature flows of the same fluid with the same pressure. In both cases, the velocity can be as low as it is compatible with a fully developed turbulent regime.

The first class of variable density fluid motions includes three distinct configurations where the density changes are associated with :

− *Boundary effects*, when the volume of a confined mass of gas is modified in response to changes in the boundary conditions.
− *Mach number effects*, in high speed free flows past obstacles, where the compressibility of the fluid leads to velocity induced pressure effects.
− *Dilatation effects*, as it is observed when the volume of the fluid element changes due to the thermal expansion of the fluid. This situation includes as a limit with respect to weak density variation, buoyant flows.

2.2 SOME SIMPLIFICATIONS TO VARIABLE DENSITY FLUID MOTION

Mainly due to the complexity of the Navier-Stokes equations of a variable density fluid motion, it is tempting to get more tractable models, using weak perturbations analysis, linearization and decoupling of pressure effects.

With the pioneering work of Kovasnay[23] in 1953, these ideas were first applied to compressible supersonic flows. Close to a reference state of the fluid motion, Kovasnay decomposed the velocity fluctuation into three modes (vorticity, acoustic and entropy modes) which are decoupled from each other to first order in the fluctuation amplitude.

Considering that the disturbances apply about potential flows round arbitrary obstacles, Goldstein [19] decomposed the velocity fluctuation into a "vortical" and "acoustic" part. This decomposition has been extended by Debiève [8] and Dussauge et al. [11] to flows with mean entropy gradients orthogonal to the mean velocity.

Some years later, in 1962, the same type of analysis was applied to the Boussinesq's approximations by Ogura et al. [27]. Eliminating acoustic wave propagation, and according to the anelastic approximation of these authors, the pressure and buoyancy terms are taken linear. As noticed by Lilly [26], these equations apply to oceanography and deep convective mesoscale atmospheric dynamics provided that the reference state is assumed to be in static neutral stability. The "pseudo-incompressible" approximation introduced by Durran [10] in 1989 permits to refer to an arbitrary temperature profile while maintaining energetic consistency for such small Mach number flows.

More recently, the same ideas have been used to derive the equations of nearly incompressible fluid motion at low Mach number by means of singular expansion techniques : Zank et al. [34], for various incompressible limits an ideal gas flow, and Bayly et al. [3], for a real fluid with a general equation of state. In both cases, temperature and density fluctuations compete with pressure fluctuations as deviations about the Boussinesq's reference state.

The reviewed studies bring out fundamental results about the theoretical distinction between the mathematical models applying to compressible flows and low speed variable density fluid motions. Some peculiarities of each type of flows can be found for instance in Lele [25] and Fulachier et al. [18] respectively. Focusing on this last situation we shall now introduce a peculiar class of variable density flow.

2.3 DIVERGENCE OF THE VELOCITY IN VARIABLE DENSITY FLUID MOTION

In a fluid motion — see Batchelor [1] page 75 for instance —, the fractional rate of change of the volume V of a material element is given by :

$$\frac{1}{V}\frac{dV}{dt}\left(\equiv -\frac{1}{\rho}\frac{d\rho}{dt}\right) = div\vec{U} \tag{1}$$

where d/dt stands for the material derivative and $\vec{U} - (U_i, i = 1,2,3)-$, is the velocity of the fluid element. Thus from eq. (1), constant density fluid flows are divergence free (solenoidal) motions, but the reciprocal is not true : the stratified stationary laminar flow with density gradients orthogonal to the streamlines is a well known example of such a situation. Let us detail this reciprocal for a binary mixture of non reactive gases. Introducing the mass fraction (C) of the component A, the equation of state is $P = \rho T(\alpha C + \beta)$, where P, ρ, T stand for the pressure, density and temperature of the mixture respectively. The two coefficients α and β are related to the ideal gas constants r_A and r_B of the pure species according to $\alpha = r_A - r_B$ and $\beta = r_B$. Taking the logarithmic material derivative of this equation directly yields in eq. (1) :

$$div\vec{U} = -\frac{1}{P}\frac{dP}{dt} + \frac{1}{T}\frac{dT}{dt} + \frac{\alpha}{\alpha C + \beta}\frac{dC}{dt} \tag{2}$$

Thus, in general, three additive effects (pressure, temperature and composition) are contributing to the divergence of the velocity vector in such fluids motions. We shall now detail separately (i) the constant pressure and temperature binary mixture, and (ii) the single component ideal gas flows.

2.3.1 The constant pressure and temperature binary mixture.
In this case, the equation of state is merely : $\rho = a\rho C + b$ \hfill (3)

with $a = -\alpha/\beta \equiv 1 - \rho_B^*/\rho_A^*$ and $b = \rho_B^* \equiv P^*/\beta T^*$, where ρ_A^* and ρ_B^* are the densities of the pure species corresponding to the constant pressure $P = P^*$ and temperature $T = T^*$. Consequently, eq. (2) reduces to :

$$div\vec{U} = \frac{a}{1 - aC}\frac{dC}{dt} \tag{4}$$

Now the mass conservation equation of species A gives : $\rho\frac{dC}{dt} = \frac{\partial m_j}{\partial x_j}$ \hfill (5)

where m_j, $j = 1,2,3$ is the molecular mass diffusion flux of the considered species. Combining the previous results, it is easy to obtain :

$$div\vec{U} = \frac{a}{b}\frac{\partial m_j}{\partial x_j} \tag{6}$$

Since a and b are two constants in this situation, the previous equation states that the divergence of the velocity vector only depends on the molecular diffusion mass flux.

2.3.2 The single component ideal gas.

In this case ($C \equiv 0$) the divergence of the velocity vector reduces to the two additive contributions :

$$div\vec{U} = -\frac{1}{P}\frac{dP}{dt} + \frac{1}{T}\frac{dT}{dt} \qquad (7)$$

Now, pressure and temperature variations are linked with, according to the energy transport equation, which can be written for an ideal gas (see Bayly et al. [3]) :

$$\rho C'_p \frac{dT}{dt} = \frac{dP}{dt} + 2\mu S_{ij}S_{ij} - \frac{2}{3}\mu\sigma^2 + \frac{\partial}{\partial x_j}\left(\lambda\frac{\partial T}{\partial x_j}\right) \qquad (8)$$

where C'_p, μ and λ are respectively the specific heat at constant pressure, the dynamic viscosity and the thermal conductivity. The strain rate tensor is denoted by S_{ij} with $\sigma = S_{ii} \equiv div\vec{U}$. Eliminating the temperature from eqs. (7) and (8) yields :

$$div\vec{U} = -\frac{1}{\gamma P}\frac{dP}{dt} + 2\nu\frac{\gamma-1}{a^2}\left[S_{ij}S_{ij} - \frac{\sigma^2}{3}\right] + \frac{\gamma-1}{\rho a^2}\frac{\partial}{\partial x_j}\left(\lambda\frac{\partial T}{\partial x_j}\right) \qquad (9)$$

introducing the celerity of sound $a^2 = \gamma r T$, the isentropic coefficient $\gamma = C'_p/C'_v$ and the kinematic viscosity $\nu = \mu/\rho$. Taking $\mathcal{L}$ and $\mathcal{U}$ as characteristic advective length and velocity scales respectively, and Θ as a temperature scale, the non-dimensional form of eq. (9) for low speed motions where the scaling of the pressure is given by $\rho\mathcal{U}^2$, is :

$$\widehat{div\vec{U}} = -M^2\frac{\widehat{dP}}{dt} + 2(\gamma-1)\frac{M^2}{R_e}\left(\widehat{S_{ij}S_{ij}} - \frac{\widehat{\sigma^2}}{3}\right) + \frac{1}{P_r \times R_e}\frac{\widehat{\partial^2 T}}{\partial x_j^2} \qquad (10)$$

In eq. (10), the wide hat denotes non dimensional quantities and M, R_e, P_r are respectively the Mach, Reynolds and Prandtl numbers based on the characteristic scales. For low Mach number flows, the last term is predominant so that the divergence of the velocity vector is proportional to the thermal molecular diffusion :

$$div\vec{U} \overset{M\to 0}{\simeq} \frac{\nu}{P_r\Theta}\frac{\partial^2 T}{\partial x_j^2} \qquad (11)$$

2.4 THE MEAN SOLENOIDAL CONDITION

In turbulent flows where molecular diffusion can be discarded as compared with turbulent transfer, it is obtained, from eqs. (6) and (11) that :

$$\partial\overline{U}_j/\partial x_j = 0 \qquad (12)$$

where $\overline{U}_j$, $j = 1, 2, 3$ denotes the statistical or ensemble average of the velocity vector. As we shall see later on, the solenoidal property of the mean velocity field — obviously satisfied by incompressible or isovolume turbulent motions —, is not restricted to weak density variations. This is the reason why mean solenoidal turbulent (M.S.T) flows should be considered as a specific class of low-speed, variable density fluid flows.

Basically, this class of mean divergence free variable density turbulent flows is concerned with *mass or thermal fluctuations dominated motions*. In other words, the concentration **or** temperature fluctuations **and** the density fluctuations are significantly more dominant than the pressure ones[1].

---

[1]   According to Zank et al. [34], the M.S.T. condition (12) can be extended to flows where the thermal transfer equation is modified by acoustic effects, provided that the density and pressure fluctuations should be linked by $p' = a^2\rho'$, where $a = \sqrt{\gamma r T}$ is the *local* celerity of sound.

3 Correlations with density fluctuation

Using a statistical or ensemble averaging operator, any instantaneous value $F(x_j,t)$ can be taken as :

$$F(x_j,t) = \overline{F}(x_j,t) + f'(x_j,t) \tag{13}$$

where $f'(x_j,t)$ is the random centered fluctuation ($\overline{f'}(x_j,t) = 0$) with respect to the mean value $\overline{F}(x_j,t)$. When averaging non linear terms involving the density, correlations with the density fluctuations (d.f.c.) are introduced. The generic form of such d.f.c. terms is :

$$\overline{\rho' f_1' f_2' \cdots f_n'} \tag{14}$$

where f_α', $\alpha = 1, \ldots n$ stands for any random scalar fluctuating function, including velocity components.

3.1 MASS-FRACTION – DENSITY CORRELATIONS

For the situation introduced in §2.3.1., the equation of state (3) can be rewritten as $\overline{p} + \rho' = a(\overline{\rho C} + \rho' \overline{C} + \overline{\rho \gamma'}) + b$ with $C = \overline{C} + \gamma'$. Substraction of the mean value gives :

$$\rho' = \frac{a}{1 - a\overline{C}} \left(\rho \gamma' - \overline{\rho \gamma'} \right) \tag{15}$$

Then, multiplying by $f_1' f_2' \cdots f_n'$ and averaging, it is obtained :

$$\overline{\rho' f_1' f_2' \cdots f_n'} = \frac{a}{1 - a\overline{C}} \left(\overline{\rho \gamma' f_1' f_2' \cdots f_n'} - \overline{\rho \gamma'} \times \overline{f_1' f_2' \cdots f_n'} \right) \tag{16}$$

This general expression simplifies for any *second order* d.f.c. since the last term in the right hand side is zero due to the centered fluctuation $\overline{f_\alpha'} = 0$. Then, it merely comes :

$$\overline{\rho' f'} = \frac{a}{1 - a\overline{C}} \overline{\rho \gamma' f'} \tag{17}$$

This result clearly states that the second order d.f.c. with any scalar fluctuation is directly linked with the correlation of the considered scalar fluctuation with the mass-fraction fluctuation times the *instantaneous* value of the density.

3.2 TEMPERATURE – DENSITY CORRELATIONS

The same analysis can be applied to the variable temperature situation described in §2.3.2 to give $(1 - \omega \gamma)\rho' = -\left(\rho \theta' - \overline{\rho \theta'} \right)/\overline{T}$, where $\omega = p'/\gamma r \overline{T} \rho'$. Thus an explicit expression similar to eq. (15) is obtained when $1 - \omega \gamma = \pm 1$. The first condition ($\omega = 0$) is nothing but the temperature dominated assumption. The second one is $\omega = 2/\gamma$ ($\omega \approx 1.43$ for air). It is approached by the Zank et al. [34] pseudosound assumption for nearly incompressible fluids ($\omega = 1$). We shall include all these situations in a single expression :

$$\rho' = -\left(\rho \theta' - \overline{\rho \theta'} \right)/\delta \overline{T} \tag{18}$$

where δ is a given parameter equal to 1, -1, $1 - \gamma$, according to which one of the three situations is considered. Hence, the general expression of density fluctuation correlations is :

$$\overline{\rho' f_1' f_2' \cdots f_n'} = -\left(\overline{\rho\theta' f_1' f_2' \cdots f_n'} - \overline{\rho\theta'} \times \overline{f_1' f_2' \cdots f_n'}\right)/\delta\overline{T} \qquad (19)$$

Any second order d.f.c. is simply deduced from density cross moment $\overline{\rho\theta' f'}$ by :

$$\overline{\rho' f'} = -\overline{\rho\theta' f'}/\delta\overline{T} \qquad (20)$$

As a partial conclusion, we can keep in mind that for low speed variable concentration or temperature flows, the following (exact) expressions apply :

$$\frac{\overline{\rho\gamma'}}{\overline{\rho\gamma'^2}} = \frac{\overline{\rho u_i'}}{\overline{\rho\gamma' u_i'}} = \frac{a}{1 - aC} \qquad (21a) \qquad\qquad \frac{\overline{\rho\theta'}}{\overline{\rho\theta'^2}} = \frac{\overline{\rho u_i'}}{\overline{\rho\theta' u_i'}} = -\frac{1}{\delta\overline{T}} \qquad (21b)$$

3.3 SOME d.f.c. PROPERTIES IN A TURBULENT FREE JET

A turbulent free jet (ρ_{e0} at the exit) is discharging into a quiescent homogeneous (ρ_∞) atmosphere under the concentration and heat fluctuations dominating assumption ($\delta = 1$). The inlet density ratio $S = \rho_{e0}/\rho_\infty$ is equal to $S = T_\infty/T_{e0}$, where T_{e0} is the temperature at the exit and T_∞ the temperature of the atmosphere, for variable temperature jets. For variable concentration isothermal jets with $C = 1$ at the exit, it is easy to see that $S = 1/1 - a$.

With these notations, eqs. (21aa and b) can be transformed to give :

$Var.\,temp.\ Hot\ Jet$: $(0 \leqslant S < 1)$ $\dfrac{\overline{\rho\theta'}}{\overline{\rho\theta'^2}} \equiv \Gamma^S(\overline{\Theta}) = \dfrac{S - 1}{S + (1 - S)\overline{\Theta}}$ $\qquad (22a)$

$Var.\,temp.\ Cold\ Jet$: $(S > 1)$ $\dfrac{\overline{\rho\theta'}}{\overline{\rho\theta'^2}} \equiv \Gamma^S(\overline{\Theta}) = \dfrac{1 - S}{1 + (S - 1)\overline{\Theta}}$ $\qquad (22b)$

$Var.\,conc.\ Light\ and\ Heavy\ Jet$: $(\forall S)$ $\dfrac{\overline{\rho\gamma'}}{\overline{\rho\gamma'^2}} \equiv \Gamma^S(\overline{C}) = \dfrac{S - 1}{S + (1 - S)\overline{C}}$ $\qquad (23)$

with $\overline{\Theta} = \overline{T} - T_{min}/(T_{max} - T_{min})$. Since $0 \leqslant \overline{\Theta}$ or $\overline{C} \leqslant 1$, it can be concluded that in a concentration fluctuations dominated *light* jet, density and mass-fraction fluctuations are *negatively* correlated, *similarly* to temperature and density ones in *heated* jets. They are *positively* correlated in concentration fluctuations dominated *heavy* jets, as *opposed* to variable temperature *cold* jets, where the correlation is still negative.

This departure between heavy and cold jets can be analyzed in more details when including the signs of the density velocity fluctuations correlations (see Chassaing et al. [5]). They are given in the following table, along with those of the mean concentration and temperature gradients $G_C \equiv \partial\overline{C}/\partial x_j$ and $G_T \equiv \partial\overline{T}/\partial x_j$, $j = 1$ and 2, and the associated transport terms :

$$H_C \equiv \overline{\rho u_j' \partial\overline{C}/\partial x_j} \quad \text{and} \quad H_T \equiv \overline{\rho u_j' \partial\overline{T}/\partial x_j} \qquad (24)$$

Some of these results were first obtained experimentally. The direct measurements by Driscoll et al. [9] in a turbulent non premixed flame give negative values of both

$\overline{\rho u'}$ and $\overline{\rho v'}$. The same negative values are obtained in a variable concentration jet by So et al. [32]. In heated jets, Corrsin et al. [7], and Chevray et al. [6] found positive velocity-temperature correlations $\overline{\theta'u'}$ and $\overline{\theta'v'}$. According to eq. (21ab) and with the approximation $\overline{\rho\theta'u'_i} \approx \overline{p}\,\overline{\theta'u'_i}$, this result confirms the negative values of the turbulent mass fluxes.

		$\overline{\rho\gamma'}$	$\overline{\rho u'}$	$\overline{\rho v'}$	$\overline{\rho\gamma'u'}$	$\overline{\rho\gamma'v'}$	G_C	H_C
Variable Concen- tration	Heavy jet	+	+	+	+	+	-	-
	Light jet	-	-	-	+	+	-	+

		$\overline{\rho\theta'}$	$\overline{\rho u'}$	$\overline{\rho v'}$	$\overline{\rho\theta'u'}$	$\overline{\rho\theta'v'}$	G_T	H_T
Variable Tempera- ture	Cold jet	-	+	+	-	-	+	+
	Hot jet	-	-	-	+	+	-	+

Table 1 Signs of d.f.c., mean gradients and associated mean transport in a variable density jets

To summarize, the following statements emerge for d.f.c. in M.S.T. turbulent jets :

– The signs of density / mass-fraction (temperature) correlation can be obtained exactly;

– Such d.f.c. explain analytical differences between heavy / cold and light / hot jets respectively :

– As far as second order d.f.c. are concerned, the situations of both light (variable composition) and hot (variable temperature) jets are qualitatively identical ;

– In heavy and cold jets the d.f.c. do not exhibit similar trends.

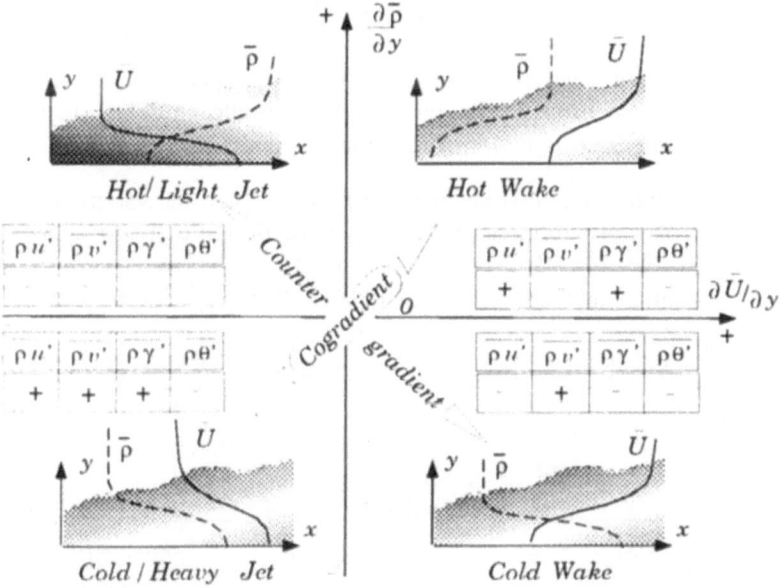

Figure 1 Signs of second order d.f.c. in free shear flows.

In fig. 1, the conclusions for jets flows have been generalized to other shear flows, such as wakes and mixing layers (see Joly [22]). If density and temperature fluctuations are obviously always negatively correlated, the signs of the three other correlations are inverted when the sign of the mean density gradient is changed. Finally, the two different cogradient (resp. countergradient) situations only differ according to the sign of $\overline{\rho v'}$.

Although qualitative, these results clearly enhance the interest of the d.f.c. to explain some density effects in turbulent shear flows. However one can still wonder about the quantitative influence of the d.f.c., which is likely to depend upon the amount of the density variation. This explains why shall now try to introduce some quantitative classification in M.S.T. flows.

3.4 TOWARDS A QUANTITATIVE CLASSIFICATION OF M.S.T. FLOWS

In M.S.T. flows, the turbulent density intensity $I_\rho = \sqrt{\overline{\rho'^2}}/\overline{\rho}$ can be markedly different from other turbulent intensities. In a turbulent wall jet of helium for instance, Harion [21] measured a value of about 40% for I_ρ at a downstream location where the turbulent longitudinal velocity intensity ($I_u = \sqrt{\overline{u'^2}}/\overline{U}_\infty$) is roughly half this value (17%). Similarly, in a pure hydrogen/air jet, it can be deduced from the measurements of Sautet [30] that $I_\rho \approx 0.52$ on the jet axis at $x/D = 12$, as compared with a concentration intensity $I_\gamma = \sqrt{\overline{\gamma'^2}}/\overline{C}$ of about 71%.

Such differences directly result form eqs. (21a). To a first order approximation in density fluctuation, it can be deduced from eq. (15) for instance, that :

$$I_\rho \approx \left| \Gamma^S(\overline{C}) \right| \times \sqrt{\overline{\gamma'^2}} \qquad (25)$$

This relation suggests to introduce directly $\Gamma^S(\overline{C})$ — resp. $\Gamma^S(\overline{\Theta})$ — to discuss the quantitative effects of density fluctuation in M.S.T. turbulence. Thus, these functions will be called "density effects indicator functions" and are plotted in the next figure.

From the analytical expressions of $\Gamma^S(\overline{\Theta})$ and $\Gamma^S(\overline{C})$, the variation range of the density ratio can be split into three parts, as sketched in figure 2, corresponding to :
a) Small indicator function ($\left|\Gamma^S\right|_{max} < 1/2$), for $2/3 \leqslant S \leqslant 3/2$;
b) Moderate indicator function ($\left|\Gamma^S\right|_{min} > 1/3$ and $\left|\Gamma^S\right|_{max} < 1$), for $1/2 \leqslant S \leqslant 2/3$ or $3/2 \leqslant S \leqslant 2$;
c) High indicator function ($\left|\Gamma^S\right|_{min} > 1/2$), for $S \leqslant 1/2$ or $2 \geqslant S$.

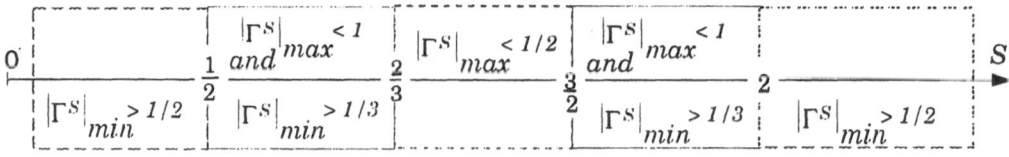

Figure 2 Quantitative gradation of density effects for turbulent jet flows.

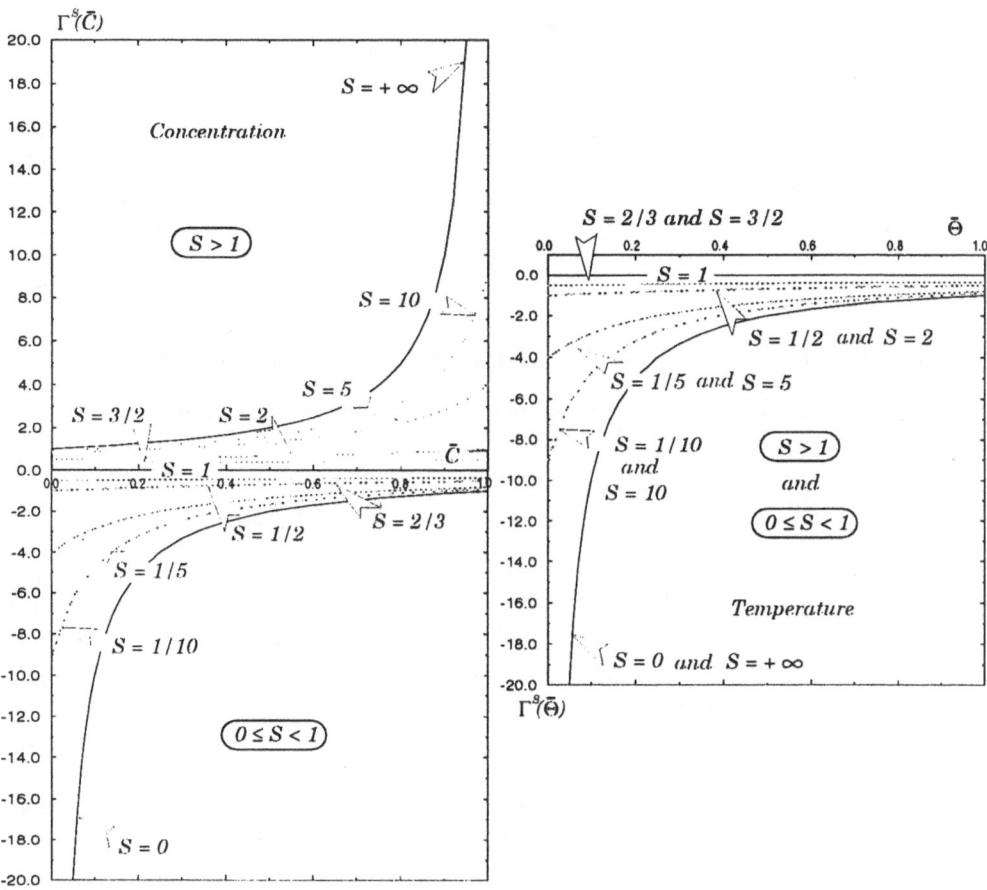

Figure 3 Density effects indicator functions for variable concentration or temperature free jets.

4 The open sets of one point statistical equations

4.1 STATISTICAL AVERAGING OF VARIABLE DENSITY TURBULENT FLOWS

So far, at least four types of method have been proposed to cope with the statistical averaging of variable density turbulent flows. The first one is the mass-weighted or Favre's averaging [12], [13],[14], [15], [16], [17]. The second is the mixed-weighted averaging introduced by Bauer et al. [2] and developed by Ha Minh et al. [20]. The third is nothing more than the conventional (Reynolds) formulation, without or with approximations to the density fluctuation, as suggested for instance by Shih et al. [31] for the mass mixing, and by Rey et al. [29] and Rey [28] for thermal mixing. The last one was originally proposed by the author in 1985 [4].

When using mass-weighted averaging, any second order d.f.c term is included into a *new* macroscopic mean value as, for instance $\overline{\rho}\,\overline{U}_i + \overline{\rho u'_i} \equiv \overline{\rho}\tilde{U}_i$. Hence a **binary** *regrouping* (B.R.) of any convective term is obtained since :

$$\overline{A}_j(F) \equiv \overline{\rho F U_j} = \overline{\rho}\,\tilde{F}\tilde{U}_j + \overline{\rho f u_j} \qquad (26)$$

An alternating to the previous method (Chassaing [4]) consists in taking apart the second order d.f.c. terms and introducing the instantaneous value of the density into triple and higher order density correlations, leading to *Reynolds' fluctuations mass-weighted second and higher order moments*. A *ternary regrouping* (T.R.) is suggested for any convective term :

$$\overline{A}_j(F) = \overline{\rho}\,\overline{F}\,\overline{U}_j + \overline{\rho f' u'_j} + \overline{\rho f'}\,\overline{U}_j + \overline{\rho u'_j}\,\overline{F} \qquad (27)$$

4.2 PHYSICAL INTERPRETATION OF B.R. AND T.R. IN M.S.T. FLOWS

Depending upon the type of regrouping which is adopted, the mean continuity equation in M.S.T. flows can be written :

Binary regrouping :
$$\frac{\partial \overline{\rho}}{\partial t} + \tilde{U}_k \frac{\partial \overline{\rho}}{\partial x_k} = -\overline{\rho}\frac{\partial \tilde{U}_k}{\partial x_k} \equiv +\overline{\rho}\frac{\partial \overline{u}_k}{\partial x_k} \qquad (28a)$$

Ternary regrouping :
$$\frac{\partial \overline{\rho}}{\partial t} + \overline{U}_k \frac{\partial \overline{\rho}}{\partial x_k} = -\frac{\partial \overline{\rho u'_k}}{\partial x_k} \qquad (28b)$$

As shown by these equations, sink / source terms are present in both right hand side, which physically correspond to (i) a mean volume variation during a macroscopically constant mass evolution, or (ii) a turbulent mass flux through the surface of a constant mean volume.

The same kind of analysis can be applied to the momentum equation, which can be written as follows in a free turbulent jet without gravity :

B.R.
$$\underbrace{\tilde{U}\frac{\partial \overline{\rho}\tilde{U}}{\partial x} + \tilde{V}\frac{\partial \overline{\rho}\tilde{U}}{\partial y}}_{(\tilde{A})} + \underbrace{\overline{\rho}\tilde{U}\left(\frac{\partial \tilde{U}}{\partial x} + \frac{1}{y^n}\frac{\partial y^n \tilde{V}}{\partial y}\right)}_{(V.V.E.)} + \underbrace{\frac{1}{y^n}\frac{\partial}{\partial y}(y^n \overline{\rho u v})}_{(T.D.)} = 0 \qquad (29)$$

T.R.
$$\underbrace{\overline{U}\frac{\partial \overline{\rho}\,\overline{U}}{\partial x} + \overline{V}\frac{\partial \overline{\rho}\,\overline{U}}{\partial y}}_{(\overline{A})} + \underbrace{\frac{1}{y^n}\frac{\partial}{\partial y}\left(y^n \overline{\rho v'}\,\overline{U}\right)}_{(T.M.F)} + \underbrace{\frac{1}{y^n}\frac{\partial}{\partial y}\left(y^n \overline{\rho u'v'}\right)}_{(T.D.)} = 0 \qquad (30)$$

Based on the predictions of the model detailed in Chassaing et al. [5], the different contributions in the previous equations can be evaluated, as plotted in the next figures. So long as the density ratio is close to one — here $S = 0.53$ and $S = 1.42$ for the light and heavy jets respectively —, the mean advection terms are quite close, so that the macroscopic volume variation and the turbulent mass flux contribution are equivalent. However, as conjectured in the thermodynamical analysis of the d.f.c., these terms change between the heavy and light situations.

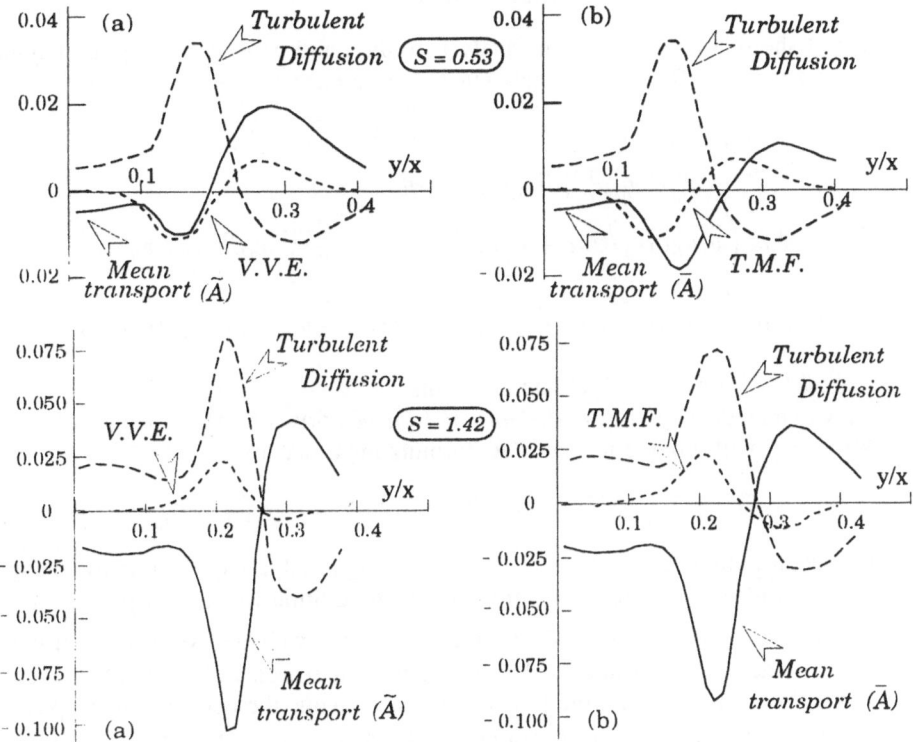

Figure 4   Mean momentum balance in a M.S.T. round free jet at $x/R_0 = 4$, according to (a) Mass-weighted velocity transport, (b) Conventional mean velocity transport. Top, light jet, bottom heavy jet.

5 Some aspects of turbulence modeling in variable density fluid motion

5.1 THE PRESSURE-VELOCITY FLUCTUATIONS CORRELATION

One key point of second order closure is the modeling of the correlation with the pressure fluctuation which appears in the "Reynolds stress" transport equations. Depending upon which formulation is adopted, the exact expression of this terms is :

$$B.R.: \quad \tilde{\Pi}_{ij} = -\left(\overline{u_i \frac{\partial p'}{\partial x_j}} + \overline{u_j \frac{\partial p'}{\partial x_i}}\right) \qquad T.R.: \quad \overline{\Pi}_{ij} = -\left(\overline{u'_i \frac{\partial p'}{\partial x_j}} + \overline{u'_j \frac{\partial p'}{\partial x_i}}\right) \quad (31)$$

Now, it can be easily seen that, since $u'_i = u_i - \overline{u}_i$:

$$\overline{u'_i \frac{\partial p'}{\partial x_j}} = \overline{u_i \frac{\partial p'}{\partial x_j}} - \overline{u}_i \times \overline{\frac{\partial p'}{\partial x_j}} \equiv \overline{u_i \frac{\partial p'}{\partial x_j}} \qquad (32)$$

Thus, the *exact expressions of the pressure fluctuation correlation terms with binary or ternary regrouping are identical*, and will be denoted simply by Π_{ij}. Consequently, the expression of the closure schemes should be independent of the formulation which is adopted.

5.2 TRANSPOSING THE INCOMPRESSIBLE SCHEMES

For sake of shortness, we restrict the analysis to the rapid part only, using the Launder, Reece and Rodi (L.R.R.) [24] formulation as a reference for the incompressible flow, i.e. :

$$\phi_{ij}^{R0} \equiv \overline{p'^{r0}\left(\partial u_i'/\partial x_j + \partial u_j'/\partial x_i\right)} = \overline{A}_{ij}^0 + \overline{B}_{ij}^0 + \overline{C}_{ij}^0 \tag{33}$$

with

$$\overline{A}_{ij}^0 = a^0 \rho^0 \overline{u_m' u_m'}(\partial \overline{U}_i/\partial x_j + \partial \overline{U}_j/\partial x_i) \tag{34}$$

$$\overline{B}_{ij}^0 = b^0 \rho^0 \left(\overline{u_i' u_k'}\partial \overline{U}_j/\partial x_k + \overline{u_j' u_k'}\partial \overline{U}_i/\partial x_k - \frac{2}{3}\overline{u_m' u_n'}\partial \overline{U}_m/\partial x_n \delta_{ij}\right) \tag{35}$$

$$\overline{C}_{ij}^0 = c^0 \rho^0 \left(\overline{u_i' u_k'}\partial \overline{U}_k/\partial x_j + \overline{u_j' u_k'}\partial \overline{U}_k/\partial x_i - \frac{2}{3}\overline{u_m' u_n'}\partial \overline{U}_m/\partial x_n \delta_{ij}\right) \tag{36}$$

where a^0, b^0, c^0 are three algebraic constants.

In variable density flows, the *formal transposition* of the previous scheme is straightforward with both B.R. and T.R. formulations, that is :

$$\overline{\phi}_{ij}^R = \overline{A}_{ij} + \overline{B}_{ij} + \overline{C}_{ij} \quad (T.R.) \quad and \quad \tilde{\phi}_{ij}^R = \tilde{A}_{ij} + \tilde{B}_{ij} + \tilde{C}_{ij} \tag{37}$$

The T.R. contributions are deduced from the incompressible expressions by changing $\rho^0 \overline{u_\alpha' u_\beta'}$ into $\overline{\rho u_\alpha' u_\beta'}$. For the B.R. formulation, in addition to the transposition of the turbulent stresses $\rho^0 \overline{u_\alpha' u_\beta'} \ \longrightarrow \ \overline{\rho u_\alpha u_\beta}$, the mean mass-weighted velocity is of course substituted to the conventional one. This last method has been widely adopted, even for high speed flows, see Vandromme [33] for instance. Now, the question of the *physical justification* of such formal transpositions is not easy to answer. Let us just observe that the specific redistributive feature of the term to be modelled is lost with the binary formulation, since $\tilde{B}_{ii} = \tilde{C}_{ii} = 0$, but $\tilde{A}_{ii} \neq 0$, so that $\tilde{\phi}_{ii}^R \neq 0$, a result contrasting with the T.R. expression, where $\overline{\phi}_{ii}^R$ is still zero for M.S.T. variable density flows.

5.3 B.R. AND T.R. REDISTRIBUTIVE CONTRIBUTIONS IN A ROUND JET

The predictions of the model [5] can be used for a quantitative comparison of B.R. and T.R. closure transposition. Considering the Reynolds and Favrian turbulent stresses :

$$\overline{\rho u_i u_j} = \overline{\rho u_i' u_j'} - \overline{\rho u_i'} \times \overline{\rho u_j'}/\overline{\rho} \tag{38}$$

the form of the computed profiles for the light ($S = 1/50$) and heavy jet ($S = 50$) have been found markedly different. However, it is worthwhile noticing that, even for such very important density contrasts, no significant differences exist between the turbulent stresses according to binary ($\overline{\rho u_i u_j}$) and ternary ($\overline{\rho u_i' u_j'}$) formulations. Of course such a conclusion does not apply to cross-correlations including mass fraction fluctuation ($\overline{\rho \gamma u_i}$ and $\overline{\rho \gamma' u_i'}$ for instance).

Now, the general expression eq. (37T.R) can be applied to the normal radial component $\overline{\rho u_r' u_r'}$, to give for this type of flow :

$$\overline{\phi}_{rr}^R = a_0 \frac{\overline{\rho u_r' u_r'}}{2}\frac{\partial \overline{V}}{\partial r} + \frac{4a_2 - 2a_1}{3}\overline{\rho u' v'}\frac{\partial \overline{U}}{\partial r} + \overline{T}_{rr} \tag{39}$$

where $\overline{T}_{rr}$ denotes additional terms that are generally considered as negligible due to the thin shear layer approximations :

$$T_{rr} = \frac{4a_1 + 4a_2}{3}\overline{\rho v'^2}\frac{\partial \overline{V}}{\partial r} - \frac{2a_1 + 2a_2}{3}\overline{\rho u'^2}\frac{\partial \overline{U}}{\partial x} - \frac{2a_1 + 2a_2}{3}\overline{\rho w'^2}\frac{\overline{V}}{r} \qquad (40)$$

The three contributions $\overline{R}_{rr} = a_0 \frac{\overline{\rho u_i' u_i'}}{2} \partial \overline{V}/\partial r$, $\overline{S}_{rr} = \frac{4a_2 - 2a_1}{3}\overline{\rho u' v'}\,\partial\overline{U}/\partial r$ and $\overline{T}_{rr}$ are plotted in the next figure.

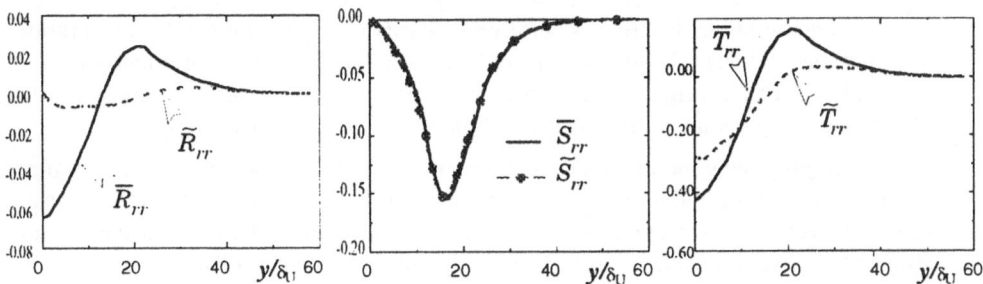

Figure 5   Modelled contributions of the rapid pressure-velocity correlation according to B.R. and T.R.

For such a light jet ($S = 1/50$), the most significant contribution results from the "second order" terms of the thin shear layer approximations (T_{rr}). If only the "main" shear contribution S_{rr} is considered, no significant differences are observed between B.R. and T.R. predictions. Finally, the first contributions are markedly different, and it can be verified that the mass-weighted expression is not divergence free (Figure 6). In other words, non redistributive effects, depending on the density ratio, are introduced when transposing divergence free incompressible schemes with the B.R formulation. This is not the case with the T.R. one, so that the models of the pressure-velocity correlation are not identical, as opposed to the exact terms.

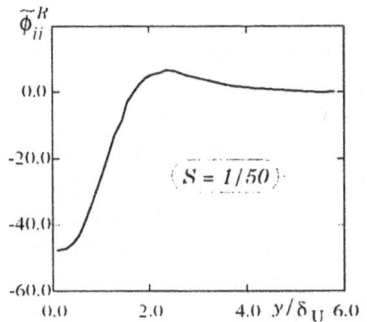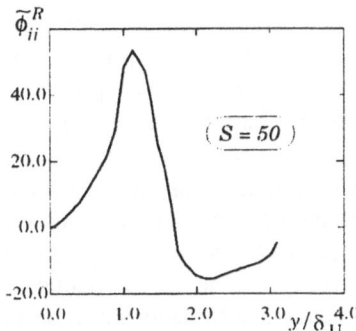

Figure 6   Non redistributive action of the B.R. transposed scheme.

6 Conclusion

1) In low speed turbulent motion of variable density fluids, a specific class of flows can

be introduced when molecular diffusion effects are negligeable. The *conventional* mean velocity field is then *divergence free*, independently of the origin of the density variation.

2) Any d.f.c. term $(\overline{\rho f'})$ is *exactly linked* with a density weighted second order moment of centered fluctuations $(\overline{\rho f' \theta'}$ or $\overline{\rho f' \gamma'})$. This conclusion, based on the equation of state, is associated with a thermodynamical interpretation of the extra correlation terms resulting from the density fluctuation.

Such an algebraic linkage gives qualitative explanation of density effects and support the idea of quantitative differences between (i) variable temperature and concentration fluid motions and (ii) co and counter gradients situations. In free turbulent jets, and independently of the origin of the density variation, the d.f.c. variation range can be split into three parts according to the turbulent density intensity level.

3) When averaging the general balance equations, the d.f.c. can be considered as "mute variables" which are incorporated in both mean an higher orders moments of mass-weighted averages. This gives rise to a *binary regrouping* of the extra correlation terms which are induced by the density variation. Such a mean picture results from a mechanical interpretation which refers to a "macroscopic" mass conservative evolution were the d.f.c. effects are now part of a new macroscopic advection. Alternatively, with the thermodynamical interpretation, all d.f.c. terms can be taken explicitly, referring to a mean open system evolution. This yields to the *ternary formulation* of the mean motion equations.

In M.S.T. flows, a *mean transport effect* of d.f.c. can be pointed out, which is associated with a macroscopic volume variation of a constant mass of moving fluid (B.R.) or a turbulent mass transfer during a mean isovolume evolution of a fluid particle (T.R.). Both interpretations are only equivalent for small density contrasts.

Acting through transportable quantities, the d.f.c. are the source of an additional *equations coupling*, as compared with the constant density flow, which can be analyzed in details when the T.R. formulation is used.

4) When dealing with second order closure of the B.R. equations, all d.f.c. are hidden, except in the mean pressure terms of the transport equation of a Favrian second order moment. Consequently, it could seem at least questionable to use closure schemes based on a functional dependence of mass-weighted arguments exclusively $(\tilde{U}_i, \ \widetilde{u_i u_j}, \ \tilde{F}, \ \widetilde{f u_i} \ ...)$. Such a peculiarity is not present with the T.R. formulation where all d.f.c. can be considered as independent arguments of the closure procedure. When incorporating the density correlation with the macroscopic mean value (B.R.), d.f.c. terms are actually acting as "forces" driving the "fluxes" in any gradient type closure scheme. In particular, such a conclusion applies to any scheme including kinematic arguments since, despite the formal analogy with the incompressible situation, the kinematic description of the mean flow field is changed when the density varies. As an example, the redistributive contribution of the pressure fluctuation correlation is not achieved when transposing the corresponding incompressible scheme.

Bibliography

[1] Batchelor G.K . *An introduction to Fluid Dynamics.* Cambridge University Press, reprinted edition, 1970.

[2] Bauer P.T. Zumwalt G.W & Fila L.J . A numerical method and an extension of the Korst jet mixing theory for multispecie turbulent jet mixing. In *AIAA Paper*, volume 68(112), New-Jork, 1968. 6th Aerospace Sciences Meeting.

[3] Bayly B.J. Levermore C.D. & Passot T . Density variations in weakly compressible flows. *Phys. Fluids*, A-4(5):945–954, 1992.

[4] Chassaing P . Une alternative à la formulation des équations du mouvement turbulent d'un fluide à masse volumique variable. *J. Méc. Théo. et Appl.*, 4:375–389, 1985.

[5] Chassaing P. Harran G. & Joly L . Density fluctuation correlations in free turbulent binary mixing. *J. Fluid Mech.*, 279:239–278, 1994.

[6] Chevray R. & Tutu N.K . Intermittency and preferential transport of heat in a round jet. *J. Fluid Mech.*, 88:133, 1978.

[7] Corrsin S. & Uberoi M.S . Further experiments on the flow and heat transfer in a heated turbulent air jet. Technical Report TN-1865, NACA, 1949.

[8] Debiève J.F . Problèmes de distorsion rapide en écoulements compressibles. In *Ecoulements turbulents compressibles.*, Poitiers-Marseille, 1986. ONERA/DRET.

[9] Driscoll J.F. Scheffer R.W. & Dibble R.W . Mass fluxes $\overline{\rho'u'}$ and $\overline{\rho'v'}$ measured in a turbulent nonpremixed flame. In *Ninth Symp. on Combustion*, pages 477–485, The Combustion Institute, 1982.

[10] Durran D.R . Improving the anelastic approximation. *J. Atmos. Sci.*, 46:1453–1461, 1989.

[11] Dussauge J.P. Debiève J.F. & Smits A.J . Rapidly distorted compressible boundary layers. *AGARDOGRAPH*, AG315:2–1, 2–11, 1988.

[12] Favre A . Equations statistiques des gaz turbulents. *C.R. Acad. Sci. Paris*, 246:2573–3216, 1958.

[13] Favre A . Equations des gaz turbulents compressibles. I Formes générales. *J. Méc.*, 4:361–390, 1965.

[14] Favre A . Equations des gaz turbulents compressibles. II Méthode des vitesse moyennes ; Méthode des vitesses macroscopiques pondérées par la masse volumique. *J. Méc.*, 4:390–421, 1965.

[15] Favre A . Equations statistiques aux fluctuations d'entropie, de concentration, de rotationnel dans les écoulements compressibles. *C.R. Acad. Sci. Paris*, 273:1289–1294, 1971.

[16] Favre A . Equations statistiques des fluides turbulents compressibles. In *Fifth Cong. Can. de Méc. Appl.*, pages G3–G34. New Brunswick Univ., 1975.

[17] Favre A . Equations statistiques des fluides à masse volumique variable en écoulements turbulents. In *Secondes Journées d'Etudes Ecoulements Turbulents à Masse Volumique Variable*. Orléans, 1992.

[18] Fulachier L. Borghi R. Anselmet F. & Paranthoen P . Influence of density variations on the structure of low-speed turbulent flows : a report on Euromech 237. *J. Fluid Mech.*, 203:577–593, 1989.

[19] Goldstein M.E . Unsteady vortical and entropic distorsions of potential flows round arbitrary obstacles. *J. Fluid Mech.*, 89:443–468, 1978.

[20] Ha Minh H. Launder B.E. & Mac Innes H . A new approach to the analysis of turbulent mixing in variable density flows. In *3rd Symp. on Turbulent Shear Flows*, pages 19.619–19.625, Davis, 1981.

[21] Harion Jean-Luc . *Influence de différences de densité importante sur les propriétés de transfert d'une couche limite turbulente.* Thèse, I.N.P., Grenoble, 1994.

[22] Joly L . *Ecoulements cisaillés libres à masse volumique variable : Analyse physique et modélisation.* Thèse n° 944, Inst. Nat. Polytech. Toulouse, E.N.S.I.C.A., 1994.

[23] Kovasnay L.S.G . Turbulence in supersonic flow. *J. Aero. Sci*, 20:657–674, 1953.

[24] Launder B.E. Reece G.J. and RODI W. Progress in the development of a Reynolds stress turbulence closure. *J. Fluid Mech*, 68(3):537–566, 1975.

[25] Lele S.K . Notes on the effects of compressibility on turbulence. *CTR Manuscript*, 145, 1993. Stanford University.

[26] Lilly K . A comparison of incompressible, anelastic, and boussinesq dynamics. *Private Com. to be published in "Atmospheric Research".*, 1995.

[27] Ogura Y. & N.A. Phillips . Scale analysis of deep and shallow convection in the atmosphere. *J. Atmos. Sci.*, 19:173–179, 1962.

[28] Rey C . Ecoulements turbulents compressibles et variables aléatoires centrées. In *Dixième Cong. Franc. de Méc.*, Paris, 1991.

[29] Rey C. & Rosant J.M . Influence of density variations on small turbulent structures of temperature in strongly heated flows. In *Ninth Int. Heat Trans. Conf.*, pages 405–409, Israel, 1990.

[30] Sautet J.C . *Effets des différences de densité sur le développement scalaire et dynamique des jets turbulents.* PhD thesis, Univ. de Rouen, 1992.

[31] Shih T.H. Lumley J.L & Janicka J . Second-order modelling of a variable-density mixing layer. *J. Fluid Mech.*, 180:93–116, 1987.

[32] So R.M.C. Zhu J.Y Otügen M.V. & Hwang B.C . Some measurements in a binary gas jet. *Exps. Fluids*, 9:237–284, 1990.

[33] Vandromme D . Turbulence modeling for compressible flows and implementation in navier-stokes solvers. In *Introduction to the modeling of turbulence.*, pages 1–65, Rhode-Saint-Genese, Belgium, 1991. V.K.I. Lect. Series 1991–02.

[34] Zank G.P. & W.H. Matthaeus . The equations of nearly incompressible fluids. I Hydrodynamics, Turbulence and Waves. *Phys. Fluids*, A-3(1):69–82, 1991.

Favre's intervention

This general lecture and several papers discuss about the choice of averaging method for the formulation of the statistical equations, which express the mean balances for transportable quantities defined per unit volume such as : the mass ρ, the momentum components ρu_k, the internal energy, the entropy ρs. I call again the attention to the case of the entropy and to the expression of the second principle of thermodynamics.

For the separation of a transportable quantity per unit volume ρw into a non-flutuating part $\overline{\rho}\dot{W}$ and a fluctuation $\rho\dot{w}$, I started, (1), (2), with a general form of $\dot{W}$, invariant in averages : $\dot{W}=\overline{\dot{W}}$. This keeps the possibility to make later the choice of the averaging method, and leads for the velocity components to $u_k=\dot{U}_k+\dot{u}_k$ with $\dot{U}_k=\overline{\dot{U}}_k$, for the entropy to $s=\dot{S}+\dot{s}$ with $\dot{S}=\overline{\dot{S}}$.

The rate of variation of the average entropy in a volume bounded by a closed surface $\dot{C}$ moving at the velocity $\dot{U}_k$ then reads :

$$\frac{D}{Dt}\int_{\tau}(\overline{\rho}\dot{S}+\overline{\rho\dot{s}})d\dot{\tau}-\int_{\dot{C}}(\overline{\frac{\lambda}{\theta}\frac{\partial\theta}{\partial x_k}}-\overline{\rho\dot{u}_k}\dot{S}-\overline{\rho\dot{u}_k\dot{s}})\dot{\lambda}_k d\dot{C}=\int_{\tau}[(\overline{\frac{\varphi}{\theta}})+\overline{\frac{\lambda}{\theta^2}\frac{\partial\theta}{\partial x_k}\frac{\partial\theta}{\partial x_k}}]d\dot{\tau}\geq 0$$

    (I)   (II)   (III)

where the last term (III) is not negative. Thus, the rate (I) of variation of the mean entropy $\overline{\rho s}=\overline{\rho}\,\dot{S}+\overline{\rho\dot{s}}$, taking into account the variations (II) due to the heat conduction and to the turbulent diffusions is not negative.

But we have not any physical principle concerning the rate of variation of the average fluctuation of entropy $\overline{\rho\dot{s}}$ in that volume. So, in the general form, the non-fluctuating part $\overline{\rho}\,\dot{S}$ of entropy has not the physical meaning of an entropy consistent with the second principle of thermodynamics.

It is only when the average fluctuating part of entropy vanishes, by definition $\overline{\rho\dot{s}}=0$, that the non-fluctuating part $\overline{\rho}\,\dot{S}$ has the physical meaning of an entropy, and this is the case for the mass-weighted average method that I have chosen. At the same time, for the velocity, the average mass into the volume $\dot{\tau}$ is constant if the mass is conservative, and the stream surfaces tangent to the mass-weighted average velocity are tight in average.

As for comparison between the results of the measurements and of the computations, it is obvious that it does not concern the validity of the averaging methods which are exact. The observed discrepancies are related to the accuracy of the measurements and computation procedures, and mainly to the validity of the assumptions made for the formulations of the models.

For the applications to the equations obtained by the mass-weighted-average method, it should be convenient to express the models in terms of the transportable quantities by unit volume, mass, momentum, internal energy, entropy.

(1) FAVRE A., 1964, *Statistical equations of turbulent gases*, In : SSSR Acad. Sc. NAUKA. Phys. and Math., Moscow, pp. 483-511; French-English trans. (1969) In : Problems of hydrodynamics and continuum mechanics, S.I.A.M. Philadelphia, pp. 231-266.

(2) FAVRE A., 1976, *Equations fondamentales des fluides à masse volumique variable en écoulements turbulents*, In : La turbulence en mécanique des fluides, Eds. Favre A., Kovasznay L.S.G., Dumas R., Gaviglio J., Coantic M., Gauthiers-Villars, pp. 24-75.

Chassaing's answer

The question raised by Professor Favre is a fundamental one. Beyond the particular and very important case of the entropy balance equation, the question which is addressed is that of giving a physical meaning to averaged balance equations in terms of balance equations of mean quantities, keeping the same (continuum) scale of analysis in both cases. Let me detail this point taking the entropy equation as an example. The corresponding instantaneous local equation can be written as :

$$\frac{\partial}{\partial t}(\rho S) + \frac{\partial}{\partial x_j}(\rho S U_j) + \frac{\partial}{\partial x_j}\left(\frac{q_j}{T}\right) = \frac{1}{T}\left(\tau_{ij}\frac{\partial U_i}{\partial x_j} - \frac{q_j}{T}\frac{\partial T}{\partial x_j}\right) \geq 0 \qquad (1)$$

Averaging this equation, the inequality condition of the second principle of thermodynamics obviously applies to the "averaged" equation, that is :

$$\frac{\partial}{\partial t}(\overline{\rho S}) + \frac{\partial}{\partial x_j}(\overline{\rho S U_j}) + \frac{\partial}{\partial x_j}\left(\overline{\frac{q_j}{T}}\right) = \overline{\left[\frac{1}{T}\left(\tau_{ij}\frac{\partial U_i}{\partial x_j} - \frac{q_j}{T}\frac{\partial T}{\partial x_j}\right)\right]} \geq 0 \qquad (2)$$

Now, if a *physical interpretation* in terms of "mean quantities" is to be given to eq.(2), I agree with Professor Favre's conclusion that the mass-weighted averaging is the only valuable choice. However, eq.(2) can still be *used* if the alternative ternary decomposition I proposed [1], is applied, provided one does not separate the non fluctuating part of ρS from the average of the fluctuations.

As a counterpart of the lack of clear physical interpretation in terms of "mean values" of the resulting entropy equation in this case, one obtains some interesting pieces of information which are not directly available when applying the mass-weighted procedure. It was part of my invited paper to detail some of them. Let me give here just one example. When using the mass weighted formulation in low speed motion, the gravity terms associated with the buoyancy fluctuations ($\rho' g_i$) are no longer present in the stress transport equations. However buoyancy effects are necessarily included in the equation, and it can be seen that they are "incorporated" in a "mean pressure" contribution implying the averaged mass weighted velocity fluctuation, e.g. $\overline{u_i}\partial\overline{P}/\partial x_j$. However this linkage with the buoyancy is not straightforward and in supersonic flows for instance, the mean pressure gradient term can accaount for other physical effects [2].

This may lead one to question which formulation would be the most appropriate to analyze density effects *from a modelling point of view*. For low velocity mixing situations at least, some evidence [3] was given to adopt the correlation with the density fluctuation ($\overline{\rho f'}$) as an independent argument (not to be grouped with the mean value $\overline{\rho}\overline{F}$).

However, as far as the numerical procedure is concerned, solving the mass weighted formulation of the equations is more efficient.

In conclusion, the basis on which one should choose between alternative averaging strategies for a variable density flow is not entirely clear to the author. In his opinion, it depends on the goal one is focusing on. His philosophy, as a modeller, is that one should rather try to make the best use of all possibilities aiming to achieve complementarity rather than contradiction. A possible path would be to use ternary regrouping to separate mean and fluctuating contributions to density effects, then develop new closure schemes that could finally be transformed to a more tractable mass-weighted formulation.

[1] P. CHASSAING 1985 Une alternative à la formulation des équations du mouvement turbulent d'un fluide à masse volumique variable. *Jl. Méc. Théo. Appl.* vol.4, pp:375–389.

[2] A. FAVRE 1976 Équations fondamentales des fluides à masse volumique variable en écoulement turbulent. *La Turbulence en Mécanique des Fluides* Ed. Gauthier-Villars

[3] P. CHASSAING, G. HARRAN & L. JOLY 1994 Density fluctuation correlations in free turbulent binary mixing. *Jl. Fluid Mech.* vol.279, pp:239–278.

TURBULENT KINETIC ENERGY IN VARIABLE DENSITY JETS AND THE MODELLING ISSUE

L. JOLY AND P. CHASSAING
Ecole Nat. Sup. d'Ing. de Constructions Aéronautiques
1, Place Emile Blouin, 31056 Toulouse Cedex.

1. Introduction

The aim of the present paper is to gain insight in the way density variations modify the turbulent-kinetic-energy field in shear flows such as jets and mixing layers. Some consequences on modelling are then examined.

Reynolds centered average being used, variable density extra-terms are isolated in the turbulent-kinetic-energy equation. Retaining leading order terms only, the cogradient and countergradient cases are shown to result in opposite density effects. Qualitative conclusions are outlined, concerning the non-homogeneous initial zone in low-speed variable-density jets.

A second part gives some guidelines for model refinements. The streamwise displacement of the on-axis turbulent-kinetic-energy maximum as a function of the jet density ratio is shown to be very sensitive to the value of one modelling constant. It is concluded that the adjustment of density effects in the dissipation-rate equation is the next important target.

2. Variable density effects on turbulent kinetic energy

2.1. THE SIMPLE SKETCH OF THE MEAN MOMENTUM MECHANISMS

In a former communication (Joly & Chassaing, 1994) we had stressed the four part structure of the simplified mean momentum equation:

$$\underbrace{\bar{\rho}\bar{U}\frac{\partial \bar{U}}{\partial x} + \bar{\rho}\bar{V}\frac{\partial \bar{U}}{\partial y}}_{A} + \underbrace{\bar{\rho}\frac{\partial \overline{u'v'}}{\partial y}}_{D} + \underbrace{\overline{\rho v'}\frac{\partial \bar{U}}{\partial y} + \overline{u'v'}\frac{\partial \bar{\rho}}{\partial y}}_{R} = \underbrace{(\bar{\rho} - \rho_o)\,g^{\pm}}_{B} \qquad (1)$$

A new streamwise partition of the jet was proposed which defined a close-to-exit region as the one concerned by density effects. Cogradient and

L. Fulachier et al. (eds.), IUTAM Symposium on Variable Density Low-Speed Turbulent Flows, 85–88.
© 1997 *Kluwer Academic Publishers.*

countergradient situations were seen to result in opposite inertia effects : *light jets undergo a more intense deceleration than heavier ones.* We propose to focus on the turbulent kinetic energy (TKE) with a similar approach.

2.2. THE STRUCTURE OF THE TKE EQUATION

Under the assumptions of two-dimensionnal, high turbulent Reynolds number, low turbulent Mach number and thin shear flow, the TKE transport equation reduces to the first order balance (Joly, 1994):

$$\underbrace{\bar{\rho}\bar{U}\frac{\partial \bar{k}}{\partial x} + \bar{\rho}\bar{V}\frac{\partial \bar{k}}{\partial y}}_{A} + \underbrace{\bar{\rho}\frac{\partial \overline{kv'}}{\partial y}}_{D} + \underbrace{\overline{\rho v'}\frac{\partial \bar{k}}{\partial y} + \overline{kv'}\frac{\partial \bar{\rho}}{\partial y}}_{R_1} = \underbrace{-\bar{\rho}\overline{u'v'}\frac{\partial \bar{U}}{\partial y} - \bar{\rho}\varepsilon}_{S} - \underbrace{\frac{\overline{\rho'v'}}{\bar{\rho}}\frac{\partial \bar{\rho}\overline{v'^2}}{\partial y}}_{R_2}$$

$$(2)$$

with $\bar{k} = \frac{1}{2}\overline{u_i' u_i'}$. When compared to its constant density equivalent, the transport equation for the TKE is supplemented with $-(R_1 + R_2)/\bar{\rho}$ on the right-hand side.

2.3. COGRADIENT VERSUS COUNTERGRADIENT JETS

Let us divide a half jet into two regions: region **A**, between the jet symetry axis and the line of local maximum of TKE, and region **B**, from this line to the outer edge. Besides, the more probable sign of a correlation $\overline{\phi'v'}$ is the opposite of that of the mean gradient $\partial\bar{\Phi}/\partial y$. The sign of the global density effects $-(R_1 + R_2)$ is given in table 1.

TABLE 1. The sign of variable density terms

Jet type	region **A**	region **B**
Light or countergradient	+	−
Heavy or cogradient	−	+

This also leads to a merely natural statement: *The kinetic energy (per volume unit) is turned into velocity fluctuations to the lighter fluid benefit.*

Lowering the exit density ratio $S_\rho = \rho_{jet}/\rho_{atm}$ should result in a displacement of the maxima of TKE toward the jet axis. Numerical simulations, based on second-order closure schemes similar to those detailed in Chassaing, Harran & Joly (1994), confirm that, as S_ρ is lowered, the line of the TKE maxima is inflected toward the jet axis, see figure 1.

Therefore, density effects on TKE on-axis evolutions appear as direct consequences of inertia effects taking place off-axis in the non-homogeneous region close to the jet exit, that is to say in the merging mixing-layers.

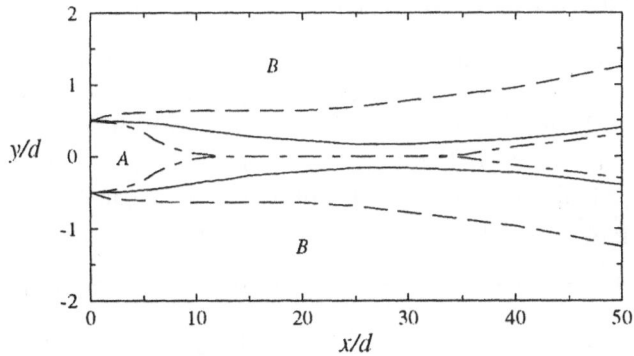

Figure 1.   Locus of the maximum of the turbulent kinetic energy in variable-density round jets: $(-\cdot-) S_\rho = 1/3$, $(—) S_\rho = 1$, $(- -) S_\rho = 3$.

3.   What are the consequences on closure schemes testing ?

3.1. CLOSURE SCHEMES FOR VARIABLE DENSITY EXTRA-TERMS

3.1.1.   *The quasi-normal approximation*
The modelling of the diffusion terms involved in the TKE equation (or in RSM) is discussed in (Joly, 1994). One key point is the way mass-weighted third-order correlations are treated. The decomposition: $\overline{\rho u'^2 v'} = \bar{\rho}\overline{u'^2 v'} + \overline{\rho' u'^2 v'}$ raises the question of modelling the fourth-order correlation. Though not experimetally supported, the widespread response appeals to the quasi-normal approximation : $\overline{\rho' u'^2 v'} = 2\overline{\rho' u'}\,\overline{u'v'} + \overline{\rho' v'}\,\overline{u'^2}$. This approximation leads to a correct restitution of density effects on the lenght of the jet potential core.

3.1.2.   *The extra source-term in the dissipation equation*
Another crucial point to variable-density closures is the way density effects are included in the dissipation-rate equation. Mac Innes (1985), reviewed by Ruffin (1994) proposed a systematic development of possible extra-terms. In constrast with such an approach yielding numerous terms, each preceded by a constant to fit, traditionnal source-term balance arguments are usually invoked to build a dissipation-rate equation in the rather simple form:

$$\bar{\rho}\bar{U}\frac{\partial \varepsilon}{\partial x}+\bar{\rho}\bar{V}\frac{\partial \varepsilon}{\partial y}+\bar{\rho}\frac{\partial \overline{\varepsilon v'}}{\partial y}+\underbrace{\overline{\rho v'}\frac{\partial \varepsilon}{\partial y}+\overline{\varepsilon v'}\frac{\partial \bar{\rho}}{\partial y}}_{R_{\varepsilon 1}}=-\frac{\bar{\rho}\varepsilon}{\tilde{k}}(c_{\varepsilon 1}\overline{u'v'}\frac{\partial \bar{U}}{\partial y}+c_{\varepsilon 2}\varepsilon+c_{\varepsilon 3}\underbrace{\frac{\overline{\rho'v'}}{\bar{\rho}^2}\frac{\partial \bar{\rho}\overline{v'^2}}{\partial y}}_{R_{\varepsilon 2}})$$

(3)

Density effetcs are of the same nature as in (2) and their intensity depends partly on the value of the constant $c_{\varepsilon 3}$, a point discussed in the next paragraph.

3.2. A PRACTICAL POINT OF VIEW

Since density effects were seen to be responsible of the spatial distribution
of TKE, the modelling issue is to get the right value at the right place. The
TKE on-axis evolution is then regarded as the result of the density-sensitive
convergence of TKE toward the jet axis. For a light pipe jet, $S_\rho = 0.14$ (the
helium-air ratio), figure 2 gives the influence of $c_{\varepsilon3}$ on the position of the
TKE on-axis peak and the inflection of the line of TKE maxima.

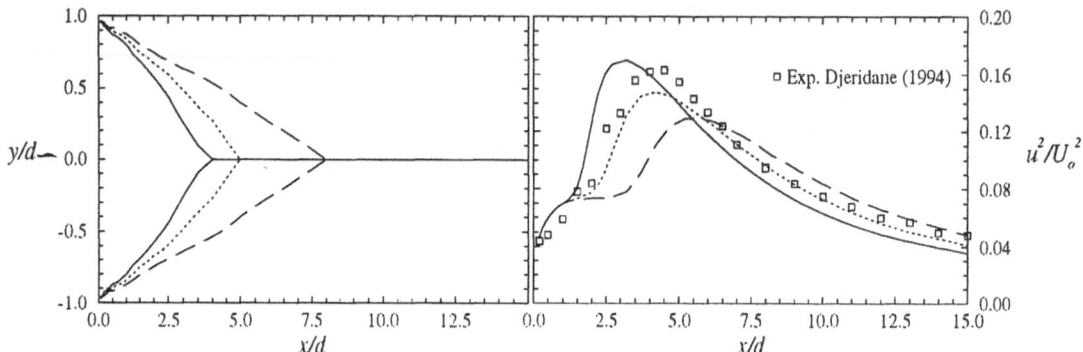

Figure 2.   Influence of $c_{\varepsilon3}$ on TKE: (left) Radi of maximum TKE, (right) on-axis
evolution of the fluctuation $\overline{u'^2}/U_o^2$. (——) $c_{\varepsilon3} = 0.$, ($\cdots$) $c_{\varepsilon3} = 1.$, (— —) $c_{\varepsilon3} = 2.$

Increasing $c_{\varepsilon3}$ damps the density effetcs on the TKE, decreasing and
shifting downstream the TKE on-axis peak. For this particular density ratio
and experimental data (Djeridane 1994) a reasonable fitting of $c_{\varepsilon3}$ could be
a little less than one. All fictive origins in self-preserving laws for mean
quantities are shown to be affected in the same way while slopes are not.
Adjusting density effects is thus important to the streamwise position of
the whole statistical field. However, the close-to-exit region being under the
influence of both initial conditions and density effetcs, it is concluded that
pipe-jets are not likely to be the proper frame to calibrate the extra-terms
of the dissipation equation.

References

Chassaing, P., Harran, G. & Joly, L. (1994) Density fluctuation correlations in free
    turbulent binary mixing, *J. Fluid Mech.* **vol. 279**, pp. 239–278.
Djeridane, T. (1994) *Contribution à l'éude expérimentale de jets turbulents
    axisymétriques à densité variable*, Thèse, Université d'Aix-Marseille II, Ref. **207-
    94-57**
Joly, L. & Chassaing, P. (1994) Analyse de l'écoulement cisaillé mince soumis à des
    gradients de masse volumique, *Conf. on Low-speed Variable-density Turbulent Flows,
    LEGI Grenoble, 15-16 Septembre.*
Joly, L. (1994) *Ecoulements turbulents cisaillés libres à masse volumique variable : analyse
    physique et modélisation*, Thèse INPT n° **944**.

SECOND-ORDER TURBULENCE MODELLING AND NUMERICAL SIMULATION OF VOLUME VARIABLE TURBULENT FLOWS

THIERRY AURIER[†], CLAUDE REY** and JEAN-FRANCOIS SINI*
[†]* LMF, URA 1217 - Ecole Centrale de Nantes
1, rue de la Noë - 44072 Nantes cedex 03
[†] present address : Ecole Polytechnique de Montréal, Départ. Génie Mécanique
CP 6079, Succ A, H3C 3A7, Montréal, Qc, Canada
** IRPHE, UMR 138
12, Avenue Général Leclerc - 13003 Marseille

1. INTRODUCTION

The analysis and the numerical simulation of turbulent volume variable flows require a mathematical description of physical variables in a statistical model. A closed set of statistical equations describing correctly the evolution of this specific flows is needed for the resolution. The variability of the mass volume increases significantly the complexity of this set of equations, enhancing a strong connection between all of them. The writing of balance laws per unit of volume for the volume variable turbulent flows using Reynolds averaging leads to an open set of equation. These equations, while exact, are unclosed; they contain correlation that are not exactly determinable, with physical signification and modelling steel complicated. Favre (1965) used density-weighted averages to reduce the mathematical complexity of this set. However, conditioned averages restrict the physical analysis and then the modelling framework. At the same time, other authors have developped other ways of investigation using hybrid decompositions like Chassaing (1985) or HaMinh and al. (1981). This work takes again the Reynolds averaging's trail and postulates that the compressible situation can be described like a deviation of the incompressible situation. Then, we can identify the different physical mechanisms and separate the incompressible terms. Consequently, we identify the "deviatoric terms" where the variation and fluctuation of volume appear. The Boussinesq assumptions are not adopted.

2. THEORETICAL APPROACH

The first step of our analysis is writing the equations of conservation for a material mass which naturally introduces Reynolds averaging (Rey, 1992). This leads to a mathematical set of equations that are well shaped to the analysis of turbulent variable density flows. Using Reynolds averaging, we identify the different physical phenomena interacting in the turbulent flows, thus enriching the analysis and the physical basis of the models. We derive a mathematical structure of the balance laws, and identify the role of the mass volume

L. Fulachier et al. (eds.), IUTAM Symposium on Variable Density Low-Speed Turbulent Flows, 89-92.

fluctuation. Per unit of mass, the general bugdet equation for a variable Ψ_i can be written as follow :

$$\Psi_{i,t} + U_k \Psi_{i,k} = A_i + \mathcal{U}_{ik,k} \tag{1}$$

This analysis per unit of mass introduces the mass volume $\mathcal{U}$ define for a perfect gas using the Reynolds averaging like :

$$\mathcal{U} = \overline{\mathcal{U}} + \upsilon \text{ with } \overline{\mathcal{U}} \approx \frac{R\overline{T}}{\overline{P}} \tag{2}$$

By decomposing volume effects into mean values and fluctuations, we identify the incompressible mechanisms and terms where fluctuation υ of volume appears. So we can write the classical budget equation (1) as follows, with $\Psi_i = \overline{\Psi}_i + \varphi_i$:

$$\underbrace{\Psi_{i,t} + \overline{U}_k \Psi_{i,k} + u_k \overline{\Psi}_{i,k} + \left(\varphi_i u_k\right)_{,k} - A_i - \overline{\mathcal{U}} J_{ik,k}}_{\text{Identical formally as incompressible flows}} = \upsilon \overline{J}_{ik,k} + \underbrace{\upsilon j_{ik,k} + \varphi_i u_{k,k}}_{\left[\Pi_{\Psi_i}\right]} \tag{3}$$

$$\text{Physicals mechanisms with volume fluctuation}$$

We proceed to establish a hierarchy of terms where volume fluctuations or deviatoric terms appear. Using Reynolds averaging, dominant deviatoric terms include the coupling between volume fluctuation and the macroscopic field of surfacic interaction $\overline{J}_{ik}$ while secondary deviatoric terms $\left[\Pi_{\Psi_i}\right]$ represent complex physical phenomena which are difficult to model and in consequence, often neglected. However, the use of a double observation Reynolds-Favre allows us to mathematically construct an "equivalent model" for secondary deviatoric terms that can be written as a function of main variables of the problem (Aurier and al., 1994). Subsequently, using the structure of (3), we propose an exactly closed set of equations with a new second-order turbulence closure keeping all the physical mechanisms with variation and fluctuation of volume. We have shown that the complementarity of the two statistical Reynolds/Favre processes can be used to model the secondary deviatoric terms. Schemes discussed in this article are single point closures and using the Reynolds averaging, we keep the possibility to write turbulent quantities in two point statistics. In (3), without any fluctuation of volume, we find an incompressible structure.

3. SECOND-ORDER TURBULENCE MODEL

A set of equations for the second-order turbulence model is generated from equation (3). As an example, the equation for heat fluxes is presented :

$$\frac{D}{Dt} \overline{u_i \theta} = \underbrace{P_{i\theta} - \frac{\partial}{\partial x_k} \overline{u_i u_k \theta} + \Phi_{i\theta}}_{Iv} - \frac{\overline{\theta^2}}{\overline{T}} g_i + \text{Sup}_{i\theta} \tag{4}$$

The term (Iv) in (4) defines respectively, the shear production, the diffusive transport and the pressure-strain correlation. The mathematical structure of (Iv) in (4) is identical to isovolume flows. Ramirez-Leon (1991) shown the influence of volume fluctuation on the modelling of triple-moment correlations and we used the approach of MacInnes (1985) to model the pressure-strain correlation. The two last terms of (4) are the buoyancy production and the deviatoric terms (dominant and secondary).

The deviatoric term can be written as follows :

$$\text{Sup}_{i\theta} = \underbrace{\frac{\overline{u_i\theta}}{\overline{T}}\left[\frac{D\overline{T}}{Dt} + \frac{\partial\overline{u_k\theta}}{\partial x_k}\right] + \frac{\overline{\theta^2}}{\overline{T}}\left[\frac{D\overline{U_i}}{Dt} + \frac{\partial\overline{u_iu_k}}{\partial x_k}\right]}_{(a)} \underbrace{- \frac{\overline{\theta^2}}{\overline{T}^2}\Phi_{i\theta} + \frac{\overline{u_i\varepsilon_\theta}}{\overline{T}}}_{(b)} \underbrace{- \mathcal{U}\overline{u_iu_k\theta}\frac{\partial}{\partial x_k}\frac{1}{\mathcal{U}}}_{(c)=\left[\Pi_{i\theta}\right]} \quad (5)$$

In the expression (5), we establish a hierarchy of the deviatoric terms in three parts. (5a) and (5b) are the dominant deviatoric terms and they represent the coupling between volume fluctuation and the macroscopic field of surfacic interaction. They are exact and don't need more modelling. The term (5c) models the secondary deviatoric terms as first introduced by Aurier and al. (1994). The diffusion terms in (5b) are modelled using the simple gradient diffusion hypothesis (Handjalic and al., 1972).

The transport equations are elliptic in space and parabolic in time and they are solved by a finite difference approximation involving three shifted rectangular grids. Centered differences for diffusion terms and upwind-weighted scheme for advection terms are considered. Non zero velocity divergence equation is directly satisfied using an extended version of artificial compressibility implicit method.

4. RESULTS

The mathematical model is used to simulate turbulent vertical plane buoyant jets and plumes vertically discharged into uniform stagnant atmosphere. The "classical" Boussinesq assumptions are not adopted.

4.1 SIMULATION CONDITIONS

The flow is governed by the relative importance of the buoyancy and the momentum fluxes at the source. Then, the definition of densimetric FROUDE number, REYNOLDS number, MACH number are respectively :

$$F_o = \frac{W_o^2}{gD_o\left[\frac{\mathcal{U}_o}{\mathcal{U}_n}-1\right]}, \quad Re_o = \frac{W_oD_o}{\mu_o\mathcal{U}_o}, \quad M_o^2 = \frac{W_o^2}{\gamma RT_o} \quad \text{where the exit width is } D_o=1m$$

The exit conditions, identified by the subcript o, are :

W_o (m/s)	ΔT_o (°K)	F_o	Re_o	M_o	$\mathcal{U}_n/\mathcal{U}_o$
5	160	4.8	$2.7\ 10^5$	0.011	0.654

4.2 INFLUENCE OF DEVIATORIC TERMS

The transversal profiles of vertical and horizontal heat fluxes are shown on Figures 1 and 2 respectively. The solid lines correspond to a situation without deviatoric terms and dashed lines indicate a situation where the deviatoric terms, that express the volumetric expansion effects, are activated. These results show that the deviatoric terms (dominant and secondary) act to spectaculary increase both the transversal and longitudinal turbulent heat fluxes (Figs 1 and 2).

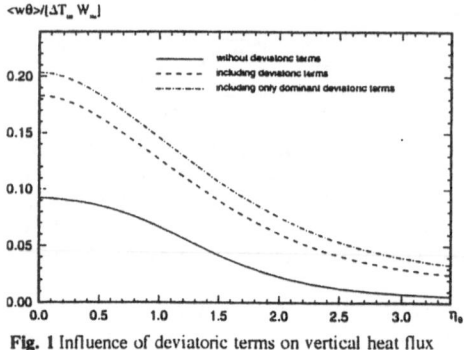 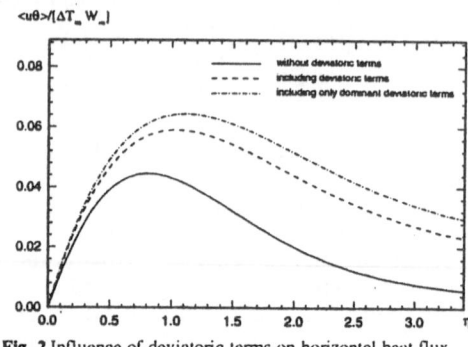

Fig. 1 Influence of deviatoric terms on vertical heat flux   **Fig. 2** Influence of deviatoric terms on horizontal heat flux

5. CONCLUDING REMARKS

A second-order turbulence model has been developed for analyzing the volume variable turbulent flows. The first results show qualitatively the influence of the deviatoric terms on the horizontal and vertical heat fluxes and more generally on the dynamical and thermal turbulent fields and on the characteristic scales (Aurier, 1995). Further work on the application of this model is actually in progress. The confrontation between numerical and experimental data (Ruffin and al., 1994) and this numerical results in plane and axisymmetric jets with the same conditions, is necessary to completely validate this new approach.

REFERENCES

Aurier T. (1995) "Modélisation au second ordre et simulation numérique des écoulements turbulents à volume variable", Thèse de Doctorat, Ecole Centrale de Nantes.
Aurier T., Rey C. (1994) "Modelling and simulation of compressible turbulent flows", Conf. on low speed variable density turbulent flows, Grenoble.
Chassaing P. (1985) "Une alternative à la formulation des équations du mouvement turbulent d'un fluide à masse volumique variable", J. de Mécanique Thèorique et Appliquée, **4**, (3), pp 375-389
HaMinh H., Launder B.E., MacInnes J. (1981), "The turbulence modelling of variable density flows, a mixed weighted decomposition", IIIrd Int. Symp. on Turb. Shear Flows, Davis
Handjalic K., Launder B.E. (1972), "A Reynolds stress model of turbulence and its application to thin shear flows", JFM, **52**, (4), pp 609-638.
MacInnes J.M. (1985) "Modelling of flows with non-uniform density", PhD Thesis, faculty of technology, Univ. Manshester.
Ramirez-Leon H. (1991) "Modélisation au second ordre d'écoulements turbulents fortement chauffés", Thèse de Doctorat, Ecole Centrale de Nantes.
Rey C. (1992) "2èmes journées d'étude sur les écoulements à masse volumique variable", Orléans, à paraître dans "Ecoulements turbulents à masse volumique variable", Edition CNRS, 1995.
Ruffin E., Schiestel R., Anselmet F., Amielh M. and Fulachier L. (1994) "Investigation of caracteristic scales in variable density turbulent jets using a second-order model", Phys. Fluids, **6** (8) pp 2785-2799.

SECOND ORDER MODELLING OF VARIABLE DENSITY JETS : FAVRE AVERAGED CLOSURES VERSUS REYNOLDS AVERAGED CLOSURES

E. RUFFIN*, R. SCHIESTEL**, F. ANSELMET***

*INERIS, Parc Technologique Alata, BP N°2, 60550 Verneuil en Halatte
,*IRPHE - Institut de Recherche sur les Phénomènes Hors d'Equilibre
**1 rue Honnorat and ***12 avenue Général Leclerc, 13003 Marseille

This paper is concerned with the development of second order closures for the numerical prediction of mixing of coaxial turbulent jets. The influence of density variations on the turbulence field can lead to complex behaviours and many of the underlying mechanisms are still incompletely understood. Several fundamental studies were published in the recent years literature (Chassaing et al., Fulachier et al., Panchakapesan and Lumley). The main purpose of the present paper is to check the prediction ability of second order modelling by comparisons between Favre averaged closures, Reynolds averaged closures and the experimental data obtained at IRPHE in circular turbulent jets of differing densities evolving in a coflow of air.

1. Turbulence Models

If constant density second order modelling has been studied for a long time (Schiestel), turbulence models for variable density flows still pose many fundamental problems to the modeler and no truly specific approach has so far been developed and proved to be satisfactory.

Favre averaged second order closures are based on transport equations for Reynolds stresses and turbulent fluxes using the mass weighted quantities. The mathematical methodology developed by Favre produces equations that nicely resemble in their formal structure to the usual corresponding equations for incompressible turbulence. Many authors took advantage of this property (Borghi

93

L. Fulachier et al. (eds.), IUTAM Symposium on Variable Density Low-Speed Turbulent Flows, 93–100.
© 1997 *Kluwer Academic Publishers.*

and Dutoya, Gouldin et al., Jones, Libby, Vandromme, Vandromme et al.) to develop variable density turbulence models that are formally deduced from constant density models using analogous closure hypotheses for the corresponding terms.

The only extra terms that appear in the transport equations for the second moment are :

$$G_{ij} = -a\bar{\rho}(\widetilde{\gamma u_i}\,\bar{P},_j + \widetilde{\gamma u_j}\,\bar{P},_i)$$

in the Reynolds stresses equations, and

$$G_{\gamma j} = -a\bar{\rho}\widetilde{\gamma\gamma}\,\bar{P},_j$$

in the turbulent fluxes equations. The model used here (table 1) is thus analogous to the one of Vandromme et al..

The Reynolds averaged statistical equations have a much more complicated appearance because the density fluctuations are explicitly taken into account (Chassaing, Donaldson et al., Ha Minh et al., Janicka and Kollmann, Mc Innes, Shih and Lumley). The second moment equations now include several extra terms that are representative of the density effects. There are extra production terms, extra diffusion terms and also the pressure-strain correlation as well as the pressure-gradient of concentration correlation both include specific terms that represent the effect of density variations. The dissipation equations for the turbulent kinetic energy and for the variance of concentration also incude many additional terms related to density variations. The closure methodology used here (table 1) is inspired from the Mc Innes model. The physical mechanisms related to density effects are thus modelled separately and represented through specific terms.

The density-velocity correlation is obtained by the following approximation :

$$\overline{\rho'u_j''} = -a\bar{\rho}^2\,\overline{\gamma u_j''}$$

2. Numerical Method

The numerical method is based on a finite volume elliptic code using SIMPLE algorithm for pressure linkage. Stability is guaranted by the use of under-relaxation and apparent diffusivities in the momentum equations (Huang and Leschziner, Ruffin).

3. Reference experimental data

The experimental set-up (Djeridane) is composed of a fully turbulent vertical round jet that can be investigated as well in a free environment as in a semi-confined environment with a low speed annular coflow. The central jet can be fed with helium, air or carbon dioxyde, allowing a large set of density ratios to be experienced ($0.14<\rho_j/\rho_e<1.5$). The coflow of air is useful for proper seeding of the jet for laser-Doppler measurements and to prevent recirculation flows to happen. In order to make meaningful comparisons on the effect of density, the momentum flux has been chosen the same for each of the three gases.

The measurements of velocity and concentration include the mean velocity, the statistical moments of velocity, the mean concentration of helium and carbon dioxyde. Several techniques are used : Pitot tube, hot wire anemometry, laser-Doppler anemometry for the velocities and oxygen chemical analyser with a zircone probe for the concentration field.

4. Comparisons between experiments, Favre averaged modelling and Reynolds averaged modelling

Emphasis will be given to the initial region of the jet which is more difficult to predict because of a stronger influence of density variations together with non equilibrium turbulence field far from similarity. As far as the mean velocity field is concerned the two models give similar results (fig 1 and 2) indicating that the axial development strongly depends on the density ratio. In these figures, the Reynolds averaged velocities are calculated from the Favre averaged velocities using the relation :

$$\overline{U_j} = \widetilde{U}_j + a\overline{\rho}^{-2}\widetilde{\gamma u}_j$$

However, different predictions are obtained for the radial velocities. Figures 3 and 4 for an helium-air jet make it clear that measurements are indeed Reynolds averaged quantities.

Concerning the turbulent field (Ruffin), the two models reproduce the experimental behaviour of the Reynolds stresses with a higher pick of the longitudinal normal stresses in the case of the helium-air jet that is located closer to the nozzle exit. The Reynolds averaged closure however gives a better agreement with experiments (fig 5 and 6). The variance of concentration and its dissipatrion rate are less satisfactory (fig 7 and 8). Indeed, the dissipation equation for the variance of the scalar is far more subtle to model than its dynamical counterpart. The turbulent scalar fluxes are very important for the coupling between the scalar and the dynamic fields. In this case, both models (fig 9) are satisfactory.

5. Final remarks

The two modelling approaches based on Favre averaging and on Reynolds averaging have been investigated to study the influence of the density variations on the turbulence field. The numerical results show that second-order modelling is a necessary level of closure to deal with such a complex phenomenon. The Favre averaged model, extensively used mainly in combustion, and one of the first that has been developed and used by several authors in the past, presents the advantage of simplicity. The Reynolds averaged model is based on a more complicated formalism in which the density effects are separated in specific terms. If the present results show only a modest im-provement over the Favre averaged model, it is clear that the potentialities of the Reynolds averaged model are greater. Indeed, this model can easily benefit of progress in the understanding of physical mechanisms deduced from experiments and numerical simulations. Some of the modelling assumptions used with Favre averages need to be reconsidered, and especially in the presence of strong density variations.

Financial support from EDF, GDF, INERIS and SNECMA is gratefully acknowledged.

References

BORGHI R. and DUTOYA D., 1978, 17th Int. Symp. on Combustion. The Combustion Institute.
CHASSAING P., 1985, J. de Mécanique Théorique et Appliquée, vol. 4, n°3, pp. 375-389.
CHASSAING P., HARRAN G. and JOLY L., 1994, J. Fluid Mech., vol. 279, pp. 239-252.
CHUA L.P. and ANTONIA R.A., 1990, Int. J. Heat Mass Transfer, vol. 33, n°2, pp. 331-339.
DJERIDANE T., 1994, Thèse Doctorat, Univ. Aix-Marseille II.
DONALDSON C.D., SULLIVAN R. and ROSENBAUM H., 1972, AIAA J., vol. 10, pp. 162-170.
FAVRE A., 1965, J. de Mécanique, vol. 4, pp. 361-421.
FULACHIER L., ANSELMET F. and AMIELH M., 1990, 27ème Colloque d'Aérodynamique Appliquée, Marseille.
GOULDIN F.C., SCHEFFER R.W., JOHNSON S.C. and KOLLMANN W., 1986, Prog. Energy Combustion Sciences, vol. 12, pp. 257-303.
HA MINH H., LAUNDER B.E. and MC INNES J.M., 1981, Proc. 3rd T.S.F. Symp., Univ. of California, USA.
HUANG P.G. and LESCHZINER M.A., 1985, Proc. 5th T.S.F. Symp., Cornell Univ., USA.
JANICKA J. and KOLLMANN W., 1979, AGARD Conf. Proc. CP 275.
JONES W.P., 1979, VKI Lecture Series 1979.
LIBBY P.A., 1977, In *Studies in Convection*, Ed. B.E. Launder, vol. 2, pp. 1-43, Academic Press.
MC INNES J.M., 1985, Ph. D. Thesis, Faculty of Technology, Univ. of Manchester, UK.

PANCHAPAKESAN N.R. and LUMLEY J.L., 1993, J. Fluid Mech.,vol. 246, p. 197 and vol. 246, p. 225.

RUFFIN E., 1994, Thèse Doctorat, Univ. Aix-Marseille II.

SCHIESTEL R., 1993, *Modélisation et simulation de la turbulence*, Ed. Hermès, Paris.

SHIH T.H.,and LUMLEY J.L., 1987, J. Fluid Mech., vol. 180, pp. 93-116.

VANDROMME D., 1980, Ph. D. Thesis, Free Univ. of Brussels.

VANDROMME D., HA MINH H., VIEGAS J.R., RUBESIN M.W., and KOLLMAN W., 1983, Proc. 4th T.S.F. Symp., Karlsruhe, Germany.

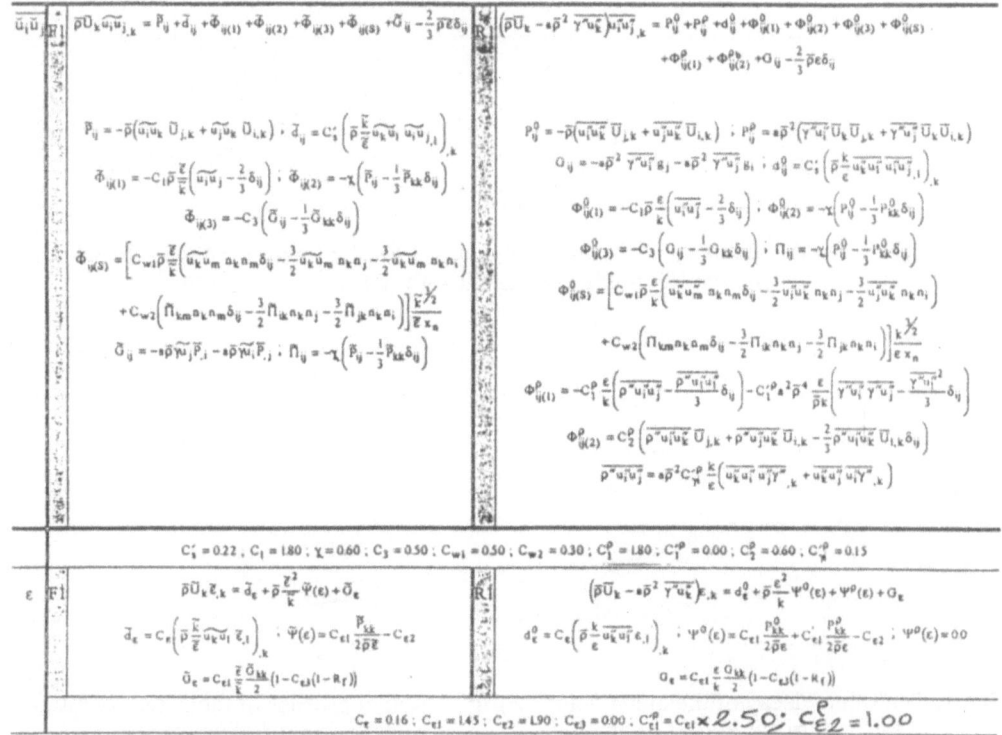

Table 1 (1st Part). Summary of the model equations for the Favre averaged closure (F) and the Reynolds average closure (R).

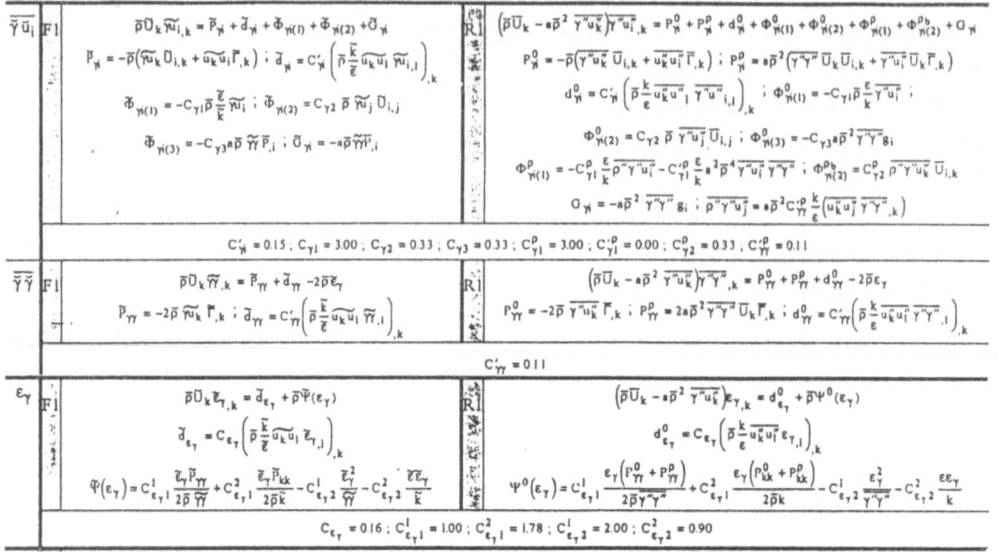

$\overline{\gamma''u_i''}$	F	$\bar{\rho}\tilde{U}_k\widetilde{u''_i}_{,k}=P_{\mathcal{H}}+d_{\mathcal{H}}+\Phi_{\mathcal{H}(1)}+\Phi_{\mathcal{H}(2)}+\tilde{O}_{\mathcal{H}}$	R	$\left(\bar{\rho}\tilde{U}_k-a\bar{\rho}^2\,\overline{\gamma''u_k''}\right)\overline{\gamma''u_i''}_{,k}=P_{\mathcal{H}}^0+P_{\mathcal{H}}^\rho+d_{\mathcal{H}}^0+\Phi_{\mathcal{H}(1)}^0+\Phi_{\mathcal{H}(2)}^0+\Phi_{\mathcal{H}(1)}^\rho+\Phi_{\mathcal{H}(2)}^\rho+O_{\mathcal{H}}$

Coefficients: $C'_{\mathcal{H}}=0.15$; $C_{\gamma1}=3.00$; $C_{\gamma2}=0.33$; $C_{\gamma3}=0.33$; $C^\rho_{\gamma1}=3.00$; $C'^\rho_{\gamma1}=0.00$; $C^\rho_{\gamma2}=0.33$; $C^\rho_{\mathcal{H}}=0.11$

$\overline{\gamma''\gamma''}$, F and R rows with coefficient $C'_{\gamma\gamma}=0.11$

ε_γ, F and R rows with coefficients $C_{\varepsilon_\gamma}=0.16$; $C^1_{\varepsilon_\gamma1}=1.00$; $C^2_{\varepsilon_\gamma1}=1.78$; $C^1_{\varepsilon_\gamma2}=2.00$; $C^2_{\varepsilon_\gamma2}=0.90$

Table 1 (2nd Part). Summary of the model equations for the Favre averaged closure (F) and the Reynolds average closure (R).

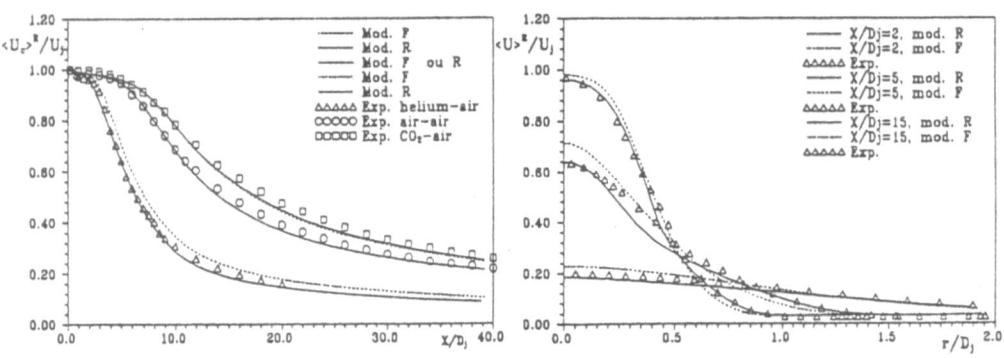

Fig. 1. Centreline decay of the mean velocity in the helium, air and CO₂ jets.　**Fig. 2.** Radial profiles of the mean longitudinal velocity in the helium-air jet.

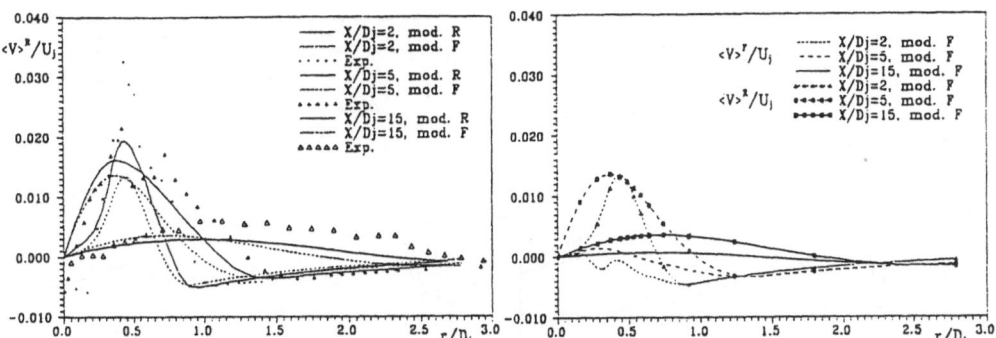

Fig. 3. Radial profiles of the radial Reynolds averaged velocity in the helium-air jet.

Fig. 4. Radial profiles of the radial Favre averaged velocity in the helium-air jet.

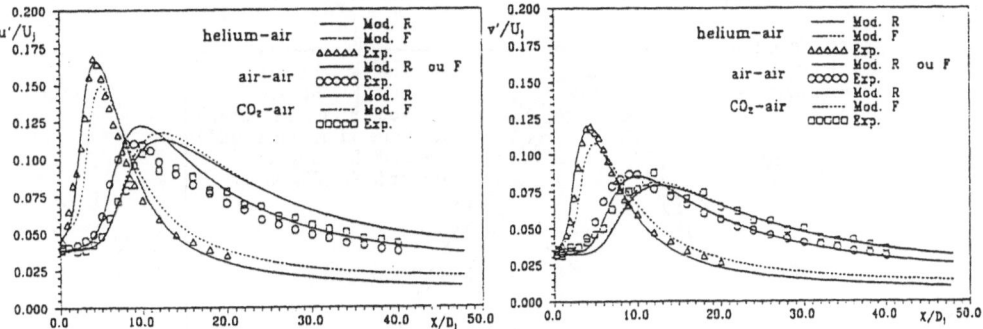

Fig. 5. Centreline evolution of the longitudinal velocity r.m.s. in the helium, air and CO_2 jets.

Fig. 6. Centreline evolution of the radial velocity r.m.s. in the helium, air and CO_2 jets.

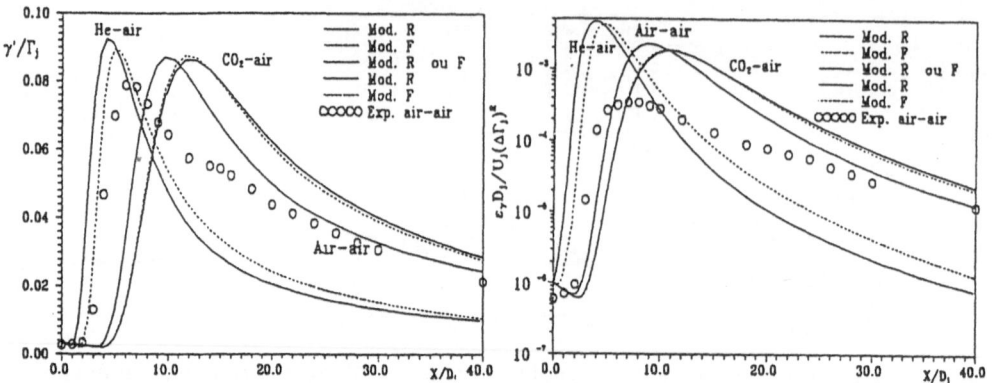

Fig. 7. Centreline evolution of the root mean square of scalar fluctuations.

Fig. 8. Centreline evolution of the scalar dissipation rate.

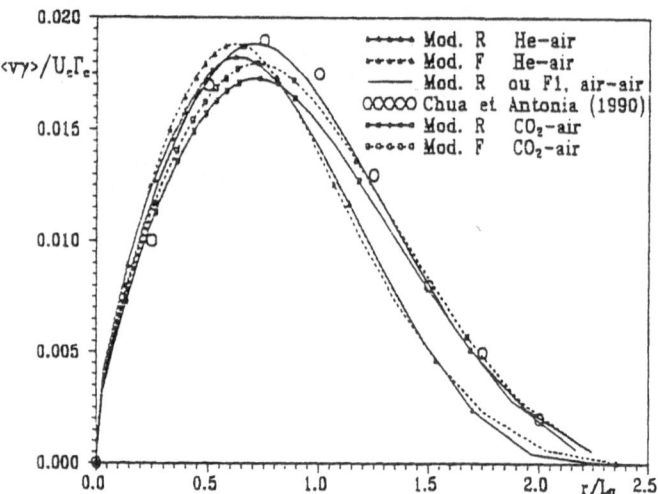

Fig. 9. Radial profile of the scalar transverse turbulent flux (at 15 diameters downstream).

AVERAGING PROCEDURES FOR THE LARGE EDDY SIMULATION OF VARIABLE DENSITY FLOWS

M. GERMANO
Politecnico di Torino
Dip. di Ing. Aeronautica e Spaziale, Torino, Italy

Abstract. Various representative methods have been adopted in the study of variable density turbulent flows. A widely used approach is that of the Favre mass weighted averaging procedure. In this paper this procedure is extended to Large Eddy Simulations in terms of the operatorial multi level formulation proposed recently by the author (Germano, 1992). This formulation does not introduce explicitly the fluctuations, and in the paper two particular points are addressed. The first point concerns the relations between the turbulent filtered quantities at different levels of resolution, both in terms of the usual averages and in terms of Favre averages. The second point regards the physical meaning of the closures based on the eddy viscosity when expressed in terms of Favre averaged quantities.

1. The multi level operatorial formulation

There are at present three different ways of computing a turbulent flow. The first is the *direct simulation* that should provide a representation resolved in all details of a generic turbulent quantity a. These simulations are limited to low values of the Reynolds number, they are very expensive and in many cases the information that they contain is highly redundant, so that a different approach is to perform the *statistical simulation* of a turbulent field that provides the statistical average $\langle a \rangle_e$ of the original quantity a

$$\langle a \rangle_e = \mathcal{E}(a) \tag{1}$$

where $\mathcal{E}$ is the *ensemble* averaging operator. These representations are produced by averaged equations, the Reynolds equations, that contain terms very difficult to model and produce only statistical informations. Thus a

101

L. Fulachier et al. (eds.), IUTAM Symposium on Variable Density Low-Speed Turbulent Flows, 101–108.

new approach has been recently developed where the simulation is conducted at an intermediate resolution level $\mathcal{F}$. This approach provides a *large eddy simulation* $\langle a \rangle_f$

$$\langle a \rangle_f = \mathcal{F}(a) \tag{2}$$

where $\mathcal{F}$ is the averaging *large eddy operator* at the $\mathcal{F}$-level and a common perception is that the averaged equations at this intermediate level of resolution are not so difficult to model, provided that the numerical resolution could capture the large turbulent structures of the flow. From a practical point of view, they could represent in future a good compromise between accuracy and computation time in the case of complex turbulent flows.

Unfortunately there is not at present a unique definition of what a Large Eddy Simulation really is and, from a basic point of view, the problem is related to the multi scale procedures. Multiple scale LES and statistical modeling has been usually developed on the basis of spectral evolution equations, (Yoshizawa, 1982), (Schiestel, 1987), but different decomposition techniques can be conceived directly in physical space. Following Leonard (Leonard, 1974) an averaging large eddy operator $\mathcal{F}$ is interpreted as a smoothing process operated by a convolution that produces a smoothed representation of the original quantity. Another different interpretation of a Large Eddy Simulation is due to Schumann (Schumann, 1975) that presents the filtering approach as a volume balance procedure, and we finally recall the operatorial formulation (Germano, 1992) due to the present author. This formulation is based on the following ordered hierarchy of filtering operators

$$\mathcal{F} \ , \ \ \mathcal{G} \ , \ \ \cdots$$

that, successively applied to the original turbulent function a, produce the following *cascade* of representations

    — The $\mathcal{F}$-*level* large eddy representation

$$\langle a \rangle_f = \mathcal{F}(a)$$

    — The $\mathcal{F}\mathcal{G}$-*level* large eddy representation

$$\langle a \rangle_{fg} = \langle \langle a \rangle_f \rangle_g = \mathcal{G}(\mathcal{F}(a))$$

and so on. We remark that a basic problem in the study of turbulent flows is to represent the $\mathcal{F}$-filtered value $\langle y(a_1, ..., a_n) \rangle_f$ of a function $y(a_1, ..., a_n)$ in terms of some modelling procedure. The simplest one is to expand $\langle y \rangle_f$ in terms of the generalized central moments associated with the turbulent arguments a_i

$$\langle y(a_1, \ldots, a_n)\rangle_f \;=\; y(\langle a_1\rangle_f, \ldots, \langle a_n\rangle_f) + \sum_{i,j} \frac{\tau_f(a_i, a_j)}{2!} y_{i,j} +$$

$$+ \sum_{i,j,k} \frac{\tau_f(a_i, a_j, a_k)}{3!} y_{i,j,k} + \cdots$$

$$(3)$$

where

$$y_{i,j} \;=\; \frac{\partial^2 y(\langle a_1\rangle_f, \ldots, \langle a_n\rangle_f)}{\partial\langle a_i\rangle_f \partial\langle a_j\rangle_f}$$

$$y_{i,j,k} \;=\; \cdots \tag{4}$$

and where

$$\tau_f(a_i, a_j) \;=\; \langle a_i a_j\rangle_f - \langle a_i\rangle_f \langle a_j\rangle_f$$

$$\tau_f(a_i, a_j, a_k) \;=\; \langle a_i a_j a_k\rangle_f - \tau_f(a_i, a_j)\langle a_k\rangle_f - \tau_f(a_j, a_k)\langle a_i\rangle_f$$
$$- \tau_f(a_k, a_i)\langle a_j\rangle_f - \langle a_i\rangle_f \langle a_j\rangle_f \langle a_k\rangle_f$$
$$\tau_f(a_i, a_j, a_k, a_l) \;=\; \cdots \tag{5}$$

are the *generalized central moments* associated with the quantities a_i. It is easy to see that when the filtering operator $\mathcal{F}$ is the ensemble average $\mathcal{E}$, the generalized central moments reduce to the usual statistical central moments. In terms of the fluctuations a_i'

$$a_i' = a_i - \langle a_i\rangle_e \tag{6}$$

they have the simple form

$$\tau_e(a_i, a_j) \;=\; \langle a_i' a_j'\rangle_e$$
$$\tau_e(a_i, a_j, a_k) \;=\; \langle a_i' a_j' a_k'\rangle_e$$
$$\tau_e(a_i, a_j, a_k, a_l) \;=\; \cdots \tag{7}$$

The operatorial formulation of the filtering approach considerably simplifies the study of turbulent flows. We notice that the generalized central moments are Galilean invariants and that in terms of the generalized central moments the filtered equations are formally invariant so that the classical formulation in terms of the *fluctuations* should be avoided when a multiscale analysis of the turbulent field is performed. The algebraic properties of the generalized central moments are very interesting in modelling and relatively unexplored. A lot of identities can be derived relating the generalized central moments at different levels, and one in particular that stands at the basis of the so called dynamic model (Germano, 1990; Germano et al., 1991) is the following

$$\tau_{fg}(a,b) = \langle \tau_f(a,b) \rangle_g + \tau_g(\langle a \rangle_f, \langle b \rangle_f) \tag{8}$$

where $\tau_g(\langle a \rangle_f, \langle b \rangle_f)$ is the resolved turbulent stress

$$\tau_g(\langle a \rangle_f, \langle b \rangle_f) = \langle \langle a \rangle_f \langle b \rangle_f \rangle_g - \langle a \rangle_{fg} \langle b \rangle_{fg} \tag{9}$$

2. The extension of Favre averages to Large Eddy Simulations

Let us now extend this formalism to variable density flows. We notice that in such cases the turbulent quantities are usually represented in terms of the Favre averaging procedure (Favre, 1965) and we recall that the Favre average $\{a\}_e$ associated with the statistical operator $\mathcal{E}$ is defined as [1]

$$\langle \varrho a \rangle_e \equiv \langle \varrho \rangle_e \{a\}_e \equiv \langle \varrho \rangle_e \langle a \rangle_e + \tau_e(\varrho, a) \tag{10}$$

where ϱ is the variable density and we can easily extend this definition to our hierarchy of filtering operators by defining the Favre average associated with $\{a\}_f$ to the $\mathcal{F}$-level as

$$\langle \varrho a \rangle_f \equiv \langle \varrho \rangle_f \{a\}_f \equiv \langle \varrho \rangle_f \langle a \rangle_f + \tau_f(\varrho, a) \tag{11}$$

and the Favre average $\{a\}_{fg}$ associated with the $\mathcal{FG}$-level as

$$\langle \varrho a \rangle_{fg} \equiv \langle \varrho \rangle_{fg} \{a\}_{fg} \equiv \langle \varrho \rangle_{fg} \langle a \rangle_{fg} + \tau_{fg}(\varrho, a) \tag{12}$$

We notice that they are related to the filtered averages by means of the generalized central moments of the second order $\tau_f(\varrho, a)$ and $\tau_{fg}(\varrho, a)$. It is interesting to list some particularities of these generalized Favre averages that are of practical use. As regards the filtered averages we obviously have

$$\langle a \rangle_{fg} = \langle \langle a \rangle_f \rangle_g \tag{13}$$

but we remark that

$$\{a\}_{fg} \neq \{\{a\}_f\}_g \tag{14}$$

[1] We introduce here this rather unusual notation representing the Favre average associated with the statistical operator $\mathcal{E}$ in order to distinguish this particular mass weighted average from the mass weighted average associated with a generic filtering operator $\mathcal{F}$.

and that

$$\{a\}_{fg} \neq \langle\{a\}_f\rangle_g \qquad (15)$$

It is easy to see that if we have to calculate the Favre average associated with the $\mathcal{FG}$-level given the values $\langle\varrho\rangle_f, \{a\}_f$ at the $\mathcal{F}$-level we have to write

$$\{a\}_{fg} = \frac{\langle\langle\varrho\rangle_f\{a\}_f\rangle_g}{\langle\varrho\rangle_{fg}} \qquad (16)$$

where obviously

$$\langle\varrho\rangle_{fg} = \langle\langle\varrho\rangle_f\rangle_g \qquad (17)$$

and we remark that some previous extensions of the dynamic model to compressible flows apparently have adopted a different formulation (Moin et al., 1991; El Hady et al., 1994).

We can extend the Favre formalism to higher order moments by defining the quantities $\vartheta_f(a,b)$ and $\vartheta_{fg}(a,b)$ as follows

$$\begin{aligned}
\langle\varrho ab\rangle_f &\equiv \langle\varrho\rangle_f\{a\}_f\{b\}_f + \langle\varrho\rangle_f\vartheta_f(a,b) \\
\langle\varrho ab\rangle_{fg} &\equiv \langle\varrho\rangle_{fg}\{a\}_{fg}\{b\}_{fg} + \langle\varrho\rangle_{fg}\vartheta_{fg}(a,b)
\end{aligned} \qquad (18)$$

and we can finally write that

$$\langle\varrho\rangle_{fg}\vartheta_{fg}(a,b) = \langle\langle\varrho\rangle_f\vartheta_f(a,b)\rangle_g + \langle\langle\varrho\rangle_f\{a\}_f\{b\}_f\rangle_g - \langle\varrho\rangle_{fg}\{a\}_{fg}\{b\}_{fg} \quad (19)$$

This identity extends to Favre averages the identity (8) and it is of practical use in the generalization of the dynamic modelling procedure to variable density flows (Germano, 1993). In this case a, b are the components of the velocity u_i, u_j and the quantities $\vartheta_f(u_i, u_j)$ and $\vartheta_{fg}(u_i, u_j)$ are represented by subgrid scale models. The identity (19) represents in such cases a consistency relation that the models should in some way respect and, in the case of the eddy viscosity model, a consistent value of the Smagorinsky constant appropriate to the particular flow could be derived.

3. The physical meaning of the eddy viscosity approximation in terms of Favre averages.

A simple extension of the eddy viscosity assumption to the subgrid scale closure of variable density turbulent flows in terms of Favre averages is given by

$$\vartheta_f^a(u_i, u_j) \sim -2\nu_f\sigma_{ij}^a \qquad (20)$$

where the superscript stands for the deviatoric part, σ_{ij} is the tensor

$$\sigma_{ij} = \frac{1}{2}\left(\frac{\partial\{u_i\}_f}{\partial x_j} + \frac{\partial\{u_j\}_f}{\partial x_i}\right)$$

and ν_f is an eddy viscosity at the filtered $\mathcal{F}$-level.

As in the case of the statistical eddy viscosity (Germano, 1996), it is interesting to explore which kind of modelling assumption stands at the basis of this closure. In order to answer to this question let us recall that

$$\langle \varrho u_i \rangle_f \;=\; \langle \varrho \rangle_f \{u_i\}_f = \langle \varrho \rangle_f \langle u_i \rangle_f + \tau_f(\varrho, u_i)$$

$$
\begin{aligned}
\langle \varrho u_i u_j \rangle_f \;=&\; \langle \varrho \rangle_f \{u_i\}_f \{u_j\}_f + \langle \varrho \rangle_f \vartheta_f(u_i, u_i) = \\
=&\; \langle \varrho \rangle_f \langle u_i \rangle_f \langle u_j \rangle_f + \langle \varrho \rangle_f \tau_f(u_i, u_j) + \langle u_i \rangle_f \tau_f(\varrho, u_j) + \\
&+\; \langle u_j \rangle_f \tau(\varrho, u_i) + \tau_f(\varrho, u_i, u_j)
\end{aligned}
\tag{21}
$$

where $\tau_f(u_i, u_j)$ are the turbulent stresses at the $\mathcal{F}$-level

$$\langle u_i u_j \rangle_f = \langle u_i \rangle_f \langle u_j \rangle_f + \tau_f(u_i, u_j)$$

and where $\tau_f(\varrho, u_i, u_j)$ are central moments of the third order. We notice that, in the first case, the description of the turbulent field is in terms of the quantities

$$\langle u_i \rangle_f \;,\;\; \tau_f(\varrho, u_i) \;,\;\; \tau_f(u_i, u_j) \;,\;\; \tau_f(\varrho, u_i, u_j)$$

while, in the case of the mass-weighted average, the turbulent field is described by the quantities

$$\{u_i\}_f \;,\;\; \vartheta_f(u_i, u_j)$$

and these quantities are related by the following expressions

$$\{u_i\}_f \;=\; \langle u_i \rangle_f + \frac{\tau_f(\varrho, u_i)}{\langle \varrho \rangle_f}$$

$$\vartheta_f(u_i, u_j) \;=\; \tau_f(u_i, u_j) - \frac{\tau_f(\varrho, u_i)\tau_f(\varrho, u_j)}{\langle \varrho \rangle_f^2} + \frac{\tau_f(\varrho, u_i, u_j)}{\langle \varrho \rangle_f} \tag{22}$$

By the use of these relations we are now ready to interpret the modelling assumption (20). With a little algebra it is easy to see that this expression is produced by assuming the following chain of modelling relations for the statistical moments

$$\tau_f^a(u_i, u_j) \;\sim\; -2\nu_f \langle s_{ij} \rangle_f^a$$

$$\tau_f(\varrho, u_i) \;\sim\; -2\nu_f \frac{\partial \langle \varrho \rangle_f}{\partial x_i}$$

$$\tau_f(\varrho, u_i, u_j) \;\sim\; -\nu_f \left(\frac{\partial \tau_f(\varrho, u_i)}{\partial x_j} + \frac{\partial \tau_f(\varrho, u_j)}{\partial x_i} \right)$$

$$\tag{23}$$

where $\langle s_{ij}\rangle_f^a$ is the deviatoric part of the tensor

$$\langle s_{ij}\rangle_f = \frac{1}{2}\left(\frac{\partial\langle u_i\rangle_f}{\partial x_j} + \frac{\partial\langle u_j\rangle_f}{\partial x_i}\right)$$

The same considerations apply to the modelling of the transport term $\langle\varrho\alpha u_i\rangle_f$ of a generic scalar α in a variable density flow. In this case we can also write the following chain of relations

$$\langle\varrho\alpha\rangle_f = \langle\varrho\rangle_f\{\alpha\}_f = \langle\varrho\rangle_f\langle\alpha\rangle_f + \tau_f(\varrho,\alpha)$$

$$\begin{aligned}\langle\varrho\alpha u_i\rangle_f &= \langle\varrho\rangle_f\{\alpha\}_f\{u_i\}_f + \langle\varrho\rangle_f\vartheta_f(\alpha,u_i) =\\ &= \langle\varrho\rangle_f\langle\alpha\rangle_f\langle u_i\rangle_f + \langle\varrho\rangle_f\tau_f(\alpha,u_i) + \langle\alpha\rangle_f\tau_f(\varrho,u_i) +\\ &+ \langle u_i\rangle_f\tau_f(\varrho,\alpha) + \tau_f(\varrho,\alpha,u_i)\end{aligned} \tag{24}$$

and we obtain

$$\{\alpha\}_f = \langle\alpha\rangle_f + \frac{\tau_f(\varrho,\alpha)}{\langle\varrho\rangle_f}$$

$$\vartheta_f(\alpha,u_i) = \tau_f(\alpha,u_i) - \frac{\tau_f(\varrho,u_i)\tau_f(\varrho,\alpha)}{\langle\varrho\rangle_f^2} + \frac{\tau_f(\varrho,\alpha,u_i)}{\langle\varrho\rangle_f} \tag{25}$$

If we now introduce the following modelling assumptions

$$\tau_f(\alpha,u_i) \sim -2\nu_f\frac{\partial\langle\alpha\rangle_f}{\partial x_i}$$

$$\tau_f(\varrho,u_i) \sim -2\nu_f\frac{\partial\langle\varrho\rangle_f}{\partial x_i}$$

$$\tau_f(\varrho,\alpha,u_i) \sim -2\nu_f\frac{\partial\tau_f(\varrho,\alpha)}{\partial x_i} \tag{26}$$

we can derive the usual modelling assumption for $\vartheta_f(\alpha,u_i)$ in terms of Favre averages

$$\vartheta_f(\alpha,u_i) \sim -2\nu_f\frac{\partial\{\alpha\}_f}{\partial x_i} = -\nu_{\alpha f}\frac{\partial\{\alpha\}_f}{\partial x_i} \tag{27}$$

We remark that no hypothesis is needed as regards $\tau_f(\varrho,\alpha)$ and that owing to the relation $\nu_{\alpha f} = 2\nu_f$ the predicted turbulent Prandtl number is 0.5. We notice finally that in the framework of the gradient diffusion hypothesis the chain of relations (26) are perfectly consistent with the identity

$$\tau_f(\varrho,\alpha,u_i) = \tau_f(\varrho\alpha,u_i) - \langle\varrho\rangle_f\tau_f(\alpha,u_i) - \langle\alpha\rangle_f\tau_f(\varrho,u_i) \tag{28}$$

and the additional closure

$$\tau_f(\varrho\alpha, u_i) \sim -2\nu_f \frac{\partial \langle \varrho\alpha \rangle_f}{\partial x_i} \tag{29}$$

4. Conclusions

In this paper we have extended the operatorial multi-level formalism to Favre averages, and some relations between the turbulent stresses at different resolution level have been provided. They can be applied to standard modelling procedures in order to produce a *dynamic* version. Another point examined in the paper has been the relation between the eddy viscosity closure and the same closure expressed in terms of mass-weighted averages. It turns out that this closure can be interpreted in terms of the gradient diffusion hypothesis.

References

N.M. El-Hady, T.H. Zang, U. Piomelli *Application of the dynamic subgrid scale model to axisymmetric transitional boundary layer at high speed*, Phys. Fluids 6, pp. 1299-1309, (1994)

A. Favre *Equations des gas turbulents compressibles* Journal de Mecanique, 4, pp. 361-390, (1965)

M. Germano *Averaging invariance of the turbulent equations and similar subgrid scale modeling* CTR Manuscript 116, Stanford, CA, (1990)

M. Germano, U. Piomelli, P. Moin, W. H. Cabot *A dynamic subgrid-scale eddy viscosity model* Phys. Fluids A 3, 1760-1765, (1991)

M. Germano *Turbulence : the filtering approach* J. Fluid Mech. 238, 325-336, (1992)

M. Germano *Multilevel subgrid modelling*, Proceedings 5th International Symposium on Refined Flow Modelling and Turbulence Measurements, pp. 81-88, Paris, (1993)

M. Germano, *The statistical meaning of the Boussinesq approximation in terms of the Favre averages*, ERCOFTAC Bulletin 28, pp. 59-60, (1996)

A. Leonard *Energy Cascade in Large-Eddy Simulations of Turbulent Fluid Flows* Adv. in Geophysics, 18A, pp. 237-248, (1974)

P. Moin, K. Squires, W. Cabot, S. Lee *A dynamic subgrid scale model for compressible turbulence and scalar transport*, Phys. Fluids A3, pp. 2746-2757, (1991)

R. Schiestel *Multiple time scale modeling of turbulent flows in one point closures*, Phys. Fluids 30, pp. 722-731, (1987)

U. Schumann *Subgrid Scale Model for Finite Difference Simulations of Turbulent Flows in Plane Channels and Annuli* J. Comput. Phys. 18, 376 (1975)

A. Yoshizawa *A statistically derived subgrid model for the large-eddy simulation of turbulence*, Phys. Fluids 25, pp. 1532-1538, (1982)

NUMERICAL PREDICTIONS OF MIXING PHENOMENA IN TURBULENT VARIABLE DENSITY FLOWS WITH SECOND ORDER CLOSURE TURBULENCE MODELS

R. ELAMRAOUI, D. GARRÉTON AND O. SIMONIN
Electricité de France - Laboratoire National d'Hydraulique
6, quai Watier, B.P. 49, 78401 Chatou Cedex, France

1. Introduction

This work is devoted to the prediction of turbulent variable density flows which is considered as an important step towards the prediction of turbulent flames. Indeed, this step is of particular interest as far as many practical devices are often characterized by a complex aerodynamics: recirculation zones, impinging jets, swirling flows, ... that determines the efficiency of the fuel-air mixing. Combustion modifies the dynamic field but in general, the flow pattern is qualitatively recovered between the case of inert mixing of the reactants and the reactive flow. In this context, tests of the turbulence models which are implemented for three dimensional computations of industrial applications [1] in non reactive flows may give precious indications for the simulation of the corresponding flames; in particular about the widely used Boussinesq assumption for the scalar turbulent flux and the assumed proportionality law between the characteristic time of the scalar dissipation and that one of the turbulent kinetic energy. These questions are addressed here through numerical predictions of two basic turbulent variable density flows. The first one is an axisymmetric turbulent jet (diameter D= 26 mm) surrounded by a low velocity (0.9 m/s) coflow of air of diameter 285 mm. The central jet is fed by slightly heated air, carbon dioxide or helium. For this latter case, density variations are of the order of magnitude of those encountered in flames. The inlet momentum is kept constant for the three experiments (see Table 1). Detailed measurements [2,3] of the dynamic and scalar fields, including probability density functions, were performed at the IRPHE (Marseille). The second study is related to a more complex turbulent flow. The flow configuration consists of a 5.4 mm diameter jet of methane (21 m/s) located in the center of a cylindrical bluff body. The air (15 m/s) is supplied through a coaxial jet surrounding the bluff body between the diameters 50 and 100 mm. The experience is a non confined case and corresponds to a blocage ratio of 25%. Fine measurements were performed in the inert and reactive cases by Sandia National Laboratories and Gaz de France - R&D Division [4]. The inert flow presents a large recirculation zone downstream the bluff body which does not enter the central jet. Along the axis, the axial mean velocity exhibits a minimum positive value (2 m/s) around 60 mm from the burner nozzle. Following the analysis of Chen et al. [5], bluff body flows can be classified into two types according as the central jet penetrates the recirculation zone (jet regime) or it does not (strongly recirculating regime). The present flow appears to be of the first type but due the very low minimum of axial velocity, it is rather close to the limit between the two regimes.

L. Fulachier et al. (eds.), IUTAM Symposium on Variable Density Low-Speed Turbulent Flows, 109–118.
© 1997 *Kluwer Academic Publishers.*

In the following, the equations and the turbulence model are briefly recalled. Numerical methods and the conditions of the computations are described. The results are then analysed, first for the dynamic field, then for the scalar one. Concluding remarks are finally given.

Table 1 : Characteristics of the three round jets -
Reynolds number is based on the central jet diameter.

Jet	density	jet bulk velocity	Reynolds number	Turbulent Reynolds number	Froude number
Air	1.16	12.	21 000	300	
Helium	0.16	32.	7 000	70	1 100
CO2	1.8	10.	32 000	500	630

Although many computations have been performed for these cases, only those which present significant differences between each other are detailed. Due to paper length limitation, we just indicate the behaviour of the other computations (with the standard k-ε model for instance) when it is of interest for the discussions.

2. Modelling

Transport equations are solved for the momentum, the Reynolds stresses, the dissipation rate of the turbulent kinetic energy, the mean scalar, its variance and the scalar turbulent flux. Some tests have been also performed with a transport equation for the scalar dissipation. The scalar, defined as the concentration of the central jet is required for the computation of the mean density in all cases except for the air jet where the computed scalar is a non dimension temperature which can be considered as a passive scalar.

Favre average is used as being the only average operator which is tractable in turbulent reactive flows. It must be said that the approach developed by Chassaing et al. [6] is very attractive for non reactive flows as i) it allows a convenient extension of the closure proposed for constant density flows, ii) additional terms due to density variations are clearly exhibited.

2.1.   GOVERNING EQUATIONS

In the following, overbar and tilda denote respectively Reynolds and Favre ageraged, while the corresponding fluctuations are ' and ". The equation set of a low Mach, variable density turbulent flow can be written as:

Equation of state (except for the constant density air jet):

$$\bar{\rho} = \frac{1}{b + a\tilde{f}} \tag{2.1}$$

f is the mass fraction of the species injected in the central jet, ρ the density, a and b are given constants.

Balance equations for mass momentum and the mean scalar mass fraction:

$$\frac{\partial}{\partial t}\bar{\rho} + \frac{\partial}{\partial x_i}\bar{\rho}\,\tilde{u}_i = 0 \tag{2.2}$$

$$\frac{\partial}{\partial t}\bar{\rho}\tilde{u}_i + \frac{\partial}{\partial x_j}(\bar{\rho}\,\tilde{u}_i\tilde{u}_j) = \frac{\partial}{\partial x_j}\left(\tau_{ij} - \bar{\rho}\,\widetilde{u_i''u_j''}\right) - \frac{\partial\bar{p}}{\partial x_i} + \bar{\rho}\,g_i \tag{2.3}$$

$$\frac{\partial}{\partial t}\, \bar{\rho}\tilde{f} + \frac{\partial}{\partial x_i}\left(\bar{\rho}\,\widetilde{u_i f}\right) = -\frac{\partial}{\partial x_i}\left[J_i + \bar{\rho}\,\widetilde{u_i'' f''}\right] \qquad (2.4)$$

where u_i are the velocity components (i=1,3), p is the pressure, g_i the i-th component of gravity and τ_{ij} the mean viscous stress tensor ij-th term. J_i is the mean scalar molecular diffusion flux i-th component.

At this level, the mean viscous stress tensor and the scalar molecular diffusion flux are closed with conventional expressions for Newtonian fluids obeying the Fick law.

A closure is also required for the turbulent correlations $\widetilde{u_i'' u_j''}$ and $\widetilde{u_i'' f''}$. In §4, results are presented when the standard k-ε model is used as described in [7] or with a second order turbulence model. In this latter case, transport equations for the turbulent correlations can be written in a very general form as:

$$\frac{\partial}{\partial t}\,\bar{\rho}\,\widetilde{u_i'' u_j''} + \frac{\partial}{\partial x_\alpha}\bar{\rho}\,\tilde{u}_\alpha\,\widetilde{u_i'' u_j''} = D_{ij} + P_{ij} + \phi_{ij} + G_{ij} + \varepsilon_{ij} \qquad (2.5)$$

and

$$\frac{\partial}{\partial t}\,\bar{\rho}\,\widetilde{u_i'' f''} + \frac{\partial}{\partial x_\alpha}\left(\bar{\rho}\,\tilde{u}_\alpha\,\widetilde{u_i'' f''}\right) = D_{if} + P_{if} + \phi_{if} + G_{if} \qquad (2.6)$$

where the right hand side terms stand for diffusion, production, pressure-strain or pressure-scalar gradient correlations, additional term due to density variations and dissipation for the last term of Equation (2.5).

2.2.   TURBULENCE MODELS

2.2.1. *Reynolds Stress (RS) And Scalar Turbulent Flux Modelling.* The turbulence model we focus on is used in 2D and 3D codes developed at LNH [8] and is denoted as the standard RS model. In order to help the discussions, the model proposed by Chassaing et al. [6] is also used for the air round jet. Both are briefly recalled here. Production term can be treated exactly at second order closure level:

$$P_{ij} = -\bar{\rho}\,\widetilde{u_i'' u_\alpha''}\frac{\partial \tilde{u}_j}{\partial x_\alpha} - \bar{\rho}\,\widetilde{u_j'' u_\alpha''}\frac{\partial \tilde{u}_i}{\partial x_\alpha} \qquad (2.7)$$

The turbulent transport is approximated by the gradient-diffusion model [9], where k and ε denote the turbulent kinetic energy and its disspation rate respectively:

$$D_{ij} = C_s \frac{\partial}{\partial x_\alpha}\left[\bar{\rho}\,\frac{k}{\varepsilon}\,\widetilde{u_\alpha'' u_\beta''}\frac{\partial \widetilde{u_i'' u_j''}}{\partial x_\beta}\right] \qquad (2.8)$$

In the standard model, following Launder et al. [10], the pressure-strain correlation is decomposed into three contributions, a return to isotropy (slow) term, an isotropization of the production (rapid) term and a wall echo term, negligible in the present studies:

$$\phi_{ij} = \phi_{ij1} + \phi_{ij2} = -C_1\,\bar{\rho}\frac{\varepsilon}{k}\left(\widetilde{u_i'' u_j''} - \frac{2}{3}\delta_{ij}\,k\right) - C_2\left(P_{ij} - \frac{1}{3}\delta_{ij}\,P_{kk}\right) \qquad (2.9)$$

Two additional terms related to the isotropization of the transposed production tensor Q and to the mean strain rate tensor T are taken into account in the model of Chassaing:

$$-A_2\left(Q_{ij} - \frac{1}{3}\delta_{ij}\,P_{kk}\right) - 2\,A_0\,\bar{\rho}\,kT_{ij}\ \text{where}\ Q_{ij} = -\bar{\rho}\,\widetilde{u_i'' u_\alpha''}\frac{\partial \bar{u}_\alpha}{\partial x_j} - \bar{\rho}\,\widetilde{u_j'' u_\alpha''}\frac{\partial \bar{u}_\alpha}{\partial x_i} \qquad (2.10)$$

The fourth term in Equation (2.5) is directly related with density variations by means of the equivalent turbulent viscosity:

$$G_{ij} = -\overline{u_i''}\frac{\partial \overline{p}}{\partial x_j} - \overline{u_j''}\frac{\partial \overline{p}}{\partial x_i}, \text{ with } \overline{u_i''} \approx -\frac{\nu_T}{\overline{\rho}}\frac{\partial \overline{p}}{\partial x_i} \quad (2.11)$$

As dissipation occurs in small scale turbulence which is assumed to be isotropic at high Reynolds number, $\varepsilon_{ij} = -\frac{2}{3}\overline{\rho}\delta_{ij}\varepsilon$, where ε is given by a transport equation:

$$\frac{\partial}{\partial t}(\overline{\rho}\varepsilon) + \frac{\partial}{\partial x_\alpha}(\overline{\rho}\tilde{u}_\alpha\varepsilon) = C_\varepsilon\frac{\partial}{\partial x_\alpha}\left[\overline{\rho}\frac{k}{\varepsilon}\widetilde{u_\alpha''u_\beta''}\frac{\partial\varepsilon}{\partial x_\beta}\right] + C_{\varepsilon 1}\ \overline{\rho}\ \frac{\varepsilon}{k}\frac{P_{ii}+G_{ii}}{2} - C_{\varepsilon 2}\ \overline{\rho}\frac{\varepsilon^2}{k} \quad (2.12)$$

with $C_\varepsilon = 0.18$; $C_{\varepsilon 1} = 1.44$ and $C_{\varepsilon 2} = 1.92$.

In a similar way, the turbulent scalar flux equations (2.6) are closed as follows:

$$P_{if} = -\overline{\rho}\ \widetilde{u_\alpha''f''}\frac{\partial \tilde{u}_i}{\partial x_\alpha} - \overline{\rho}\ \widetilde{u_i''u_\alpha''}\frac{\partial \tilde{f}}{\partial x_\alpha} \text{ and } D_{if} = C_{sf}\frac{\partial}{\partial x_\alpha}\left[\overline{\rho}\frac{k}{\varepsilon}\widetilde{u_\alpha''u_\beta''}\frac{\partial\widetilde{u_i''f''}}{\partial x_\beta}\right] \quad (2.13)$$

$$\Phi_{if} = \Phi_{if1} + \Phi_{if2} = -C_{f1}\ \overline{\rho}\ \frac{\varepsilon}{k}\widetilde{u_i''f''} + C_{f2}\ \overline{\rho}\ \widetilde{u_\alpha''f''}\frac{\partial \tilde{u}_i}{\partial x_\alpha} \text{ and } G_{if} = -\overline{f''}\frac{\partial\overline{p}}{\partial x_i} = -a\overline{\rho}\ \widetilde{f''^2}\frac{\partial\overline{p}}{\partial x_i} \quad (2.14)$$

A term related to the transposed production is introduced by Chassaing et al. [6]:

$$\Phi_{if3} = -C_{f3}\overline{\rho}\ \widetilde{u_\alpha''f''}\frac{\partial \tilde{u}_\alpha}{\partial x_i} \quad (2.15)$$

Table 2: Values of the constants for the standard RS model and the model proposed by Chassaing et al. [6]

RS model	C_1	C_2	A_0	A_2	C_s	C_{f1}	C_{f2}	C_{f3}	C_{sf}
Standard	1.8	0.6	0.	0.	0.22	3.	0.5	0.	0.18
Chassaing	2.8	0.76	0.16	0.08	0.21	6.	0.8	0.2	0.15

2.2.2. *Scalar Variance and Dissipation Modelling.* Finally, the scalar variance is computed with:

$$\frac{\partial}{\partial t}\overline{\rho}\ \widetilde{f''^2} + \frac{\partial}{\partial x_i}\overline{\rho}\ \tilde{u}_i\ \widetilde{f''^2} = C_f\frac{\partial}{\partial x_i}\left(\overline{\rho}\frac{k}{\varepsilon}\widetilde{u_i''u_j''}\frac{\partial\widetilde{f''^2}}{\partial x_j}\right) - 2\overline{\rho}\ \widetilde{u_i''f''}\frac{\partial\tilde{f}}{\partial x_i} - 2\overline{\rho}\ \varepsilon_f \quad (2.16)$$

where

$$\varepsilon_f = K\frac{\overline{\partial f''}}{\partial x_i}\frac{\partial f''}{\partial x_i} \quad (2.17)$$

Two closures are used for the dissipation rate of the scalar fluctuations: an algebraic conventional expression, when the characteristic times of scalar dissipation and turbulent dissipation are supposed to be proportional ($C'_T = 1.6$):

$$\varepsilon_f = \frac{1}{C'_T}\frac{\varepsilon}{k}\widetilde{f''^2} \quad (2.18)$$

or the transport equation which was proposed by Jones & Musonge [11]:

$$\frac{d\varepsilon_f}{dt} = \frac{\partial}{\partial x_i}\left[C_{\varepsilon f}\frac{k}{\varepsilon}\widetilde{u_i''u_j''}\frac{\partial\varepsilon_f}{\partial x_j}\right] + P_{\varepsilon f} - C_{D2}\frac{\varepsilon}{2k}\tilde{\varepsilon}_f - C_{D1}\frac{\varepsilon_f^2}{\widetilde{f''^2}} \quad (2.29)$$

with
$$P_{\varepsilon f} = -C_{D3} \frac{\varepsilon}{k} \overline{u_i'' f''} \frac{\partial \tilde{f}}{\partial x_i} - C_{D4} \frac{\varepsilon_f}{k} \overline{u_i'' u_j''} \frac{\partial \tilde{u}_i}{\partial x_j}$$
(2.20)

and the following constants : $C_{D1} = 2$; $C_{D2} = 1.8$; $C_{D3} = 1.7$; $C_{D4} = 1.4$; $C_{\varepsilon f} = 0.22$

3. Application

3.1.   NUMERICAL METHOD

Computations are performed with a two-dimensional numerical code which is based on finite difference discretization with colocated velocity components and an incremental version of the original fractional step method. Advection is solved with the two-dimensional characteristics method, using third order space interpolation, diffusion and source term step is solved with a semi-direct implicit method after splitting into orthogonal directions. Pressure is located at the center of the meshes and determined with a Poisson equation, which is treated with a conjugate gradient method. The mean scalar and its variance are computed with a finite volume method, using a Leonard scheme for the advection. To avoid numerical oscillations, the Reynolds stress and scalar turbulent flux components are computed on the pressure half staggered subgrid. Special efforts have been made to minimize the CPU time increase and to get a numerically stable coupling between Navier-Stokes and Reynolds stress equations [8]. Conventional wall functions (linear or logarithmic laws) are used for solid wall conditions.

3.2.   INLET CONDITIONS AND COMPUTATIONAL GRID

3.2.1.   *For the round jets*. Inlet conditions are provided by measurements performed at $z/D = 0.2$, except for the dissipation rate which is determined, using the measured turbulent kinetic energy and the integral length scale corresponding to a fully developped pipe of same flow rate and diameter. Preliminary tests about the mesh show that the results are mainly sensitive to the radial refinement.
Three grids are tested: 25X84, 49X84 and 97X84, respectively radial and axial number of nodes. For all grids, the computed domain represents 22 diameters in the axial direction. Finally, as the second order model requires a finer grid than the k-ε model, computations with the RS model are performed with the 49X84 grid, while the 25X84 grid is convenient for the k-ε computations. However, the helium/air jet exhibits sharper gradients than the other jets and the simulations of this flow with second order closure should have been performed with the finest grid in order to get absolutely grid independent results. Nevertheless, deviations between the medium grid and the finest one are of minor importance.

3.2.2.   *For the Bluff Body Flow*. The determination of the inlet conditions in this case is more complex as no measurement is available. The length of the upstream methane pipe is long enough to allow fully developed turbulence pipe conditions for the central jet, but first computations have shown a strong dependence on the characteristics of the boundary layer of the air inlet [12]. Therefore, the air flow inside the burner has been computed to provide realistic conditions at the air exit.

The computations are performed with a grid of 72X125 nodes, representing respectively 175 mm radially and 270 mm axially. The grid independence has been checked with a grid of 116X214 [13].

4. Results and Discussions

4.1.   MAIN FEATURES OF THE COMPUTATIONS

4.1.1.   *For the round jets.* First computations are performed with the standard k-ε model, the standard RS model and the model of Chassaing, and compared with the measurements for the air jet: with the standard model, as well as for the k-ε one, the mean axial velocity is underpredicted by about 20%, which corresponds to an overestimation of the radial jet expansion (figure 1); with the model of Chassaing, the predictions are clearly in a better agreement with the experimental data, except for the r.m.s. value of the radial velocity which is overestimated by 15%. Actually, these results, which are very similar to those obtained with the model of Rodi [14], where C_μ and $C_{\epsilon 2}$ are varying, led us to perform computations of the three jets with the standard RS model where the values of C_1 and C_{f1} are those proposed by Chassaing. The result of this test is the following: i)- for the air and CO2 jets, a significant improvement of the predictions is observed for dynamic and scalar quantities, they are even very close to the results obtained with the model of Chassaing in the case of the air jet; a deeper analysis shows that the slight differences which can be noticed on the mean scalar field between the modified standard model and the model of Chassaing are more related to differences of the other constants (in particular C_2 and C_{f2}) than to the additional terms which are found to be negligible; ii)- such an improvement with the modified model is not recovered for the helium jet; when comparisons are done in a similar region relatively to the development of the three jets, the agreement with the experiments is found much better for the helium jet than for the others with the standard RS model while the modified model leads to the opposite conclusion.

4.1.2.   *For the bluff body flow.* As opposed to the previous cases, second order closure for the Reynolds stresses allows a better prediction of the flow than the standard k-ε model. In particular, the location of the axial velocity minimum along the axis is rather well recovered with the standard RS model while the central jet penetration is strongly underestimated with the k-ε model (figure 2). As far as the bluff body flow consists of the mixing of a light jet in air, the dependency of the results with the constants of the model has been tested only for C_{f1}. When using the value of 6 for C_{f1} instead of 3, the penetration of methane is significantly increased but the comparison with the experiment does not show a real improvement of the prediction for the mean scalar field. The analysis of the different terms in the scalar turbulent flux equation confirms this sensitivity with C_{f1}: for the radial turbulent flux, major terms are the production by the mean scalar gradient and the slow term Φ_{if1}(figure 3). If equilibrium between these two terms was perfectly ensured, one should recover a gradient law for the radial component of the scalar turbulent flux as:

$$\overline{u_r'' f''} = - \frac{1}{C_{f1}} \frac{k}{\varepsilon} \overline{u_r'' u_r''} \frac{\partial \tilde{f}}{\partial x_r}$$

where the subscript r denotes the radial direction. For the axial component, the production term by the mean velocity gradient is also an important contribution which prevents the axial turbulent flux following a gradient law. Counter gradient evolution occurs inside the recirculation zone.

Finally, as far as the radial derivative of the radial turbulent flux is much more important than the similar term in the axial direction, the evolution of the mean scalar field is mainly controlled by that of the radial turbulent flux. This explains the sensitivity of the mean scalar field with the value of C_{fl}. A similar analysis can be made for the scalar turbulent fluxes in the round jets, except that no counter gradient effect is predicted. An equivalent Schmidt number can be deduced from the radial turbulent flux: it is found rather constant around 0.9 for air and CO2 jets, and 0.75 for helium jet, while it presents larger variations from 0.45 to 0.6 for the bluff body flow. It should be noted that in this case as well as for the round jets, the additional terms related to density variations in the second order correlation equations (Equations 2.11 and 2.15) are quite negligible.

However, experimentally, the overall behaviour of the scalar field for the bluff body flow differs significantly from that of the round jets: in particular, a sharp drop of the axial profile can be observed around the height x= 60 mm. This phenomena, which does not occur for the jets, is studied by Namazian et al. [4], who relates it to the presence of large scale structures and intermittent mixing: planar images show either a continuous central jet passing through the downstream end of the recirculation zone, or a jet that is stagnated by upstream flowing air and dispersed in the recirculation zone. Such a behaviour cannot be taken into account in the numerical simulations like those presented here.

4.2.   CHARACTERISTIC TIME OF THE SCALAR DISSIPATION

Computations have been performed using the two closures defined respectively by Equations (2.18) and (2.29) for the air jet and the bluff body flow as measurements of the scalar fluctuations are only available in these cases. For the air jet, the model of Jones&Musonge predicts more realistic but still too large fluctuation levels than the algebraic closure with a time scale ratio of 0.8 between the scalar dissipation characteristic time and that of turbulent dissipation (figure 4). With the non algebraic closure, this time scale ratio is found almost constant within the flow around 0.4 (figure 6). Considering the experimental data, one should think that a value of 0.35 should be more convenient. This rather good behaviour of the model of Jones&Musonge obtained for the air jet is not recovered for the bluff body flow, where the time scale ratio varies between 0.35 and 0.6 but the comparison with the experiments is not improved relatively to the algebraic closure.

5. Concluding Remarks

Computations have been performed with second order closure for the Reynolds stresses and the scalar turbulent fluxes in two types of density varying turbulent flows: roud jets and a bluff body flow. The radial expansion of the air and CO_2 jets is overestimated with standard RS model but rater well recovered when the constants related to the return to isotropy terms are taken as proposed by Chassaing et al. [6]. On the contrary, some features are reasonably recovered with the standard RS model for the helium jet and the bluff body flow, particularly in the latter case concerning the central jet penetration. The study of the scalar turbulent flux components show that the departure from Boussinesq assumption which is observed for the axial flux plays a minor role for the evolution of the mean scalar. Second order closure in the present cases allows a moderate improvement, but one should keep in mind its interest for three dimensional turbulent flows when the aerodynamics is particularly complex [15] as well as in cases where counter gradient effects act on other phenomena like combustion [16]. The last point is related to the time scale ratio between the scalar dissipation characteristic time and that of turbulent dissipation. When using the model of Jones&Musonge, this time scale ratio is found almost constant for the round jets and presents variations of $\pm 25\%$ for the bluff body flow. Further developments are required for the modelling of the scalar dissipation in recirculating turbulent flows as shown by the comparison between predictions and measurements of the scalar variance.

Finally, the influence of large scale structures in the bluff body flow is experimentally found of importance on the jet penetration and future works will have to deal with this through Large Eddy simulations.

This work is done in collaboration with Gaz de France - R&D Division and CORIA-Université de Rouen.

References

[1] Méchitoua, N., Mattéi, J.D., Garréton, D. and Chaumeton, B. (1994) Three dimensional flow and combustion modelling of a laboratory gas turbine combustor, International Symposium on Turbulence, Heat and Mass Transfer, Lisbon, Portugal.
[2] Djeridane , T., Amielh, M., Anselmet, F. and Fulachier, L. (1996) Turbulence in the near-field region of axisymmetric variable density jets, Phys. Fluids 8, 1614-1630.
[3] Anselmet, F., Djeridi, H. and Fulachier, L. (1994) Joint statistics of a passive scalar and its dissipation in turbulent flows, J. Fluid Mech. 280, 173-197.
[4] Namazian, N., Kelly, J., Schefer, R.W. (1992) Concentration imaging measurements in turbulent concentric jet flows, AIAA Journal, 30(2), 384-394.
[5] Chen, R.H., Driscoll, J.F., Kelly, J., Namazian, M. and Schefer, R.W. (1990) A comparison of bluff body and swirl-stabilized flames, Combust. Sci. and Tech. 71, 197-217.
[6] Chassaing, P., Harran, G. and Joly, L. (1994) Density fluctuation correlations in free turbulent binary mixing, J. Fluid Mech. 279, 239-278.
[7] Launder, B.E. and Spalding, D.B. (1974) The numerical computation of turbulent flows, Computer Method in Applied Mechanics and Engineering 3, 537-566.
[8] Bel Hassan, M. and Simonin, O. (1993) Second-Moment Predictions of Confined Turbulent Swirling Flows, in Presses de l'ENPC (eds), Proc. of the 5th International Symposium on Refined Flow Modelling and Turbulence Measurements, pp. 537-544.
[9] Daly, B.J. and Harlow, F.H. (1970) Transport equations in turbulence, Physics of Fluids 13(11), 2634-2649.
[10] Launder, B.E., Reece, G.J. and Rodi, W. (1975) Progress in the development of a Reynolds stress turbulence closure, J. Fluid Mech. 68(3), 537-566.
[11] Jones, W.P. and Musonge, P. (1988) Closure of the Reynolds stress and scalar flux equations, Phys. Fluids 31, 3589-3604.
[12] Garréton, D. and Simonin, O. (1995) ASCF: Report of Activities, ERCOFTAC Bulletin, June, 29-35.
[13] Elamraoui, R. and Garréton, D. (1996) Simulation numérique au second ordre du mélange turbulent de deux écoulements turbulents en aval d'un obstacle, EDF internal report HE-44/96/012.
[14] Rodi, W. (1972) The prediction of free turbulent boundary layers by use of a two-equation model of turbulence, Ph. Thesis, Imperial College.
[15] Deutsch, E., Méchitoua, N. and Mattéi, J.D. (1996) Flow Simulation in piping system dead legs using second moment closure and k-epsilon model, Proc. of the 6th International Symposium on Refined Flow Modelling and Turbulence Measurements, Sept. Thallahasse, USA.
[16] Bailly, P., Champion, M. and Garréton, D. (1995) Numerical study of a combustion zone stabilized by a rectangular section cylinder, in 10th Symposium on Turbulent Shear Flows, 22(19)-22(24).

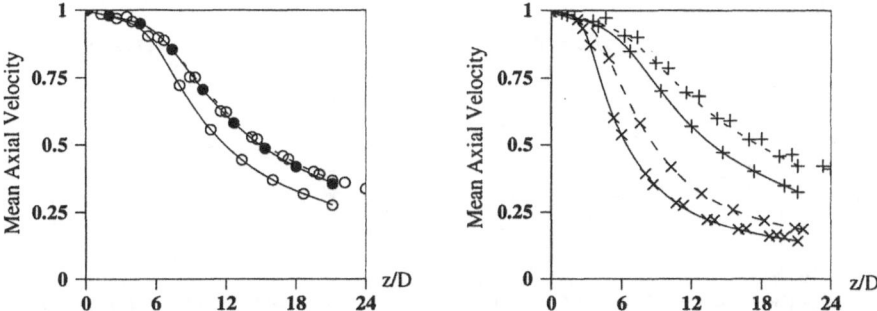

Figure 1: Non dimension mean axial velocity profiles along the centreline on the left for the air (o) jet, on the right for the CO2 (+) and helium (x) jets. Solid line: standard RS model, dashed line: standard RS model with modifed values of C1 and Cf1, •-•: model of Chassaing, isolated symbols: measurements.

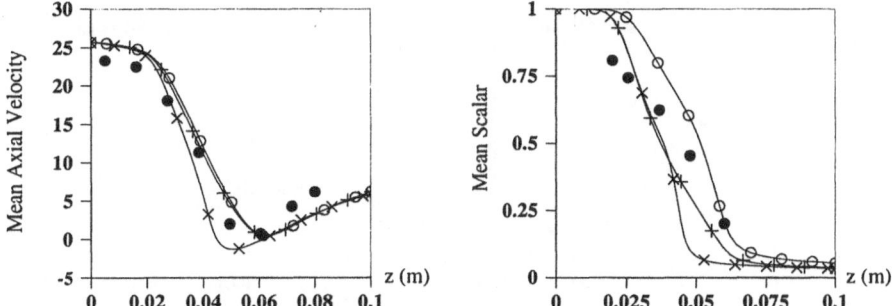

Figure 2: axial profiles for the bluff body flow of the mean axial velocity (on the left) and the mean scalar (on the right), +-+: standard RS model, x-x: standard k-ε model, o-o: standard RS model with modifed value of Cf1 ; •: measurements.

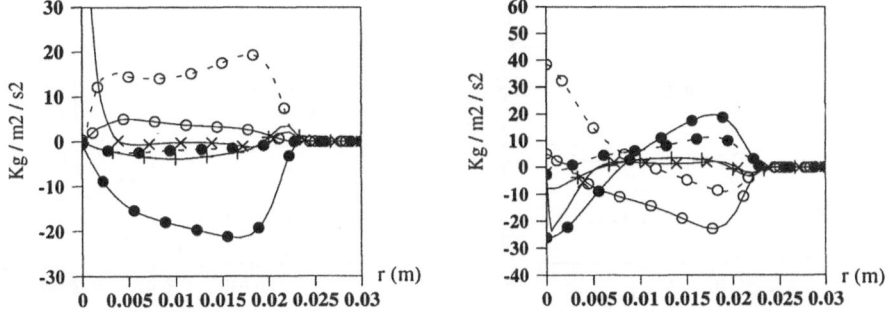

Figure 3: Radial profiles at 60 mm of the budgets for the radial (on the left) and axial (on the right) scalar flux components for the bluff body flow; +-+: convection, x-x: turbulent diffusion, o-o: production by mean velocity gradients, o- -o: production by mean scalar gradients, •-•: slow part of the pressure-scalar correlation, •- -•: rapid part.

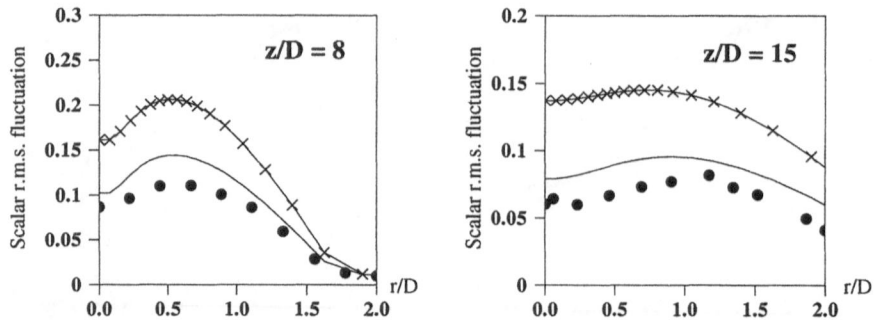

Figure 4: Radial profiles of the scalar r.m.s. fluctuation at 8 and 15 diameters for the air jet, x-x: time scale ratio of 0.8, solid line: model of Jones & Musonge, •: measurements.

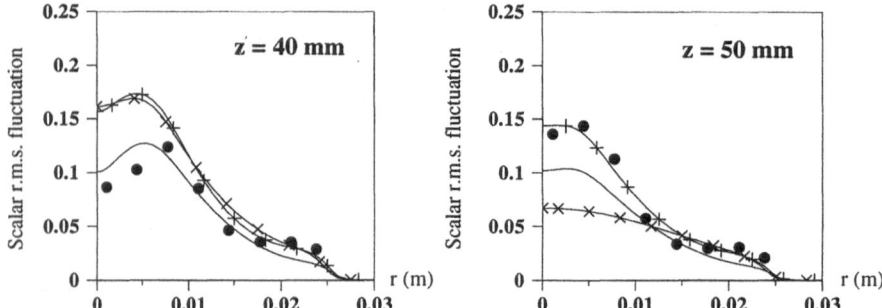

Figure 5: Radial profiles of the scalar r.m.s. fluctuation at 40 and 50 mm for the bluff body flow, solid line with symbols: time scale ratio of 0.8, + for RS model, x for k-ε model, solid line: model of Jones & Musonge, •: measurements.

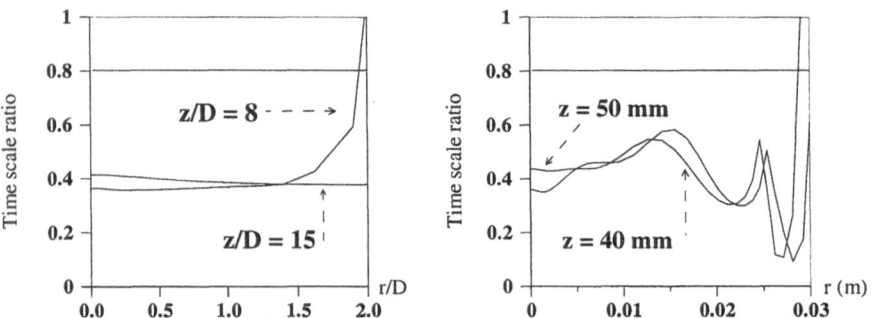

Figure 6: Radial profiles of the time scale ratio for the air jet (on the left) and for the bluff body flow (on the right).

VARIATION OF THE EFFECTIVE NOZZLE DIAMETER IN ROUND TURBULENT JETS WITH VARIABLE DENSITY

JC. SAUTET and D. STEPOWSKI
CORIA URA CNRS 230 - Université de Rouen
76821 Mont Saint Aignan - FRANCE

1. INTRODUCTION

The dynamic behaviour of round turbulent jets with variable density has been the subject of a great deal of studies (Abramovich [1] (1963), Pitts et al [2] (1984), Dowling and Dimotakis [3] (1988), Pitts [4] [5] (1991)). Most investigations are assuming low Mach number, constant pressure (remote from walls) and high Reynolds number (negligible molecular viscosity). Such turbulent flows may be classified according to the relative value of two forces : the momentum flux, $M(x) = \pi \int_0^\infty \rho U^2 d(r^2)$, and the buoyancy flux,

$B(x) = \pi g \int_0^x \int_0^\infty (\rho_\infty - \rho) d(r^2) dx$, where x and r are the axial (vertical) and radial coordinates respectively, ρ_∞ is the density of the ambient fluid. In the near development field the inertial forces dominate the buoyancy forces and the flow is a pure jet over which the density varies as the centerline velocity decays. Chassaing [6] and Chen [7] have shown that the axial distance before which buoyancy effects are negligible is

$x_M = \dfrac{d_0}{2} Fr^{1/2} R_\rho^{1/4}$ where d_0 is the nozzle diameter, Fr is the Froude number and R_ρ is the initial density ratio between injected and ambient fluids.

The development of a constant density jet jet may be self-similar if one length scale, d_0, is sufficient to render its relative time-averaged quantities dimensionless functions of one geometrical variable only. In the first development stages of a variable density jet over which buoyancy forces are still negligible, the jet density varies from ρ_0 to ρ_∞ and a pseudo-similar behaviour can be found if a variable scale $d_{eff}(x)$ is introduced to account for the variation of the momentum effective jet density. We have derived the effective nozzle diameter, $d_{eff}(x)$, from laser Doppler velocity measurements over different section of several jets with different density ratios, $R_\rho = \rho_0/\rho_\infty$. With such a variable length scale the rate of the hyperbolic decay of the centerline velocity is constant and independent of R_ρ. In this paper we propose a general law for the axial evolution of $d_{eff}(x)$ as a function of R_ρ. Beyond the problem of variable density jets, the relative evolution of this length scale is related to the decay of the global strength of the nozzle fluid as the jet develops.

L. Fulachier et al. (eds.), IUTAM Symposium on Variable Density Low-Speed Turbulent Flows, 119–126.

2. EXPERIMENTAL CONFIGURATION (figure 1)

We have investigated the near field ($5 \leq x/d_0 \leq 20$) of four round turbulent jets with density ratio R_ρ ranging from 0.07 to 1 (Sautet and Stepowski [8]). The turbulent jets ($U_0 = 45$ m/s; $u'_0/U_0 = 9\%$) issuing from a vertical pipe ($d_0 = 10$ mm, $L = 1$ m) are fed with various mixtures of hydrogen and nitrogen. Table 1 lists the input parameters and the distance x_M before which buoyancy effects are negligible. The jets are discharging into a slow coflow air stream ($U_\infty = 4$ m/s) which allows the seeding of the ambient with particles as required for careful laser Doppler velocimetry. Particular attention was paid to the seeding conditions and averaging procedures in order to minimize the different biases which can be very significant in the near field. For each density ratio both the jet and the coflow were equitably seeded in terms of mass rate. Then time averaged data were obtained by low frequency resampling of the real time registered data.

On the other hand the coflow makes possible to derive the mixture fraction profiles from laser Mie scattering measurements with either the jet or the coflow being seeded (Sautet [8].).

TABLE 1. Properties of turbulent jets investigated

Jet / coflow	$R_\rho = \dfrac{\rho_0}{\rho_\infty}$	Re	x_M/d_0
N_2 / air	1	27000	1250
50% H_2 - 50% N_2 / air	0.53	16400	210
80% H_2 - 20% N_2 / air	0.25	8700	30
H_2 / air	0.07	4100	10

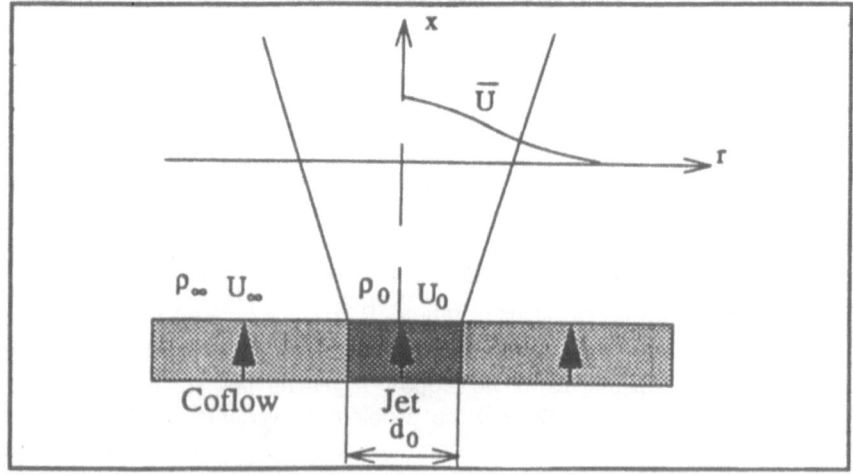

Fig 1. Experimental configuration

3. DIMENSIONAL ANALYSIS

As far as buoyancy effects are negligible ($x<x_M$), the integral conservation equations for mass and momentum fluxes write (Sautet and Stepowski [9] , 1995)

$$\frac{\partial}{\partial x}\int_0^{\infty}\overline{\rho\,U}\,d(r^2)+2\left[r\,\overline{\rho\,V}\right]_0^{\infty}=0 \tag{1}$$

$$\frac{\partial}{\partial x}\int_0^{\infty}\overline{\rho\,U^2}\,d(r^2)+2\left[r\,\overline{\rho\,U\,V}\right]_0^{\infty}=0 \tag{2}$$

where V is the radial jet velocity. Equations (1) and (2) yields to

$$\int_0^{\infty}\overline{\rho U(U-U_\infty)}\,d(r^2)=\rho_0 U_0(U_0-U_\infty)\frac{d_0^2}{4} \tag{3}$$

In order to simplify, the coflow velocity, U_∞, will be formally neglected in the following presentation. This simplification comes first to deal with momentum flux instead of excess momentum flux, second to deal with normalized radial profiles of velocity instead of excess velocity profiles and to replace $\sqrt{U_{cl}^2-U_\infty^2}\Big/U_0(U_0-U_\infty)$ by the simple decay of the centerline velocity $U_{cl}\Big/U_0$. However all measurements and data processings have been actually performed taking into account the coflow, in terms of excess quantities.

The concept of effective nozzle diameter has been introduced by Thring and Newby [10] (1961) to account for density effects in the far field region where the mixing degree is so high that the local density, ρ, can be approximated by the ambient density, ρ_∞. They defined an effective nozzle diameter $d_{eff\infty}=d_0\sqrt{\rho_0/\rho_\infty}$, which reduces the far field jet decay to that of a constant density jet. Physically, $d_{eff\infty}$, is the diameter of the pipe fed with ambient that would induce the same far field momentum flux as that really flowing. However buoyancy effects are not accounted for in this approach. This concept has been extended by Sautet et al [9] (1995) to the near field region (where buoyancy forces are negligible) over which the jet density cannot be approximated by the ambient density. In order to reduce the momentum flux conservation (Eq (3)) to that of a pseudo-constant density flow over a given section, we define the momentum effective jet density :

$$\rho_{eff}(x)=\frac{\int_0^{\infty}\overline{\rho\,U^2}\,d(r^2)}{\int_0^{\infty}\overline{U^2}\,d(r^2)}=\frac{\rho_0 U_0^2\,d_0^2}{4\int_0^{\infty}\overline{U^2}\,d(r^2)} \tag{4}$$

and the effective nozzle diameter,

$$d_{eff}(x) = d_0 \sqrt{\frac{\rho_0}{\rho_{eff}(x)}} \tag{5}$$

so that the momentum flux conservation reduces to

$$\int_0^{\overline{\;}} \overline{U}^2 \, d(r^2) = \frac{U_0^2 \, d_{eff}^2(x)}{4} \tag{6}$$

4. PSEUDO SELF-SIMILARITY WITH $d_{eff}(x)$ SCALING

The momentum effective jet density, $\rho_{eff}(x)$, has been derived (through Eq (4)) from LDV measurements over different sections x of the investigated jets. Figure 2 shows the axial evolution of $\rho_{eff}(x)/\rho_\infty$ in the near development field of the turbulent jets. The growth rate of the effective jet density is higher for the low density jets.

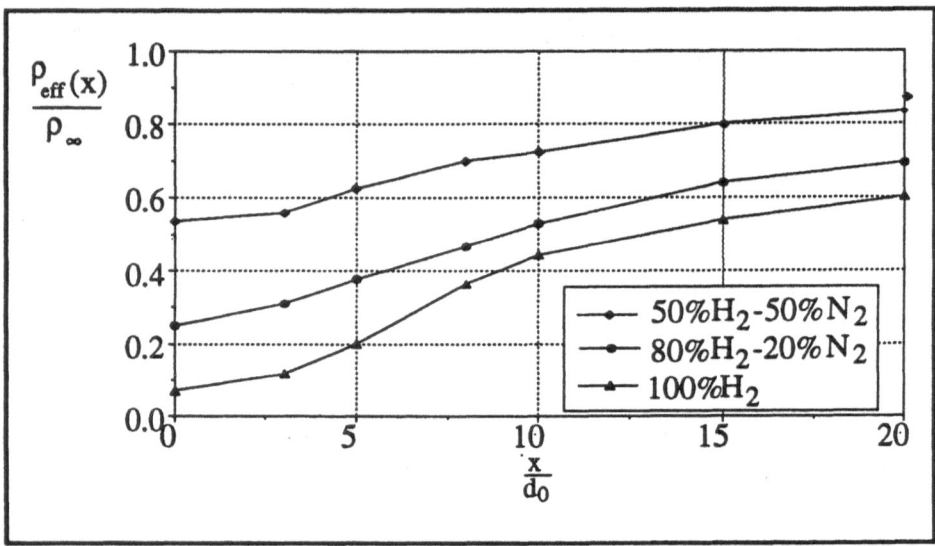

Fig 2. Axial evolution of the normalized effective jet density, $\rho_{eff}(x)/\rho_\infty$

Using the effective nozzle diameter as a variable length scale should reduce the jet decay to the hyperbolic decay of a constant density jet. Accordingly the centerline velocity decays have been plotted in figure 3 as a function of the reduced variable $x/d_{eff}(x)$. This plot shows a remarkable collapse of the data along a unique straight line for all jets according to

$$\frac{U_0}{\overline{U}_{cl}} = K \, \frac{x - x_0}{d_{eff}(x)} \tag{7}$$

where $x_0 = d_0 \left[-6.5 + 5.4 \sqrt{R_\rho} \right]$ is the virtual origin of the similarity (Sautet [9]). The decay rate $K = 0.118 \pm 0.003$ is in good agreement with reported data (Green and Whitelaw [11]). Recent results (Sautet and Stepowski [9]) have shown that the centerline growth of the relative fluctuations are also collapsing along a unique function of $x/d_{eff}(x)$. The later result underlines the effectiveness of the scaling by $d_{eff}(x)$.

To avoid tricky measurements of the momentum effective density over different sections of the jet we have investigated a general law for the evolution of $d_{eff}(x)$. Such a law would allow to predict the near velocity field of any jet with variable density.

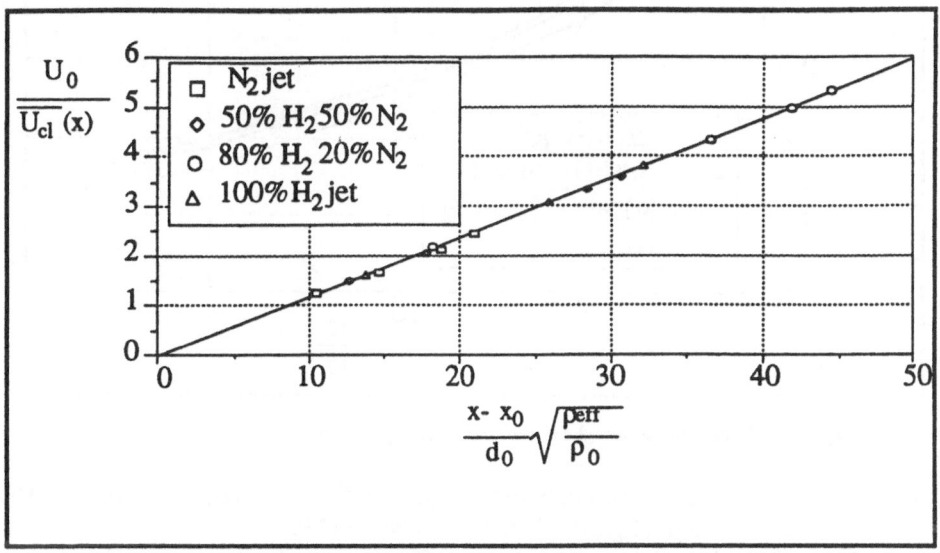

Fig 3. Centerline velocity decay as a function of the reduced variable, $x/d_{eff}(x)$

5. EVOLUTION OF THE EFFECTIVE NOZZLE DIAMETER

The curves shown in figure 2, where the growth rates of $\rho(x)$ is higher for the low density jets, suggest that a proper normalization of the data could collapse the evolutions into a unique function decaying from 1 at the nozzle, where $\rho = \rho_0$ and $d_{eff} = d_0$, down to 0 in the far field, where $\rho \to \rho_\infty$ and $d_{eff} \to d_0 \sqrt{\rho_0 / \rho_\infty}$. After different tries, it appears that our effective density data collapse (within 10%) along a unique curve (see figure 4) when plotted as

$$\beta(x) = \frac{\sqrt{\rho_{eff}} - \sqrt{\rho_\infty}}{\sqrt{\rho_0} - \sqrt{\rho_\infty}} = \frac{d_{eff}^{-1} - d_{eff\infty}^{-1}}{d_0^{-1} - d_{eff\infty}^{-1}} \tag{8}$$

Beyond an inertial region where $\beta(x)$ keeps its initial value ($\beta(x)=1$ when $x<5\ d_0$), the decay of b can be approximated by

$$\beta(x) = \frac{4.67\ d_0}{x} \qquad (9)$$

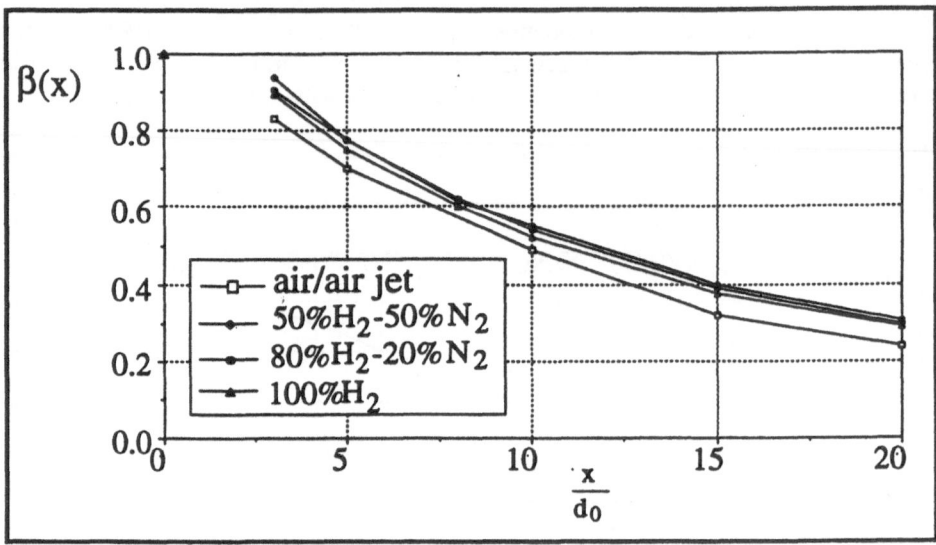

Fig 4. Axial evolution of the global unmixedness degree, $\beta(x)$.

In the air/air jet in which β is formally indeterminated, β has been derived from the decay of the nozzle fluid strength, $Z_{eff}(x)$, given from relation (13).

Then, when $x>5\ d_0$, the axial evolution of the effective nozzle diameter can be described by

$$\frac{d_{eff}(x)}{d_0} = R_\rho^{\frac{1}{2}} \left[1 + \left(R_\rho^{\frac{1}{2}} - 1 \right) \frac{4.67\ d_0}{x} \right]^{-1} \qquad (10)$$

6. INTERPRETATION OF $\beta(x)$

$\beta(x)$ can be considered as a global unmixedness degree over the jet based on dynamic mixing. In the literature (Wignanski and Fiedler [12], Birch et al [13]), $x/d_0=45$ is often given as a typical distance for approaching the far field regime of complete similarity in which turbulence is only due to transport by the mean and the turbulent motion (so that turbulent viscosity concept can be used). This estimate is consistent with our results showing that the global mixing level $(1-\beta)$ reaches 90% when $x=45\ d_0$.

Although the formulation of $\beta(x)$ (Eq. 8) leads to an apparent indetermination for constant density jets. The definition of the global dynamic unmixedness degree, $\beta(x)$, applies to any free jet and may be related to the global strength of the nozzle fluid over a given section. The normalized strength of the nozzle fluid, noted $Z_{eff}(x)$, writes

$$Z_{eff}(x) = \frac{\int_0^\infty \overline{\rho Z U^2} \, d(r^2)}{\int_0^\infty \overline{\rho U^2} \, d(r^2)} \qquad (11)$$

where the mixture fraction, Z, is the local mass fraction of fluid which originated in the jet. Under constant temperature and pressure conditions, $Z = (\rho^{-1} - \rho_-^{-1})/(\rho_0^{-1} - \rho_-^{-1})$ so that

$$Z_{eff} = \frac{d_{eff}^2(x) - d_{eff-}^2}{d_0^2 - d_{eff-}^2} \qquad (12)$$

In the limit of constant density jets, $Z_{eff}(x) \equiv \beta(x)$, the global strength of the nozzle fluid should decay as the global unmixedness degree does, according to

$$\left[\frac{\int_0^\infty \overline{Z U^2} \, d(r^2)}{\int_0^\infty \overline{U}^2 \, d(r^2)} \right]_{\rho=Cste} = \beta(x) \qquad (13)$$

Joint laser measurements of mean mixture fraction and velocity have been used to estimate the evolution of the nozzle fluid strength in constant density jet. In figure 4 one can verify that the decay of $[Z_{eff}(x)]_{\rho=Cste}$ fits with the decay of $\beta(x)$ as soon as the contribution of the fluctuating terms can be neglected ($x>7$ d_0).
More generally, in any pure jet the decay of the nozzle fluid strength can be described by

$$Z_{eff}(x) = \frac{R_\rho}{1-R_\rho} \left[\left(1 + \left(\sqrt{R_\rho} - 1 \right) \beta(x) \right)^{-2} - 1 \right] \qquad (14)$$

using the approximation $\beta(x)=1$ when $x/d_0<5$ and $\beta(x) = \dfrac{4.67 \, d_0}{x}$ when $x/d_0>5$.

7. CONCLUSION

The near development field of turbulent, variable density jets can be investigated through a pseudo-self-similar approach using a variable length scale. This length scale is provided by the momentm effective nozzle diameter which varies from its bulk value , d_0, towards its far field value, $d_0\sqrt{\rho_0/\rho_-}$. Based on laser measurements in four turbulent jets with different density ratio, we have proposed a general law for the near field evolution of $d_{eff}(x)$ which makes it possible to predict the near velocity field in jets with variable density. Beyond the effects of density variation on the jet development, the relative normalized evolution of the effective diameter measures the decay of global unmixedness as the jet develops. The decay of this dynamic unmixedness degree is related to that of the nozzle fluid strength which reduces to a simple hyperbolic decay in constant density jets.

REFERENCES

[1] Abramovich, G.N. (1963) The theory of turbulent jets, Cambridge, MA: The MIT Press

[2] Pitts, W.M and T. Kashiwagi, T. (1984) The application of laser-induced Rayleigh light scattering to the study of turbulent mixing, J Fluid Mech, Vol 141

[3] Dowling, D.R and Dimotakis, P.E. (1988) On mixing and structure of the concentration field of turbulent jets, Proceedings of the first national Fluid Dynamics Congress, part 2, 982-988. American Institute of Aeronautics ans Astronautics

[4] Pitts, W.M. (1991) Effects of global density ratio on the centerline mixing behavior of axisymmetric turbulent jets, Exp in Fluids, Vol 11

[5] Pitts, W.M. (1991) Reynolds number effects on the mixing behavior of axisymmetric turbulent jets", Exp in Fluids, Vol 11

[6] Chassaing, P. (1979) Mélange turbulent de gaz inertes dans un jet tube libre. Thèse de doctorat. Toulouse.- France

[7] Chen, C.J and Rodi, W. (1980) Vertical turbulent buoyant jets. A review of expérimental data. Pergamon, New YorK

[8] Sautet, JC and Stepowski, D. (1994) Single shot laser Mie scattering measurements of the scalar profiles in the near field of turbulent jets with variable densities, Experiments in Fluids, Vol 16

[9] Sautet, JC and Stepowski, D. (1995) Dynamic behavior of variable-density, turbulent jets in their near development fields, Physics of Fluids, Vol 7(11)

[10] Thring, M.W and Newby, M.P. (1953) Combustion length of enclosed turbulent jet flames, Flames and Fuel Jets

[11] Green, H.G and Whitelaw, J.H. (1985) Velocity and concentration measurements in the near field of round jets, Imperial College of Science and Technology, Mech Eng Department, London

[12] Wygnanski, L and Fiedler, H (1969) Some measurements in the self-preserving jet. J Fluid Mech Vol 38, 577

[13] Birch, A.D. and Brown, D.R. and Dodson, M.G. and Thomas, J.R (1978) the turbulent concentration fieldof a methane jet. J Fluid Mech, Vol 88, 431

MIXING IN COAXIAL JETS WITH LARGE DENSITY DIFFERENCES

M.FAVRE-MARINET[1] - E.CAMANO[2]

[1]*Laboratoire Des Ecoulements Géophysiques et Industriels.*
Institut de Mécanique de Grenoble   CNRS-UJF-INPG
B.P.53 X - 38041 Grenoble Cedex
[2]*Instituto de Pesquisas Hidraulicas, Porto Alegre, Brasil*

1. Aim of research

Coaxial jets are present in several practical applications, especially in the field of combustion, where the problem of mixing is largely controlled by the flow dynamics. This has motivated many studies on isothermal coaxial jets, principally in the homogeneous case (Champagne and Wygnanski 1971, Kwan and Ko 1976, Au and Ko 1987, Dahm et al 1992). On the contrary, very few studies have considered the effects of density on these flows. Gladnick et al (1990) investigated the velocity and concentration fields in an axisymmetric jet flow of CFC-12 issuing into a coannular jet flow of air (density ratio: 0.26). They found an increase of mixing with increasing velocity ratio (limited to 2 in their study). This was attributed to the presence of annular vortex rings originating in the mixing layer between the jet and coannular flows.

The present research is an attempt to simulate the flow conditions of liquid propellant rocket engine injectors, characterized by a low velocity-high density inner jet surrounded by a high velocity - low density outer jet. By using various gas combinations (air, helium, SF_6), it is possible to reach very large density differences in the flow field near the nozzle. The objectives of the research are to investigate the effects of density variations on the flow, focused in this paper on the problem of mixing. More precisely, the aim of the present work was to investigate the density field by using thermo-anemometry and to relate the results to the flow dynamics.

2. Experimental set-up and instrumentation

The experimental set-up consists of a pair of concentrical axisymmetric nozzles discharging into quiescent ambient air (density: ρ_∞). The inner nozzle is supplied by air and the annular one by air or helium (figure 1). The experimental conditions are:

* Inner jet (i)   Diameter: $D_i = 20$ mm   Bulk velocity: U_i density:ρ_i
* Outer jet (e)   Diameter: $D_e = 27$ mm   Bulk velocity: U_e density:ρ_e
* Diameter ratio $\beta = D_e/D_i = 1.35$ (constant in this experiment)
* Bulk velocity ratio $r_v = U_e/U_i$ $(3 < r_v < 70)$

127

L. Fulachier et al. (eds.), IUTAM Symposium on Variable Density Low-Speed Turbulent Flows, 127–134.
© 1997 *Kluwer Academic Publishers.*

* Density ratio $S = \rho_e/\rho_i$ $(0.028 < S < 1)$
* Flux momentum ratio $M = r_v^2 S$ $(1 < M < 200)$

The Reynolds number used results from the momentum conservation equation:
$Re_M = Re_{ext} [1+(1-M)/\beta^2 M]^{1/2}$, where $Re_{ext} = \rho_e U_e D_e/\mu_e$. Typical values are:
* air/air jets: $Re_{ext} = 11000$ $7300 < Re_M < 7800$,
* helium/air jets: $Re_{ext} = 3200$ $2100 < Re_M < 2400$,

To vary the velocity ratio r_v, the inner velocity was progressively changed and the outer velocity was kept constant (6 m/s for air and 16 m/s for helium). SF_6 was used as the inner fluid to investigate very low density ratio. For most experiments, the jet issued into ambient air. To test the effect of ambient fluid density, some measurements were made with the jet issuing into a chamber filled with helium (cross-section 0.66 m x 0.90 m).

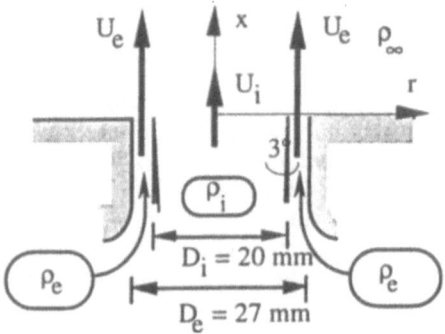

Figure 1 - Sketch of the coaxial jet.

Measurements were made with an aspirating probe, similar to that used by Brown and Rebollo (1972) for the study of the inhomogeneous mixing layer. It consists of a very thin tube (I.D. 80 μm, O.D. 300 μm) connected to a vacuum pump. When the aspirating pressure is lower than a critical value, the mass flow is blocked in the tube and a hot-wire placed inside is then insensitive to velocity fluctuations. Contrary to the arrangement of Brown & Rebollo, in the present experiment the hot-wire was located at the entrance of the suction tube in order to improve the time-response of the probe. The probe was calibrated in the inner nozzle flow of known velocity and density. It has been verified that it remains only sensitive to density variations in this configuration. The principle of the probe implies perturbations of the flow since nearly sonic conditions are maintained at the tube entrance. Fortunately, perturbation velocities decrease very fast ($\approx 1/r^2$, r: distance to the probe) and the spatial resolution is estimated theoretically to a distance of the order of 1 mm. The time-response of the probe is about 1 ms. These probe characteristics have been checked in a helium-air mixing layer and in the far-field of a helium-air jet. In the latter flow, the measured mean density and the corresponding fluctuation r.m.s are in good agreement with the results of Pitts (1986).

3. Results

3.1 ONSET OF RECIRCULATION

The initial stages in the development of the flow are governed by the growth of the inner mixing layer, generated between the inner and the coannular jets and that of the outer mixing layer between the coannular jet and the ambient fluid. An important feature of the flow is the entrainment of fluid issuing from the central nozzle into the inner mixing layer, which results in a decrease of the mass flow rate in the central region. The corresponding entrainment velocity increases with increasing velocity ratio r_v and it is expected that for sufficiently high values of r_v, a regime of recirculation occurs near the nozzle. Although the present situation is much more complex, the plane mixing layer may be considered as a reference to understand the influence of the parameters which characterize the flow. Effects of density have been considered by Brown and Roshko (1974) and more recently by Dimotakis (1986), who has modeled the asymmetric entrainment on the two sides of the mixing layer. Very recently, numerical simulations of Soteriou and Ghoniem (1995) have confirmed the observed tendency to enhancement of mixing when the slower stream becomes denser. In Dimotakis' model, the mixing layer growth is due to entrainment by the large-scale vortical structures present in the flow and is predicted by modelling the entrainment velocity v_E on each side of the mixing layer. On the low-speed side of the inner mixing layer (free stream velocity:U_i), v_E is given by:

$$v_E / (U_c - U_i) = \varepsilon$$

ε seems to be independent of r_v and S (at least for moderate values of S). The convection velocity U_c of the vortical structures is related by the model to the freestream velocities and to the density ratio S, resulting in:

$$v_E/ U_i = \varepsilon \frac{S^{1/2}(r_v - 1)}{1 + S^{1/2}} \qquad (1)$$

It can be seen that for high values of the velocity ratio r_v, v_E is directly related to the momentum ratio M, and in the limit $S \longrightarrow 0$, $v_E /U_i = \varepsilon M^{1/2}$. When applied to coaxial jets with high velocity ratios, this model suggests that the effect of density may be taken into account by considering the flux momentum ratio M to characterize the flow. However, when M is kept constant, a second-order effect of density is still present and (1) indicates an enhancement of entrainment of the low-speed stream when the high-speed stream is at low density.

These predictions are confirmed by the present experimental investigation. It has been shown that the flow may have two regimes depending on the outer to inner jet velocity ratio (Villermaux et al 1994, Camano and Favre-Marinet 1994). The effect of density has been explored by using various combinations of air, helium and SF6 and by varying S from 1 to values as low as 0.028. The onset of recirculation was detected by visualizations, laser-velocimetry and thermo-anemometric techniques (Camano 1996). These different methods are in good agreement and show that effects of density are rather well taken into account by considering the jet momentum ratio M and not r_v and S separately. As a matter of fact, the separation between the two regimes may be defined in terms of a critical ratio M_c . When M is higher than M_c (≈ 50), recirculation occurs near the nozzle and the jet dynamics are then radically modified. For $M<M_c$, regular axisymmetric structures and longitudinal vortices appear in the first three diameters of the annular jet. For $M>M_c$, this pattern of vortices becomes more disordered whereas large-scale low-frequency instability of the recirculating bubble is observed. Figure 2 presents laser-sheet visualizations of air-SF6 coaxial jets (S = 0.21) for the two regimes. For this gas combination, the boundaries of the recirculation bubble are closer to the nozzle compared to the homogeneous case. This should be due to enhancement of entrainment in agreement with equation (1).

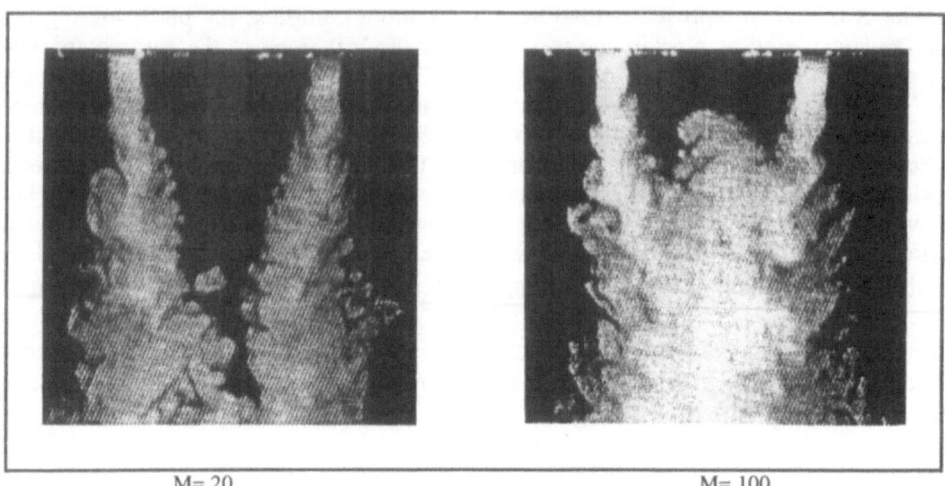

M= 20   M= 100

Figure 2- Visualizations of air-SF6 coaxial jets by laser-sheet

3.2 MIXING IN COAXIAL JETS

3.2.1. *Density measurements along the jet-axis*

Density was first measured by an aspirating probe in a helium-air jet flowing in the vertical upward direction. For this combination of gas, it has been verified that the results obtained in the near-field are not affected by viscous effects, nor by gravity effects. The first point was checked by comparing measurements for Re_M = 2520 and 1260. On the other hand, experiments were conducted with the nozzle in the upward and in the downward direction. Quite similar results were found in the different experiments.

Density effects were first tested by comparing pure helium-pure air jets (S = 0.138) to coaxial jets where a moderate value of S (S = 0.655) was obtained by using a air-helium mixture as the annular jet fluid. Figure 3 compares the evolution of the normalized density ($\rho^* = \dfrac{\rho - \rho_e}{\rho_i - \rho_e}$, here $\rho_\infty = \rho_i$) along the jet-axis for two experiments conducted with the same value of the velocity ratio r_v. It clearly demonstrates a significant effect of density on the evolution of $\overline{\rho^*}$ along the jet-axis when r_v is kept constant. The length of the potential core is considerably reduced by an increase of S, indicating an enhancement of mixing. This is in agreement with equation (1) which gives an inner entrainment velocity 65% higher for S = 0.655 than for the other case. It must be noted that M is very close to the critical value M_c for the highest value of S.

When M is kept constant (Figure 4), mixing is much less affected by effects of density. It is observed that for S = 0.655 the length of the potential core is slightly reduced and that the minimum of $\overline{\rho^*}$ is a little higher than for S = 0.138. However, the downstream evolutions of $\overline{\rho^*}$ and $\rho^{*\prime}$ ($= \sqrt{\overline{\rho^{*\prime 2}}}$) are very similar for the two cases. This result also holds for the regime of recirculation (M = 144).

The downstream evolution of the mean density and fluctuation r.m.s $\rho^{*\prime}$ is significantly affected by M (figure 5). The mean density begins to decrease on the jet-axis when helium particles issued from the outer jet reach the central region.

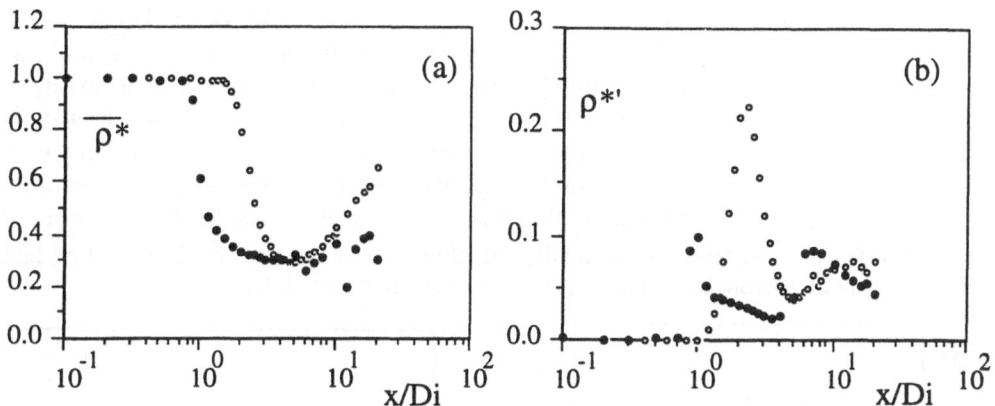

Figure 3 - Variations of density along the jet-axis for a fixed value of the velocity ratio $r_v = 8$,
(a) mean value, (b) r.m.s of fluctuations.
  o   S = 0.138 — M = 9   •   S = 0.655 — M = 42

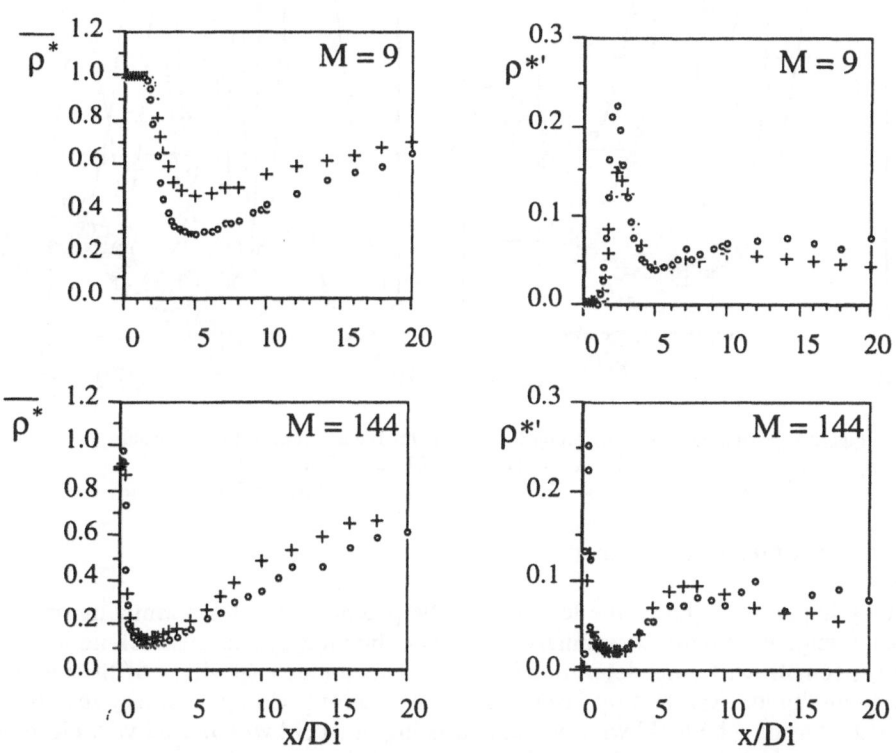

Figure 4 Variations of density along the jet-axis for fixed values of the momentum ratio M,
(a) mean value, (b) r.m.s of fluctuations. o S = 0.138   + S = 0.655

The corresponding length of the potential core L decreases with increasing M (figure 5a).
For the highest values of M ($M>M_c$), $\overline{\rho^*}$ and $\rho^{*\prime}$ reach low values in the recirculation
region ($0.6<x/D_i<1$). Intense ρ'-fluctuations take place in the upstream region, which

may be related to large oscillations of the upstream boundary of the recirculation bubble, as shown by visualizations. The low minimum values of the mean density ($\overline{\rho^*}_{min}$) and $\rho^{*\prime}$ downstream from the recirculation bubble correspond to a high degree of mixing in the initial inner region. For lower values of M, the mean density on the jet-axis is affected by mixing with outer ambient air and $\overline{\rho^*}_{min}$ increases substantially with decreasing M. $\rho^{*\prime}$ reaches a maximum value at an abscissa $x_{\rho}{\prime}$, slightly downstream from the end of the potential core (figure 5b). It is clear from these results that the onset of recirculation does not affect dramatically the global features of the evolution of $\overline{\rho^*}$ and $\rho^{*\prime}$, and cannot therefore be deduced from such measurements of density.

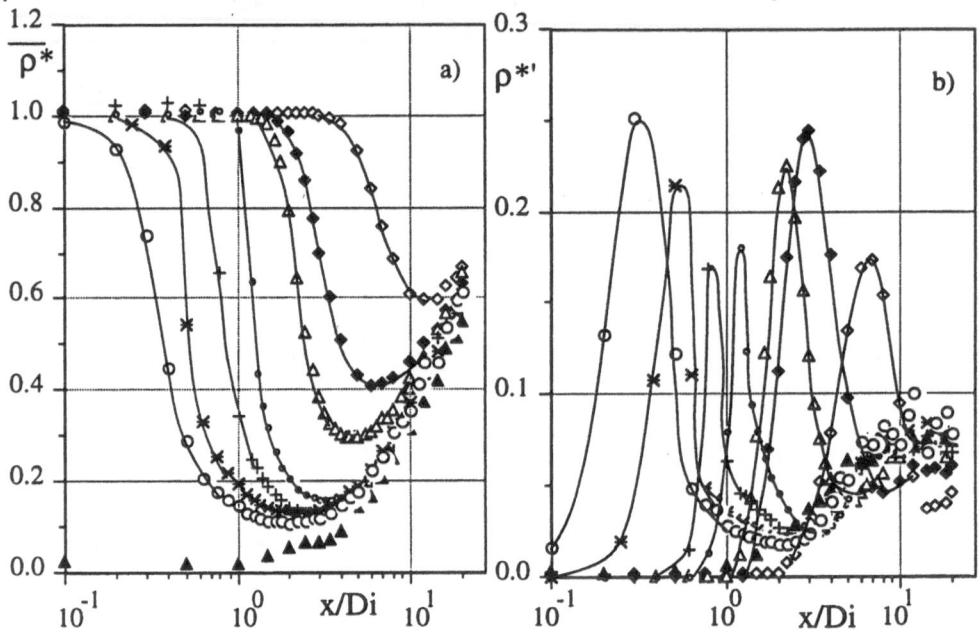

Figure 5 Variations of density along the jet-axis for fixed values of the momentum ratio M, S = 0.138 : (a) mean value, (b) r.m.s of fluctuations.
M= ◇ 1, ◆ 4, △ 9, ● 36; + 64, ✳ 100, ○ 144, ▲ ∞

3.2.2. *Length of the potential core*

It is very important to know the length L of the potential core for optimizing operation of rocket-engines. Figure 6 summarizes results obtained by various techniques in the homogeneous flow. L was determined by laser-anemometry (position of the minimum velocity on the jet-axis) and by laser-sheet visualizations. Complementary results were obtained by using a hot/cold wire probe, consisting of a cold wire placed very close from a hot-wire (typical distance: 20μm). In this method, L was determined by detecting the maximum rms fluctuations of the signal given by the cold wire. All methods are in remarkable agreement and show that L follows a power-law $M^{-1/2}$. Moreover, the present results collapse with that of Rehab et al (1996) obtained in a geometrically similar water channel. The length of the potential core was also determined from the measurements of density as the distance L_{ρ} where $\overline{\rho^*}$ departs from 1 by a given percentage (say 1%). For the moderate value of S (0.655), the behaviour of L_{ρ} is quite similar to that of L

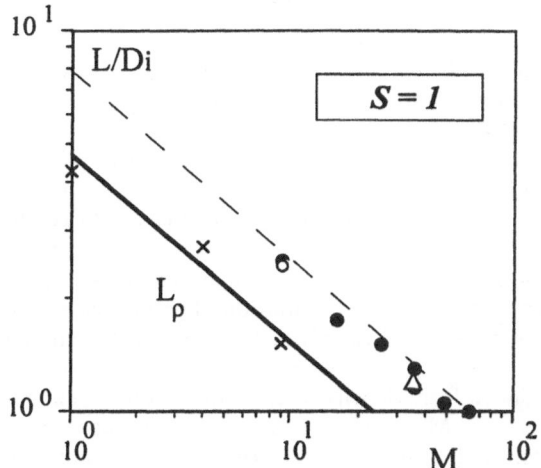

Figure 6 Length of the potential core. Homogeneous flow
● "wake" hot/cold-wire probe;   o, laser-sheet visualizations;
△ , laser anemometry; ✕ ——— , aspirating probe S=0.655
— —   l/Di = 8/M$^{1/2}$ Rehab et al (1996)

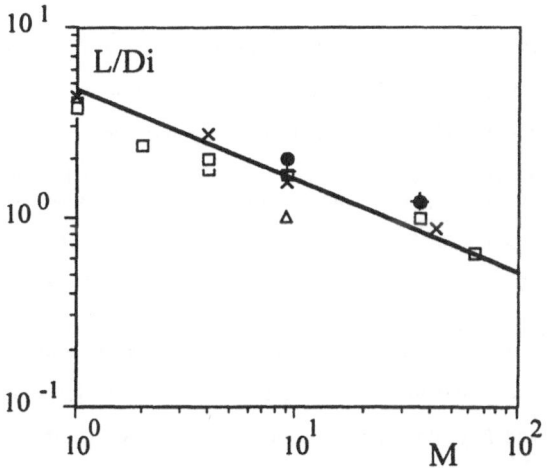

Figure 7 Influence of S on the length of the potential core
——— ✕ S = 0.655 - amb. air
□ S = 0.138 - amb. air;   △ S = 0.138 - amb. He
+ S = 0.028 - amb. air;   ● S = 0.028 - amb. He

(Figure 6). However, due to its definition, L_ρ is much smaller than the other determinations of L.

Further measurements for different values of S show that L_ρ is not considerably affected by density effects (Figure 7). By using pressure arguments based on continuity of pressure, Rehab et al (1996) related the entrainment velocity to the turbulent fluctuation velocity in the inner mixing layer and to S and obtained an estimation of v_E similar to (1) for large velocity ratios. They calculated the length of the potential core by equating the mass flow rate issuing from the central nozzle and that entrained across the surface of the potential core. The model predicts the length L as given by the relation: $L/D_i \approx 6M^{-1/2}$, which slightly underestimates the length of the potential core. Present measurements are in good agreement with this prediction and confirm the significant role played by M in the dynamics of the flow. However, small effects of density may be observed on Figure 7, especially for moderate values of M. As expected from equation (1), entrainment of inner fluid is enhanced and L_ρ is reduced when S is decreased. This effect is more pronounced, when the ambient and the outer jet fluid is the same (helium in this experiment). In this case, it is expected that the outer jet is less slowed down by the ambient fluid, resulting in an increase of entrainment of the central jet by the inner mixing layer.

The behaviour of He/SF6 coaxial jets seems paradoxical, since compared to the results for S = 0.655, the potential core length is a little increased for this combination of gas with very high density differences. Contrary to the predictions given by (1), entrainment is reduced by this further reduction of S to very low values.

4. Conclusions

This study shows that an aspirating probe with a hot-wire placed at the tube entrance is well suited to mean and fluctuating density measurements in variable density jets, except perhaps in thin layers very close to the nozzle.

The concentration field of coaxial jets is strongly affected by the velocity and density ratios of the coannular to central jets; it has been shown that effects of density are rather well taken into account by considering the flux momentum ratio M, instead of r_V and S separately.

No drastic difference was found for the mean and r.m.s fluctuation density evolution on the jet-axis between the cases with and without recirculation. The curves are only shifted towards the nozzle when M is increased and the onset of recirculation cannot be estimated from such measurements.

The length of the potential core deduced from density measurements on the jet axis is principally governed by the momentum ratio M and evolves as $M^{-1/2}$ in agreement with rather simple calculations. A reduction of this length is observed for low values of S ($S \approx 0.14$). This effect is in agreement with the behaviour of the mixing layer with density differences. The opposite behaviour occurs for extremely low values of S and this unexpected result has not been explained so far.

5. References

Au H. and Ko N.W.M. (1987) Coaxial jets of different mean velocity ratios, *Journal of Sound and Vibration* 116 (3), 427-443.
Brown G.L. and Rebollo M.R. (1972) A small fast-response probe to measure composition of a binary mixture, *AIAA J.* 10, 5 , 649-652.
Brown G.L. and Roshko A. (1974) On density effects and large structure in turbulent mixing layers, *Journal of Fluid Mechanics* 64, 4,775-816
Camano E.B (1996) Etude expérimentale des jets coaxiaux avec différences de densité. Thèse INP Grenoble
Camano E.B. and Favre-Marinet M. (1994) On the initial region of inhomogeneous coaxial jets, Advances in Turbulence V 58-62, R.Benzi Ed., Kluwer Academic Publishers.
Champagne F.H. and Wygnanski I.J. (1971) An experimental investigation of coaxial jets, *Int.Journal of Heat and Mass Transfer* 14, 1445-1461
Dahm W.J.A., Frieler C.E., Tryggvason (1992) Vortex stucture and dynamics in the near field of a coaxial jet, *Journal of Fluid Mechanics* 241, 371-402.
Dimotakis P.E. (1986) Two-dimensional shear-layer entrainment, *AIAA Journal* 24,11 1791-1796
Gladnick P., Enotiadis J., LaRue J., Samuelsen G. (1990) Near-field characteristics of a turbulent coflowing jet, *AIAA Journal* 28,8 1405-1414
Kwan A.S.H. and Ko N.W.M. (1976) Coherent structures in subsonic coaxial jets, *Journal of Sound and Vibration* 48 (2), 203-219.
Pitts W.M. (1986) Effects of global density and Reynolds number variations on mixing in turbulent, axisymmetric jets, Nat. Bur. Stand., NBSIR 86-3340.
Rehab H., Villermaux E., Hopfinger E.J. (1996) Flow regimes of large velocity ratio coaxial jets, submitted to *Journal of Fluid Mechanics*
Soteriou M.C. and Ghoniem A.F. (1995) Effects of the free-stream density ratio on free and forced spatially developing shear layers, *Phys.Fluids* 7, 8, 2036-2051
Villermaux E., Rehab H., Hopfinger E.J. (1994) Breakup regimes and self-sustained pulsations in coaxial jets, *Meccanica* 29, 4 , 393-401.

The authors would like to acknowledge the Société Européenne de Propulsion for financial support of this work.

FORCES ON BODIES MOVING IN A WEAK DENSITY GRADIENT WITHOUT BUOYANCY EFFECTS

I. EAMES AND J.C.R. HUNT

D.A.M.T.P.
Cambridge University
Silver Street
Cambridge
UK

1. Introduction

Particle laden flows and variable density gas streams passing through cool environments or near cool boundariés, are just two examples of environments where density gradients occur and solid particles or lumps of fluid with different densities move relative to their surrounding flow. In these situations the effect on the flow of density gradients and the flow induced by pressure gradients may be greater than buoyancy effects, and can materially alter the nature of the flow, such as the eddy structure and entrainment into plumes (*e.g.* Rooney & Linden 1996). Previous studies of particles and lumps have focussed on how the flow and force depend on external factors such as the gradient of local velocity field (Auton *et al.* 1987) and on differences between the local fluid and body density. Current models of turbulence and two phase flows tend to consider the effect of the movement of control volumes having different densities from their surroundings (Hunt, Perkins & Fung 1995), but do not generally represent the effect of the local gradients of density on the forces acting on the particles or eddies, although Chassaing *et al.* (1994) have proposed a model that includes these effects. In this paper we examine the flow generated by a body moving with constant velocity $\mathbf{U}$ in constant density gradient, $\nabla \rho_0$, and provide an insight into the forces and movement of eddies or particulate in flows with variable density. We focus on flows where inertia force dominates viscous and buoyancy forces, so that they are characterised by high Reynolds and Froude numbers.

135

L. Fulachier et al. (eds.), IUTAM Symposium on Variable Density Low-Speed Turbulent Flows, 135–142.
© *1997 Kluwer Academic Publishers.*

Our problem is to analyse the incompressible inviscid flow around a rigid body moving from rest with a constant velocity $\mathbf{U}$ into a region $\mathcal{D}$. The unperturbed density field, $\rho_0(\mathbf{x}')$, is assumed to be weak in the sense that $\epsilon = a|\nabla\rho_0|/\rho_0 \ll 1$, where a is the length-scale of the body. The solution to this problem $\mathbf{u}(\mathbf{x}, t)$, $\rho(\mathbf{x}, t)$ satisfies the conservation of momentum and mass, and the incompressiblity condition

$$\rho\frac{D\mathbf{u}}{Dt} = -\nabla p, \qquad \nabla.\mathbf{u} = 0, \qquad \frac{D\rho}{Dt} = 0. \tag{1}$$

We introduce the coordinate $\mathbf{x}'$ relative to the body, and a relative velocity field $\mathbf{v}(= \mathbf{u} - \mathbf{U})$. Now the velocity and density fields are subject to the boundary conditions that far upstream of the body, as $\mathbf{x}'.\mathbf{U} \to \infty$, $\mathbf{v} \to -\mathbf{U}$ and $\rho \to \rho_0(\mathbf{x})$, and on the surface of the body $\mathbf{v}.\hat{\mathbf{n}} = 0$, where $\hat{\mathbf{n}}$ is a unit vector normal to the body surface.

We write the velocity and density fields (w.l.o.g.) as

$$\mathbf{v}(\mathbf{x}', t) = \mathbf{v}_1(\mathbf{x}') + \mathbf{v}_2(\mathbf{x}', t), \qquad \rho(\mathbf{x}', t) = \rho_1(\mathbf{x}', t) + \rho_2(\mathbf{x}', t). \tag{2}$$

Here the velocity field, $\mathbf{v}_1$, represents the irrotational flow in the absence of a density gradient ($\epsilon = 0$), and ρ_1 represents the density field generated by the irrotational flow, $\mathbf{v}_1$ advecting the initial density field $\rho_0(\mathbf{x})$; $\mathbf{v}_2$ and ρ_2 are the residual terms. In the absence of a density gradient, the fluid is irrotational and steady in the frame moving with the body $\mathbf{v}_1 = \nabla\phi_1$, where the velocity potential, ϕ_1, satisfies Laplaces equation ($\nabla^2\phi_1 = 0$ the boundary condition $\nabla\phi_1.\hat{\mathbf{n}} = 0$. The velocity field $\mathbf{v}_2$ satisfies $\mathbf{v}_2.\hat{\mathbf{n}} = 0$ on the surface of the body. By definition, the leading order density field, $\rho_1(\mathbf{x}, t)$, is derived from (1) namely

$$\frac{\partial\rho_1}{\partial t} + \mathbf{v}_1.\nabla\rho_1 = 0, \tag{3}$$

subject to the boundary condition that the perturbation density gradient equals the given value $\nabla\rho_0$ far upstream of body: $\rho_1(\mathbf{x}, t) \to \rho_0 + \mathbf{x}'.\nabla\rho_0 + \mathbf{U}t.\nabla\rho_0$ as $x' \to \infty$. The residual component, ρ_2, must satisfy, $D\rho_2/Dt + \mathbf{v}_2.\nabla\rho_1 = 0$. Since $\mathbf{v}_1$ is irrotational, the vorticity $\nabla \times \mathbf{v} = \nabla \times \mathbf{v}_2 = \omega_2$ is determined by the perturbation flow. When $0 < \epsilon \ll 1$, the generation of vorticity is governed by

$$\frac{D(\rho_0\omega_2)}{Dt} = -\frac{D\mathbf{v}_1}{Dt} \times \nabla\rho_1 + (\rho_0\omega_2.\nabla)\mathbf{v}_1, \tag{4}$$

where ρ_1 satisfies (3).

When the body moves perpendicular to the density gradient, ($\mathbf{U}.\nabla\rho_0 = 0$), the density field $\rho_1(\mathbf{x}')$ is steady because the isopycnal surfaces are not permanently displaced forward and is weak in the sense its gradient of the same order as the initial weak density gradient. Therefore the vorticity distribution and velocity field is steady everywhere in $\mathcal{D}$, *i.e.* in the sub-region $\mathcal{D}_1$. In $\mathcal{D}_1$ the perturbed velocity field is $|\mathbf{v}_2| = O(U\epsilon)$ and so $\rho_2 = O(\rho_0\epsilon^2)$.

When the body moves parallel to the density gradient, $(\mathbf{U} \times \nabla\rho_0 = 0)$, the density field ρ is unsteady in the body frame of reference because isopycnal surfaces are permanently displaced forward. These surfaces "pile-up" on the body and are stretched as they are convected downstream resulting in a density field $\rho_1(\mathbf{x}, t)$ which is singular on the surface of the body and downstream streamline(s). However, when $Ut/a \ll 1/\epsilon$, the density perturbation ρ_2 can be neglected, and when $Ut/a \gg 1$, the density gradient $\nabla\rho_1$ is steady. In region $\mathcal{D}_1$ where $|\nabla\rho_1|/\rho_0 \ll 1/a$, $|\mathbf{v}_2| \ll |\mathbf{v}_1 + \mathbf{U}|$, $|\nabla\rho_2| \ll |\nabla\rho_1|$ the vorticity is small compared to the irrotational strain and is steady. However in the thin region $\mathcal{D}_2$ of size $O(a\epsilon)$ which is adjacent to the surface of the body and the attached streamline(s) which do not come from upstream, the density gradients are large enough that the perturbation velocity $|\mathbf{v}_2|$ is comparable to $|\mathbf{v}_1 + \mathbf{U}|$. The associated nonlinear unsteady effects are neglected from our analysis which excludes the effects in the narrow region, $\mathcal{D}_2$.

2. Density field perturbed by the irrotational flow, ρ_1

The flow around the body advects density from one place to another, and so the perturbed density field is determined by (3). We determine the perturbed density field by finding the mapping between the Lagrangian (X, Y) and Eulerian frames (x, y) of reference. In the material frame, the density gradient is $\rho_1(X, Y) = \rho_0 + \nabla\rho_0.\mathbf{X} = \rho_0 + \frac{d\rho_0}{dx}X + \frac{d\rho_0}{dy}Y$. The Lagrangian and Eulerian coordinates are related by

$$(X, Y) = (x, y) - \int_0^t (v_{1,x}, v_{1,y})dt. \tag{5}$$

The solution to (3) is $\rho_1(x, y) = \rho_1(X, Y)$ (Hunter 1986). When the flow is irrotational, $\mathbf{v}_1 = \nabla\phi_1$, and $\mathbf{v}_1 \to -\mathbf{U}$ as $|\mathbf{x}'| \to \infty$, Eames *et al.* (1994) showed that $X(x, y) = 2x' + \frac{\phi_1}{U} - X_d + Ut$, where Darwin's drift function X_d is defined by $X_d = \int_0^t \frac{|\nabla\phi_1 - \mathbf{U}|^2}{U}dt$. Asymptotic expressions for Darwin's drift downstream of the body have been calculated specifically for a sphere (Lighthill 1956) and cylinder (Darwin 1953). The drift function downstream has the property, $\int_{-\infty}^\infty \lim_{x' \to -\infty} X_d dy = C_M V$, where C_M is the added-mass coefficient of the body, and V the body volume (Darwin's proposition).

The Y coordinate may be determined from the streamfunction. In two dimensions, $Y = -\frac{\psi_1}{U}$, where ψ_1 is the streamfunction corresponding to the velocity potential ϕ_1. In three-dimensional axisymmetric flow, $Y = y\sqrt{-2\psi_1/U}/R$, where R is the distance from the centreline.

3. Flow around body moving normal to the density gradient, $\mathbf{U}.\nabla\rho_0 = 0$

The flow around a sphere moving normal to the density gradient was first studied by Hawthorne & Martin (1955). Later, Drazin (1961) considered the steady vorticity field generated by two- and three-dimensional bodies moving in weakly sheared flows, and perpendicular to density gradients.

3.1. 2-D ANALYSIS

The vorticity field can be calculated from (4)

$$\omega_2(\mathbf{x}').\hat{\mathbf{z}} = -\frac{1}{2U\rho_0(Y)}\frac{\mathrm{d}\rho_0(Y)}{\mathrm{d}y}\left(\mathbf{v}_1^2 - U^2\right), \tag{6}$$

showing that the vorticity field is localised, decaying from the body.

3.2. 3-D ANALYSIS

A significant difference between two- and three-dimensional flows is that vortex stretching is present which generates a streamwise component of vorticity. The streamwise component can be calculated by considering the circulation around a vortex element, which is proportional to $\omega_2.\hat{\mathbf{s}}/v$, where $\hat{\mathbf{s}}$ is the unit vector parallel to the streamline (Hawthorne & Martin 1955. Note there is a sign error in eqn. (26), which occurs throughout the paper). The generation of circulation is not affected by longitudinal stretching and is governed by

$$\frac{\mathrm{D}}{\mathrm{D}t}\left(\frac{\omega_2.\hat{\mathbf{s}}}{v}\right) = \frac{2\kappa}{\rho v}\nabla\left(\frac{1}{2}\rho v^2\right).\hat{\mathbf{b}} \tag{7}$$

(Scorer 1978, p.83) where $1/\kappa$ is the radius of curvature of the streamlines. The binormal and normal unit vectors are $\hat{\mathbf{b}}$ and $\hat{\mathbf{n}}$, and satisfy the relation $\hat{\mathbf{s}} \times \hat{\mathbf{n}} = \hat{\mathbf{b}}$. By applying Bernoulli's theorem, and Hawthorne et al. (1955) were able to show that the the downstream vorticity distribution is

$$\omega_2.\hat{\mathbf{s}} = \frac{v_1}{2\rho_0}\frac{\mathrm{d}\rho_0}{\mathrm{d}y}\frac{z}{R}\int\limits_{-\infty}^{\infty}\frac{U^2}{v_1^4}\frac{\partial v_1^2}{\partial n}\frac{\sqrt{-2\psi_1/U}}{R}\mathrm{d}s. \tag{8}$$

The trailing vorticity field is not localised and consists of a trailing horse-shoe vortices (see figure).

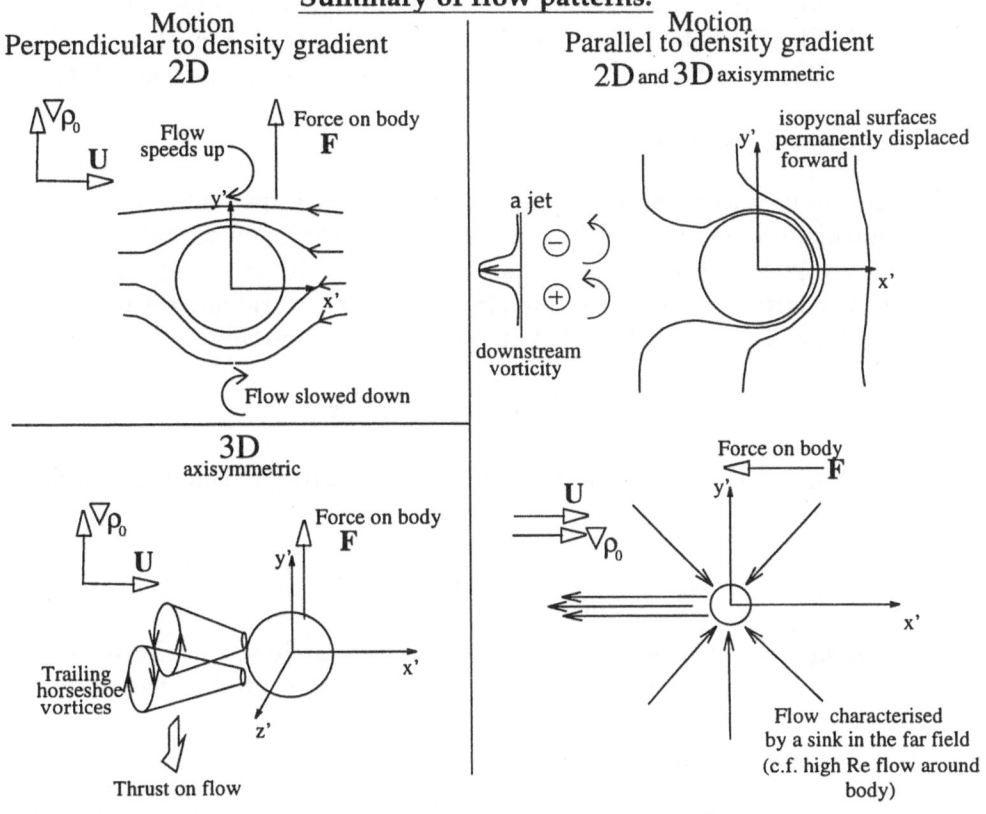

Summary of flow patterns.

4. Flow when the body moves parallel to the density gradient, $U \times \nabla\rho_0 = 0$

4.1. 2-D ANALYSIS

In a two dimensional flow there is no vortex stretching $(\boldsymbol{\omega}.\nabla)\mathbf{v} = \mathbf{0}$, so

$$\frac{D(\rho_0\boldsymbol{\omega}_2.\hat{\mathbf{z}})}{Dt} = \hat{\mathbf{z}}.\frac{D\mathbf{v}_1}{Dt} \times \nabla\rho_1. \tag{9}$$

By integrating the baroclinic torque along the streamline, we can show that

$$\boldsymbol{\omega}_2.\hat{\mathbf{z}} = \frac{1}{\rho_0}\frac{d\rho_0}{dx}\left(\frac{1}{2}v_1\frac{\partial X_d}{\partial n} - v_{1,y}\right), \tag{10}$$

so that the vorticity distribution is not localised. Far downstream $(x' \rightarrow -\infty)$ the velocity field on the attached streamline is

$$\lim_{x'\to-\infty} v_{2,x}(x,y) = -\frac{1}{2}\frac{U}{\rho_0}\frac{d\rho_0}{dx}X_d, \tag{11}$$

which describes a localised jet-like flow along the centreline (see figure). As the body moves into denser fluid, a jet in the opposite direction to U is generated. The far field flow is characterised by a sink strength located at the origin. The direction of the jet is reversed when $d\rho_0/dx$ is negative.

4.2. 3-D AXISYMMETRIC BODY

When the body is axisymmetric, the only non-zero component of vorticity is the azmuthial component, where $\hat{\varphi}$ is the unit vector in the azmuthial direction. As a vortex ring is advected past the body, it is stretched thereby increasing vorticity, whilst preserving circulation. The stretching is described by $\hat{\varphi}.(\omega_2.\nabla)\mathbf{v}_1 = v_{1,R}\omega.\hat{\varphi}/R$, so that the generation circulation around the vortex is

$$\frac{D}{Dt}\left(\rho_1\frac{\omega_2.\hat{\varphi}}{R}\right) = \frac{\hat{\varphi}}{R}.\frac{D\mathbf{u}_1}{Dt} \times \nabla\rho_1. \tag{12}$$

By manipulating the baroclinic torque, and integrating (12) along a streamline, we can show

$$\omega_2.\hat{\varphi} = \frac{1}{\rho_0}\frac{d\rho_0}{dx}\left(\frac{1}{2}v_1\frac{\partial X_d}{\partial n} - v_{1,R}\right). \tag{13}$$

The secondary flow is similar to the two dimensional flow, and characterised by a downstream jet, and in the far field by a sink located at the origin.

5. Force on the body

The force on a body moving in an inviscid fluid is a result of the pressure variation over the body surface and is defined by $\mathbf{F} = \int_{S_B} p\hat{\mathbf{n}}dS$ where S_B is the surface of the body, and $\hat{\mathbf{n}}$ is the unit normal pointing out of the fluid (Batchelor 1967, p.138). The momentum-integral theorem is applied to the steady flow in the control volume $\mathcal{D}_1\cup\mathcal{D}_2$ surrounding the body. The control volume is arbitrary - we choose a cylinder length N, and radius M, and dimensions $1 \ll (M/a, N/a) \ll 1/\epsilon$. Gauss's Theorem can be applied to show

$$\mathbf{F} = -\int_{S_\infty} p\hat{\mathbf{n}}dS - \int_{S_\infty} \rho(\mathbf{v}.\hat{\mathbf{n}})\mathbf{v}dS + \int_{\mathcal{D}_1\cup\mathcal{D}_2} (\mathbf{v}.\nabla\rho)\mathbf{v}dV. \tag{14}$$

Note that the density is non-uniform in (14).

5.1. FORCE WHEN THE BODY MOVES PERPENDICULAR TO THE DENSITY GRADIENT, $\mathbf{U}.\nabla\rho_0 = 0$

In two dimensions, the fluid is accelerated over the top of the body by the rotational component of the body, and the fluid beneath the body is decelerated. As a result, there is a pressure drop across the body, and the body experiences a lift force, pushing it towards the denser fluid. In three dimensions, the trailing horseshoe vortices impart a thrust on the flow and the body experiences a force pushing it towards the denser fluid. The magnitude of the force can be evaluated using expressions for the far field to evaluate the momentum flux far from the body (14). The force on a two-dimensional body is

$$\mathbf{F} = \frac{1}{2}(C_M + 1)V\frac{\mathrm{d}(\rho_0 U^2)}{\mathrm{d}y}\hat{\mathbf{y}}, \tag{15}$$

and on a three-dimensional axisymmetric body is

$$\mathbf{F} = \frac{1}{2}C_M V\frac{\mathrm{d}(\rho_0 U^2)}{\mathrm{d}y}\hat{\mathbf{y}}, \tag{16}$$

The force can be written vectorially as

$$\mathbf{F} = C_L V(\mathbf{U} \times \nabla\rho_0) \times \mathbf{U},$$

where the lift coefficient takes the value $C_L = C_M/2$ or $(C_M+1)/2$ for two- or three-dimensional bodies respectively. It is also relevant to note that (15) and (16) include the force on a body moving in a shear, where $U = U(y)$.

5.2. FORCE WHEN THE BODY MOVES PARALLEL TO DENSITY GRADIENT, $\mathbf{U} \times \nabla\rho_0 = 0$

The force on the body moving parallel to the density gradient can be shown to be

$$\mathbf{F} = -C_M VU\mathbf{U}\frac{\mathrm{d}\rho_0}{\mathrm{d}x} + \int_A \rho_0 \mathbf{U}\mathbf{v}_2.\hat{\mathbf{n}}\mathrm{d}A, \tag{17}$$

The first term arises from the momentum deficit caused by the *density variation* on the downstream attached streamline caused by the distortion of the isopycnal surfaces by the *irrotational* component of the flow. The second term arises from the momentum deficit caused by the *velocity variation* in the wake caused by the *rotational* component of the flow.

We have shown that $\mathbf{v}_2.\hat{\mathbf{n}}$ is negative downstream of the body (11), and that the volume flux in the wake for a two- and three-dimensional axisymmetric body, is (from (11) and Darwin's proposition) $\int_A \mathbf{v}_2.\hat{\mathbf{n}}\mathrm{d}A = \frac{1}{2\rho_0}C_M VU\frac{\mathrm{d}\rho_0}{\mathrm{d}x}$. The contribution to the force from the density variation in the wake is twice as large as the contribution from the velocity variation. However, the density variation exerts a drag on the body whereas the velocity variation exerts a thrust on the body. The total force can be written vectorially as

$$\mathbf{F} = -C_D V(\mathbf{U}.\nabla\rho_0)\mathbf{U}, \tag{18}$$

where the drag coefficient, $C_D = C_M/2$ for two dimensional bodies and three dimensional bodies axisymmetric about $\mathbf{U}$. The direction of the force depends on the sign of the density gradient: a body moving into denser/lighter fluid experiences a drag/accelerating force.

6. General Remarks

In this paper, we have studied the inviscid flow generated by a body moving at a constant vleocity through a non-uniform density field, and the force acting on the body. We have shown that two and three dimensional axisymmetric bodies moving perpendicularly to a density gradient, experience a lift force tending to drive them towards the denser fluid. When the body moves parallel to the density gradient, the body experiences a drag or thrust depending on whether the body is moving into denser or lighter fluid. The results show that density gradients must be taken into account when considering the dynamics of eddies and solid bodies. For instance, "lumps" of fluid moving in density gradients are pushed towards the denser fluid, increasing the spreading rate of light jets and positively buoyant non-Boussineq plumes, but decreasing the spreading angle of heavy jets or negatively buoyant plumes (Rooney *et al.* 1996, Chassaing *et al.* 1995).

References

Auton, T. R. 1987 The lift force on a spherical body in a rotational flow. *J. Fluid Mech.* **183**, 199–218.

Auton, T. R., Hunt, J. C. R. & Prud'homme, M. 1988 The force exerted on a body in an inviscid unsteady non-uniform rotational flow. *J. Fluid Mech.*, **197** 241–257.

Chassaing, P., Harran, G. & Joly, L. 1995 Density fluctuation correlation in free turbulent binary mixing. *J. Fluid Mech.* **279**, 239–278.

Darwin, C. 1953 A Note on hydrodynamics *Proc. Cam. Phil. Soc.* **49**, 342–354.

Drazin P. G. 1961 On the steady flow of fluid of variable density past an obstacle. *Tellus* **13**, 239–251.

Eames, I., Belcher, S.E. & Hunt, J.C.R. 1994 Drift, partial drift and Darwin's proposition. *J. Fluid Mech.* **275**, 201–223.

Hawthorne, W. R. & Martin, M. E. 1955 The generation of secondary vorticity in the flow over a hemisphere due to density gradient and shear. *Proc. Roy. Soc. London (A)*. **232**, 184–195.

Hunt, J. C. R., Perkins, R. J. & Fung, J. C. H. 1994 Problems in modelling disperse two-phase flows. *A.M.S.E.* **47**, 50–60.

Hunter, S. C. 1983 *Mechanics of a continuous media*. Eliswood.

Lighthill, M. J. 1956 Drift. *J. Fluid Mech.* **1** 31–53.

Rooney, G.G. & Linden, P.F. 1996 Similarity considerations for non-Boussinesq plumes in an unstratified environment. *J. Fluid Mech.* **318**, 237–250.

Scorer, R.S. 1978 *Environmental aerodynamics*. Ellis Horwood.

EXPERIMENTAL INVESTIGATION OF THE INTERACTION BETWEEN SCALAR DISSIPATION AND STRAIN RATE IN A COUNTERFLOW GEOMETRY

K. SARDI, A.M.K.P. TAYLOR AND J.H. WHITELAW

Imperial College of Science Technology and Medicine
Mechanical Engineering Department
Exhibition Road, London, SW7 2BX, U.K.

1. Introduction

Combustion is an important practical application of low speed, variable density flows and information on the statistical properties of a scalar field and the interaction between scalar fluctuations, θ, and their dissipation rate, χ, is needed to quantify the mixing processes. For example, it has been shown by Bilger (1980) that the reaction rate of a turbulent diffusion flame is a function of the joint probability density function, *pdf*, between θ and χ. Also, Liew *et al.* (1984) have shown that flame extinction in the laminar flamelet regime may be modelled in terms of the cumulative probability density function, *cdf*, of the values of χ that correspond to a family of strained flamelets, which exceeds a quenching value χ_q.

$$P_q = 1 - \int_0^{\chi_q} P(\chi)d\chi \qquad (1)$$

The aim of this work is to determine experimentally the distributions of a passive scalar, *i.e.* temperature, and the dissipation rate of its fluctuations in a non-combusting counterflow system of low turbulent Reynolds number. The current flow arrangement overcomes many of the difficulties associated with scalar dissipation measurements in the presence of combustion while retaining the qualitative characteristics of the latter. In addition, the counterflow geometry renders the results directly applicable to the laminar flamelet regime because the spatial thickness of the reaction zone is smaller than the smallest scales of turbulence, Mastorakos (1993).

L. Fulachier et al. (eds.), IUTAM Symposium on Variable Density Low-Speed Turbulent Flows, 143–150.
© 1997 Kluwer Academic Publishers.

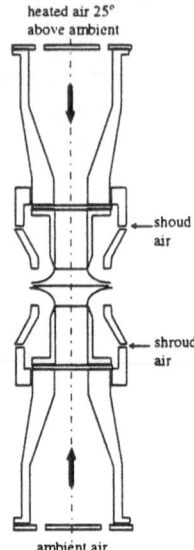

Figure 1. The experimental arrangement

2. Experimental Arrangement and Instrumentation

2.1. FLOW CONFIGURATION

The measurements were conducted in the flow arrangement shown in figure 1 which comprised two identical, vertically opposed nozzles with contraction ratio 9 separated by a distance, H, equal to one nozzle diameter, D. To enhance turbulence, perforated plates with solidity of 45% and 4 mm hole diameter were fitted after the contraction and before the straight section of each assembly. Air from a shop compressor was filtered to remove particles larger than 0.1 μm, metered by a pair of calibrated sonic nozzles and supplied to the lower and upper flows. The temperature of upper stream was increased by an electric heater (4.8 kW) to 25°C above ambient with a tolerance of ±1°C. In order to obtain a temperature profile, uniform to ±2°C at the nozzle exit, the upper stream was insulated by two layers of 50 mm thick insulating cord and the opposed jets were shielded from temperature and velocity fluctuations of the ambient air by a secondary, annular counterflow with velocity equal to one tenth of the velocity of the primary streams.

2.2. INSTRUMENTATION

Simultaneous measurements of temperature fluctuations and the axial and radial components of the scalar dissipation rate, χ_z and χ_r, were obtained along the centreline between the opposed jets using pairs of parallel cold wires made of platinum and operated in constant current mode. The heating current was 0.1 mA and the velocity sensitivity of the sensors was of the order of 1%, Wyngaard (1971a). The prongs were tapered to a tip diameter of 75 μm and the sensors were of 0.5μm dimeter and 0.6 mm long resulting in a length to diameter ratio of 1200. A 15% spatial attenuation of the measured value of χ was estimated from Wyngaard's analysis (1971b). The optimum separation between the two sensors was fixed at 0.3 mm and was selected after measuring the squared gradient of the temperature fluctuations at each point in the flowfield by parallel probes of spacing in the range of 0.2 to 0.4 mm, according to the procedure proposed by Anselmet et al. (1994). The time constant of the sensors in the vicinity of the stagnation plane, evaluated from the relation proposed by Collis & Williams (1959), was 20 μs resulting in a cut-off frequency of 8 kHz, approximately a factor of two higher than the estimated Kolmogorov frequency of 4.7 kHz so that compensation was not required.

3. Initial Conditions and Related Scales

In order to quantify the effect of bulk velocity, U_b, and the bulk strain rate, $S_b = 2U_b/H$, on the evolution of the scalar dissipation, measurements were obtained for two different flow conditions. The velocities of the cold, U_c, and the heated jet, U_h, were adjusted so that the momenta of the two jets were equal and the stagnation plane was located at the half distance, $H/2$, between the two opposing assemblies. The flow conditions and the related length scales are summarised in Table 1, where $t_{RES} = 1/S_b$ is the residence time in the mixing layer. The turbulent Reynolds number, Re_t − based on the integral length scale, L_t −, the Kolmogorov microscale, ν_k, and the eddy-turnover time, t_{ov}, were estimated from the measurements of Mastorakos (1993) in a similar configuration.

TABLE 1. Flow conditions and estimated scales

U_c (m/s)	U_h (m/s)	S_b (s^{-1})	$t_{RES}(ms)$	L_t (mm)	Re_t	ν_k (mm)	$t_{ov}(ms)$
3.8	4.1	256	3.9	2	35	0.13	6.2
5.4	5.9	362	2.7	2	52	0.10	4.3

Temperatures are reported in terms of a mixture fraction defined as $\Theta = (T - T_c)/(T_h - Tc)$, where T is the instantaneous temperature and T_c and T_h are the temperatures of the cold and hot jets at their exit planes respectively. The instantaneous dissipation rate, χ_i, of the normalised temperature fluctuations, θ, on the axial, z, or radial direction, r, is defined as, $\chi_i \equiv 2D_t(\partial\theta/\partial x_i)^2$, where D_t is the air thermal diffusivity assumed constant and having a value of 2.2×10^{-5} m^2/s at 20°C.

4. Results

The mean normalised temperature, $<\Theta>$, can be approximated by an error function, with zero and unity at the cold and hot boundaries respectively and 0.5 at the stagnation plane, as shown in figure 2 (a), in which the axial distance along the centreline has been normalised as $z^* = (z - z_s)/(H/2)$ and z_s is the location of the stagnation point. The distributions, corresponding to the two strain rates, coincide, suggesting that the mean thickness of the mixing layer is independent of S_b, in agreement with the findings of Mastorakos (1993) in variable density and combusting flows. The evolution of the r.m.s. of the normalised temperature fluctuations, θ', along the centreline between the two jets is presented in figure 2(b) and shows that the scalar fluctuations reach a maximum value at the stagnation plane and that their evolution is also independent of the bulk strain rate. From figures 3

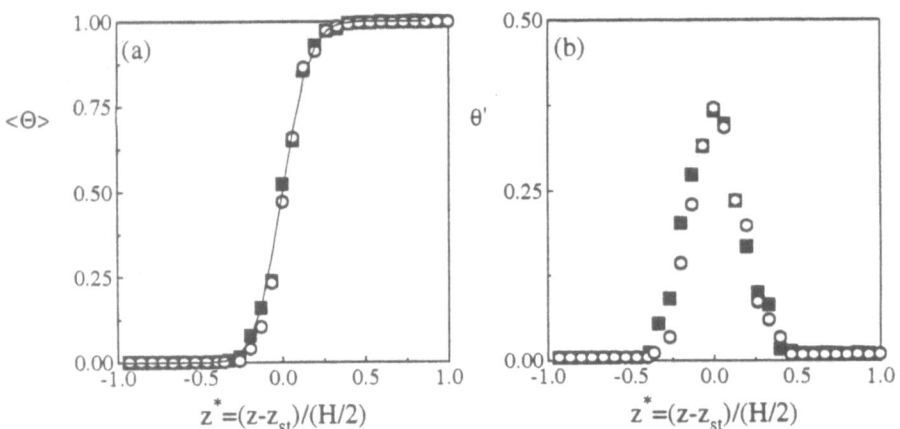

Figure 2.   Mean (a) and r.m.s. (b) normalised temperature distributions along the centreline between the opposed jets: $\bullet$ $U_c = 3.84m/s$, $\circ$ $U_c = 5.4m/s$.

and 4 it can be seen that the axial and radial mean, $<\chi_i>$, and r.m.s., χ_i', components of the dissipation rate achieve the maximum value at the stagnation plane and that, in contrast to the behaviour of the scalar field, $<\chi_i>$ and χ_i' increase proportionally with increasing bulk strain rate, by 50% and 40% respectively, in agreement with the DNS simulations of Ashurst *et al.* (1987). Due to the low values of the turbulent Reynolds number, Table 1, local isotropy is not satisfied and the ratios of the mean and r.m.s. of the axial to the radial components, $<\chi_z>/<\chi_r>$ and χ_z'/χ_r', are approximately equal to 3, for both values of the strain rate. The absence of local isotropy implies that small scales depend on large scales, Mi *et al.* (1995).   As

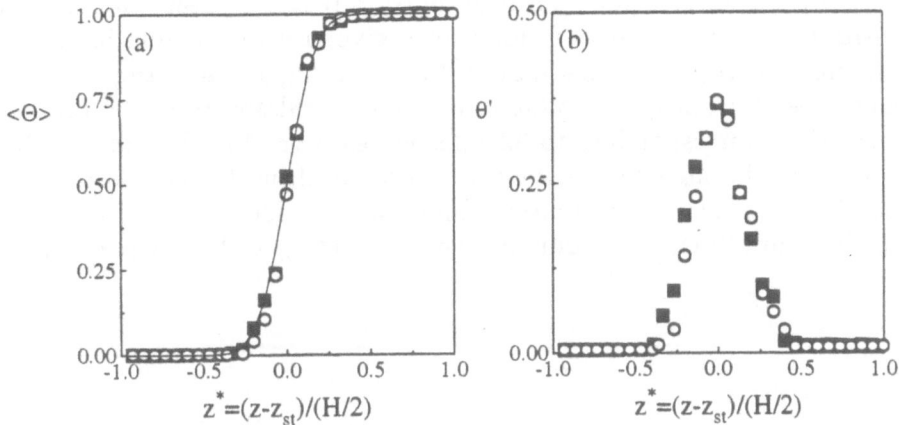

Figure 3.   Mean (a) and r.m.s. (b) of the axial scalar dissipation component, χ_z, along the centreline: • $U_c = 3.8m/s$, ∘ $U_c = 5.4m/s$.

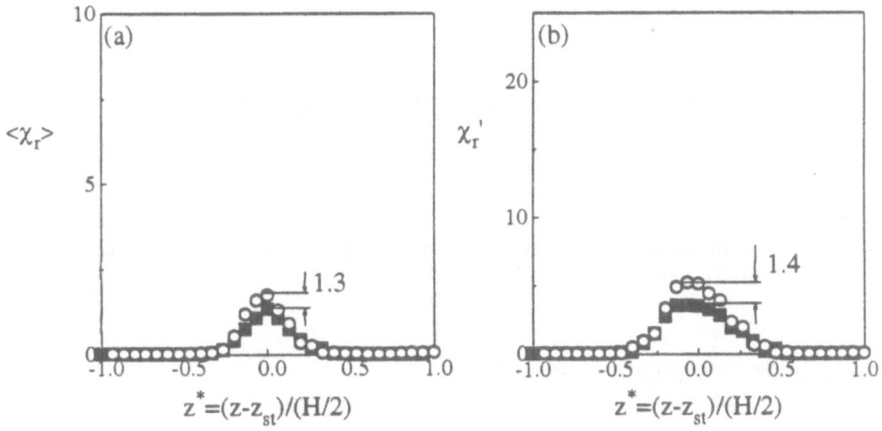

Figure 4.   Mean (a) and r.m.s. (b) of the radial scalar dissipation component, χ_r, along the centreline: symbols as in figure 3.

shown in Table 1, the eddy turnover time was longer than the residence time and this characteristic is also likely to be true in variable density flows near any flame stabilisation region. This implies that the scalar turbulence is *young* as confirmed by the bimodal scalar pdfs at the stagnation plane, figure 5, where the instantaneous scalar fluctuations have been normalised as $\theta^+ = \theta/\theta'$. It can be seen that the pdfs corresponding to the two strain rates coincide, suggesting that single point statistics are independent of turbulent Reynolds number. The respective time traces of the normalised temperature and dissipation signals on the stagnation plane are presented in figure 6 for an arbitrary duration of 0.2s. It can be seen that the sharp transitions between cold and hot values give rise to maxima in the axial component of the scalar dissipation 20 times higher than the mean. The frequencies of crossing the value of $\theta = 0.5$, calculated from the total time record of 4s, corresponding to 32768 samples, were 104 Hz and 138 Hz for the low and the high S_b cases respectively, resulting in the subsequent increase in the measured dissipation values, as also demonstrated by the pdfs of χ_z in figure 7(a). The cumulative distribution of the axial component

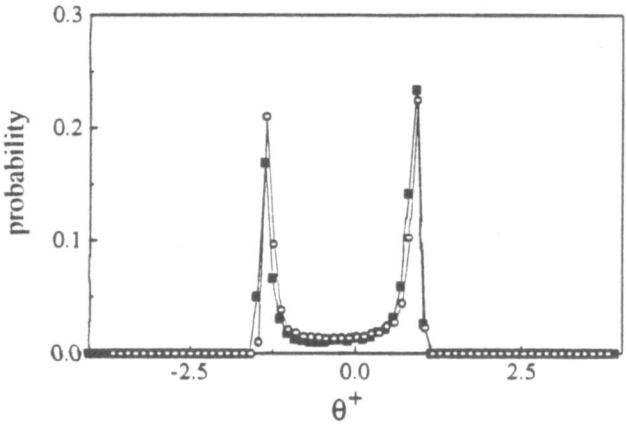

Figure 5. Probability distributions of the scalar fluctuations, θ, on the stagnation plane as a function of U_b : symbols as in figure 3.

of the scalar dissipation, defined as

$$cdf(\chi_z) = \int_0^{\chi_z} P(\chi_z)d\chi_z \qquad (2)$$

is shown in figure 7(b) as a function of S_b. Comparison of equations (1) and (2) confirms that the cdf of χ is the probability that a flamelet burns. The figure shows that by increasing the bulk strain rate by 40% the cdf of the scalar dissipation is proportionally displaced to higher values of χ_z, implying that the probability of P_c increases linearly with the velocity increment and hence extinction is approached.

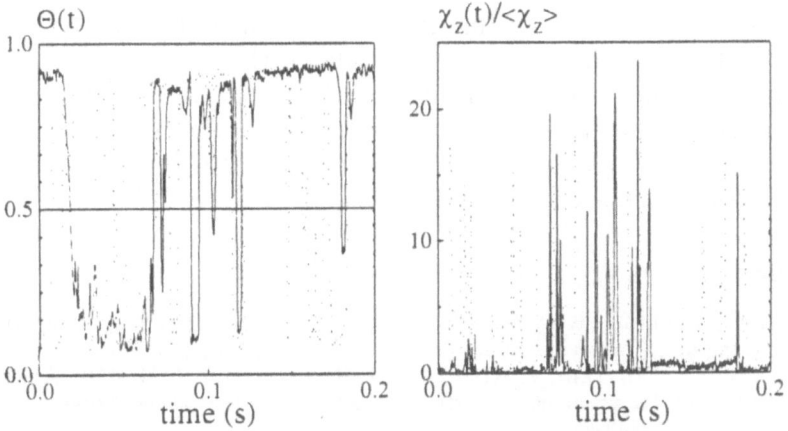

Figure 6.   Time traces of the scalar and the axial scalar dissipation component at the stagnation plane: — $U_c = 3.84m/s$, – – $U_c = 5.4m/s$.

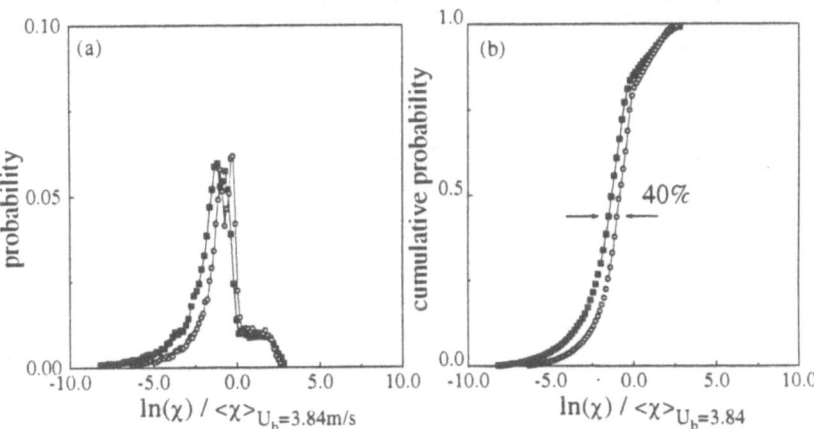

Figure 7.   Probability, (*a*), and cumulative, (*b*), distributions of χ_z : • $U_c = 3.84m/s$, ○ $U_c = 5.4m/s$.

5. Conclusions

Simultaneous measurements of a passive scalar and the dissipation rate of its fluctuations have been performed in a low turbulent Reynolds number, slightly heated counterflow system as a function of the bulk velocity and the bulk strain rate. It was found that the distributions of the mean and r.m.s. normalised temperatures are independent of U_b and that the axial and radial components of the scalar dissipation, as well as their cumulative distributions, increase proportionally with the bulk strain rate implying that, for combusting flows, extinction is approached. The results presented here are likely to be representative of scalar turbulence near the stabilisation region of many flames. Work is in progress to quantify the correlation between θ and χ and utilise the observation that the scalar turbulence is young to provide a quantitative model that predicts the relation between the scalar fluctuations and their dissipation.

References

ANSELMET, F., DJERIDI, H. & FULACHIER, L. 1994 Joint statistics of a passive scalar and its dissipation in turbulent flows. *J. Fluid Mech.* **280**, 173–197.

ASHURST, WM.T., KERSTEIN, A.R., KERR, R.M. & GIBSON, C.H. 1987 Alignment of vorticity and scalar gradient with strain rate in simulated Navier-Stokes turbulence. *Phys. Fluids* **30**, 2343-2353.

BILGER, R.W. 1980 Turbulent flames with non-premixed reactants, *In Turbulent Reacting Flows.* (ed. P.A. Libby & F.A. Williams) Springer Verlag.

COLLIS, D.C. & WILLIAMS, M.J. 1959 Two dimensional convection from heated wires at low Reynolds number. *J. Fluid Mech.* **6**, 357–384.

LIEW, S.K., BRAY, K.N.C. & MOSS, J.B. 1984 A stretched laminar flamelet model of turbulent non-premixed combustion. *Combust.Flame* **56**, 199–213.

MASTORAKOS, E. 1993 Turbulent combustion in opposed jet flows. Ph.D. Thesis, University of London.

MI, J., ANTONIA, R.A. & ANSELMET, F. 1995 Joint statistics between temperature and its dissipation rate components in a round jet. *Phys. Fluids* A **7**, 1665–1673.

WYNGAARD, J.C. 1971a The effect of velocity sensitivity on temperature derivative statistics in isotropic turbulence. *J.Fluid Mech.* **48**, 763–769.

WYNGAARD, J.C. 1971b Spatial resolution of a resistance wire temperature sensor. *Phys. Fluids* **4**, 2052–2054.

ANALYTICAL SOLUTION OF THE EQUATION FOR THE PROBABILITY DENSITY FUNCTION OF A SCALAR IN DECAYING GRID–GENERATED TURBULENCE WITH A UNIFORM MEAN SCALAR GRADIENT

V.A. SABELNIKOV
Central Aero–Hydrodynamic Institute (TsAGI),
Zhukovsky, Moscow Region, 140160, Russia

Abstract

An asymptotic self–similar, closed–form solution of the equation for the one–point probability density function (pdf) of a passive scalar in decaying grid-generated turbulence with a uniform mean cross–stream scalar gradient is obtained. The present solution generalizes the well–known asymptotic (time $t \to \infty$) closed–form solution of Y.Sinai and V.Yakhot (Phys. Rev. Lett. 63, 1962 (1989)) for the scalar pdf equation for homogeneous turbulence with no mean gradient. It is shown that when the conditional expectation of transverse velocity is a linear function of a scalar (experimental data of K.S.Venkataramany and R.Chevray (J.Fluid Mech. 86, 513(1978)) indicate that this is a reasonable assumption) the present solution is solely a function of the conditional expectation of scalar dissipation and looks exactly the same as the asymptotic solution for the case with no mean scalar gradient. This result explains why the pdf's measured by Jayesh and Z.Warhaft (Phys. Fluids A 4, 2292 (1992)) in a decaying grid-generated turbulence with a uniform scalar mean gradient are accurately represented by the asymptotic solution of Y.Sinai and V.Yakhot if the measured conditional expectation of scalar dissipation is substituted in the latter solution. An exact expression that relates the conditional expectation of molecular diffusion to the conditional expectation of transverse velocity is derived. This expression generalizes the result of L.Valiño *et al.* (Phys. Rev. Lett. 72, 3518 (1994)) for the conditional molecular diffusion in the absence of a mean scalar gradient. The case of the temperature fluctuations in grid–generated turbulence under the action of a stable (negatively buyoant), linear, mean temperature profile is briefly studied. Here, the results are less definite because the strict self–similar solution of the equation does not exist. The analysis show, however, that approximate (quasi) self–similar solution can be obtained for the above case.

L. Fulachier et al. (eds.), IUTAM Symposium on Variable Density Low-Speed Turbulent Flows, 151–158.
© 1997 *Kluwer Academic Publishers.*

1. Introduction

The impetus for this study stems from the recent experiment [1] and the subsequent theoretical work [2] which deal with the one–point probability density function (pdf) of a passive temperature fluctuations, with a uniform mean temperature gradient, in decaying grid–generated turbulence. As a solution for the one–point scalar pdf equation for a uniform mean scalar gradient case is still lacking, Jayesh and Warhaft [1] compared their measured scalar pdf and conditional expectation of scalar dissipation with the Sinai and Yakhot [3] asymptotic solution in the absence of a mean scalar gradient. This asymptotic solution has a self–similar shape (time $t \to \infty$) for the scalar pdf equation and it is a unique function of the conditional expectation of scalar dissipation. Jayesh and Warhaft [1] calculated the pdf's from the Sinai and Yakhot asymptotic solution by substituting in it the measured conditional expectation of scalar dissipation. The pdf's thus calculated, as Jayesh and Warhaft [1] stated, are in remarkably good correspondence with the measured pdf's. Jayesh and Warhaft [1] claimed that the observed fit may be somewhat fortuitous since, as they emphasized, the Sinai and Yakhot [3] asymptotic solution is for the case of no mean gradient.

The above mentioned problem was further theoretically adressed by Sahay and O'Brien [2]. They remarked that the one–point scalar pdf equation contains, in the general case, two unknown statistical functions, both conditioned on the scalar value. The first is the conditional expectation of scalar dissipation, the second quantity of equal interest is the conditional expectation of velocity. Sahay and O'Brien [2] illustrated the fundamental importance of the conditional expectation of velocity by analyizing the reverse problem. They assumed two particular self–similar scalar pdf's (gaussian and exponential) and conditional expectation of velocity (taking into account the cubic term in series expansion) and calculated directly from the pdf equation the conditional expectation of scalar dissipation. It was shown by Sahay and O'Brien that the shape of the conditional expectation of scalar dissipation is quite sensitive to the conditional expectation of velocity which was assumed.

The importance of modeling the conditional expectation of velocity to describe the evolution of the scalar pdf in shear turbulent flows was also stressed by Kuznetsov and Sabelnikov [4] and by Li and Bilger [5].

In the present work the asymptotic self–similar solution of the equation for the pdf of a scalar, with a uniform mean scalar gradient in decaying grid–generated turbulence is obtained. The present solution leads to the analytical closed–form relationship between pdf and conditional velocity and dissipation. This solution is the generalization of the Sinai and Yakhot [3] asymptotic solution for the case with no mean gradient. It sheds further light on the role played by the conditional expectation of velocity. An exact expression that relates the conditional expectation of molecular diffusion and the conditional expectation of transverse velocity is also derived. This expression generalizes the result of Valiño *et al.* [6] for the conditional molecular diffusion in the absence of a mean scalar gradient. The case of the temperature fluctuations in grid–generated turbulence under conditions of stable thermal stratification is briefly studied.

2. Passive scalar field

The transport equation for the one–point pdf of a passive scalar, in decaying homogeneous grid–generated turbulence with a uniform mean cross–stream scalar gradient, neglecting molecular diffusion in physical space and the stream–wise turbulent flux term, is given by (see, e.g. Refs. [2,4,7])

$$u_0 \frac{\partial P}{\partial x} + \frac{\partial}{\partial y} \langle v|c \rangle P = -\frac{\partial^2}{\partial c^2} \langle N|c \rangle P, \tag{1}$$

where $P(c, x, y) = F(c - \langle c \rangle, x)$ is the pdf of a passive scalar c ; u_0 is the mean stream–wise velocity; x and y are stream–wise and cross–stream coordinates respectively; $\langle v|c \rangle$ and $\langle N|c \rangle$ are the conditional expectations of transverse velocity v and of scalar dissipation $N = D(\nabla c)^2$ respectively; D is the molecular diffusivity; $\langle \ \rangle$ denotes the unconditional expectation.

The equation for the pdf can alternatively be written as (e.g. Refs. [4,7])

$$u_0 \frac{\partial P}{\partial x} + \frac{\partial}{\partial y} \langle v|c \rangle P = -\frac{\partial}{\partial c} \langle w|c \rangle P, \tag{2}$$

where $\langle w|c \rangle$ is the conditional expectation of molecular diffusion, $w = D \nabla^2 c$. The mean scalar profile for the uniform mean scalar gradient case is given by

$$\langle c \rangle = E(y - y_0), \tag{3}$$

where $E = \dfrac{d\langle c \rangle}{dy}$ and y_0 are constants.

Some insight into the nature of the scalar pdf may be obtained from analysis of the asymptotic self–similar solutions of equations (1) and (2). They have the following form

$$P(c, x, y) = \frac{1}{\sigma} g(s), \quad s = \frac{c - \langle c \rangle}{\sigma}, \tag{4}$$

where $\sigma^2 = \langle (c - \langle c \rangle)^2 \rangle$ is the scalar variance; experimental data of Jayesh and Warhaft [1] shows that σ^2 depends only on the stream–wise coordinate x and increases with x. It is assumed that g is solely a function of the self–similar variable s. Physically, the self–similar solution is a limiting ($Re_\lambda \to \infty$, $x/M \to \infty$; Re_λ is Reynolds number based on Taylor microscale, M is the mesh length) solution. It can be observed when the influence of the initial conditions on the pdf are lost.

The experimental verification of the self–similarity of the scalar pdf was done by Jayesh and Warhaft [1]. They showed that (4) provides quite close collapse of the measured pdf's onto a single curve for the range $42.4 \leq x/M \leq 152.4$ of downstream locations (the grid is located at $x = 0$). The observed departures from self–similarity can be attributed to moderate value of Re_λ in [1] ($Re_\lambda \leq 74.4$, and decays with x).

For self–similar solutions to Eq.(4) to exist conditional expectations of the scalar dissipation $\langle N|c \rangle$, molecular diffusion $\langle \omega|c \rangle$ and velocity $\langle v|c \rangle$ must have the following form

$$\langle N|c \rangle = \langle N \rangle n(s), \quad \langle v|c \rangle = \frac{q}{\sigma}V(s), \quad \langle \omega|c \rangle = \frac{\langle N \rangle}{\sigma}\Omega(s), \tag{5}$$

where $q = \langle vc \rangle$ is the cross–stream turbulent flux , $\langle N \rangle$ is the unconditional scalar dissipation; q and $\langle N \rangle$ depend only on x; functions n, V and Ω, like function g in (4), are solely dependent on variable s. Space derivatives of the self–similar pdf (4) are (Eq. (3) is used)

$$\frac{\partial P}{\partial x} = -\frac{1}{\sigma^2}\frac{d\sigma}{dx}(sg)', \quad \frac{\partial}{\partial y}\langle v|c \rangle P = -\frac{q}{\sigma^3}(Vg)'E, \tag{6}$$

where $(f(s))'$ denotes the derivative of f with respect to s.

Inserting Eqs. (4) -(6) into (1) and (2), one obtains:

$$(ng)'' + [(\alpha_1 s + \alpha_2 V)g]' = 0, \tag{7}$$

$$[(\alpha_1 s + \alpha_2 V)g]' = -(\Omega g)', \tag{8}$$

where α_1 and α_2 are constants for self–similar solution to exist. They are defined by

$$\alpha_1 = -\frac{u_0}{2\langle N \rangle}\frac{d\sigma^2}{dx}, \quad \alpha_2 = -\frac{Eq}{\langle N \rangle}. \tag{9}$$

It follows from the balance equation for the concentration variance

$$\frac{1}{2}u_0\frac{d\sigma^2}{dx} + \langle vc \rangle\frac{d\langle c \rangle}{dy} = -\langle N \rangle, \tag{10}$$

that the following equality appears

$$\alpha_1 + \alpha_2 = 1. \tag{11}$$

Values of constants α_1 and α_2 experimentally measured by Sirivat and Warhaft [8] are $\alpha_1 = -0.5, \alpha_2 = 1.5$.

Equation (8) leads to the relationship (it is assumed further that $g(s)$ decreases sufficiently rapidly as $|s| \to \infty$)

$$\Omega = -(\alpha_1 s + \alpha_2 V), \tag{12}$$

for arbitrary self–similar pdf g. It means that the second form of writing the equation for the scalar pdf (2) cannot be used for the determination of the self–similar solution.

According to Eq. (12) one may conclude that for the self–similar solution (4) to exist the conditional expectation of molecular diffusion has to be closely linked with the conditional expectation of transverse velocity. Eq. (12) is the generalization of the Valiño *et al.* result [6] for the conditional molecular diffusion $\Omega = -s$ which is valid in the absence of a mean scalar gradient.

Let us analyze now Eq. (7). First consider the case of no mean scalar gradient, i.e. $E = 0$, $\alpha_2 = 0$, $\alpha_1 = 1$ (due to Eq. (11)). Eq. (7) in this case can be integrated easily to yield

$$(ng)' + sg = 0. \tag{13}$$

The solution of Eq. (13) is:

$$g(s) = \frac{A_1}{n(s)} exp\left[- \int_0^s \frac{\zeta}{n(\zeta)} d\zeta \right], \tag{14}$$

where the constant A_1 is determined by the normalization condition $\int g(s)ds = 1$. Eq. (14) is the well-known asymptotic solution found by Sinai and Yakhot [3] using a different method.

The solution of Eq. (7) gives us the principal result of the present paper

$$g(s) = \frac{A_2}{n(s)} exp\left[- \int_0^s \frac{\alpha_1 \zeta + \alpha_2 V(\zeta)}{n(\zeta)} d\zeta \right]. \tag{15}$$

Eq. (15) is an analytical closed-form relationship between pdf and conditional velocity and dissipation.

In order to extend the analysis, one needs information concerning function V. The conditional expectation of velocity $\langle v|c \rangle$ in the grid-generated turbulence with the uniform mean cross-stream scalar gradient was measured by Venkataramany and Chevray [9]. It was shown that a linear dependence

$$\langle v|c \rangle = \frac{q}{\sigma^2}(c - \langle c \rangle) \tag{16}$$

holds reasonably well for the experimentally observed conditional expectation of velocity $\langle v|c \rangle$. In such a case, one finds using Eqs. (4), (5) and (16) that V is the linear function

$$V(s) = s \tag{17}$$

Using Eq. (17) and identity (11) it is seen that the self-similar solution (15) for the uniform mean scalar gradient case looks exactly the same as the solution (14) for the case with no mean scalar gradient.

This result explains why pdf's measured by Jayesh and Warhaft [1] for a mean scalar gradient case are well described by Sinai and Yakhot [3] asymptotic solution with no mean gradient (14). It is clear that the above agreement is a consequence of linearity of the conditional expectation of velocity, Eq. (17).

However, it should be emphasized, as the measurements of Jayesh and Warhaft [1] showed, that the physical meaning of the self-similar solutions for the above two cases is quite different. Indeed, for the case without a mean scalar gradient, the conditional expectation of scalar dissipation $\langle N|c \rangle$ is nearly constant (so that

$n \approx 1$) and the pdf's are close to Gaussian. For the mean gradient case the pdf's have exponential tails and the conditional expectation of scalar dissipation $\langle N|c \rangle$ becomes U–shaped and is quite far from being constant; in fact $n(s) \sim |s|, |s| \gg 1$.

Let us return now to expression (12) for the conditional expectation of molecular diffusion. Again using Eq. (17) with identity (11) leads to

$$\Omega = -s, \tag{18}$$

irrespective of the form of the pdf.

The success of the linear dependence (17) in describing the experimentally observed conditional expectation of velocity provides a possible explanation of why linear dependence (18) is quite a good approximation to the experimentally measured conditional expectation of molecular diffusion (see, e.g. measurements in turbulent wake [10]) even for strongly non–Gaussian pdf's. It follows also from expression (12) that deviations of the conditional expectation of molecular diffusion from linear dependence are linked with deviations of the conditional expectation of velocity from a linear function.

Finally, it should be noted that Pope and Ching [11] proposed another explanation as to why the experimental pdf's of Jayesh and Warhaft [1] are well described by the Sinai–Yakhot [3] formula (14). It is based on the use of Eq. (18). Both explanations are interrelated, since it was shown above that equations (17) and (18) are equivalent.

3. Temperature field in stably stratified turbulence

In conclusion, the case of the temperature fluctuations in grid–generated turbulence under conditions of stable thermal stratification is briefly analyzed. The mean temperature field is described by stable linear profile, so $E > 0$ in Eq. (3) (y is the vertical coordinate, gravity acts in the - y direction). Eqs. (1) and (2) continue to be valid for the case under consideration. Thus once self–similarity of the pdf is assumed and Eqs. (4) and (5) are formally used, Eqs. (7) and (8) will be obtained. Contrary to the passive case, α_1 and α_2 defined by Eq. (9) are now not constants, but they depend on stream–wise coordinate x (measurements done by Yoon and Warhaft [12] show that α_1 decreases with x). Eq. (15) can be considered for this case in a "loose" formal sense as the quasi self–similar solution, where coordinate x is a parameter. For this formal solution to be meaningful have sense the derivative of g with respect to x must be small.

Eq. (15) will give us exact self–similar solution for a thermal stratification when the conditional expectation of the vertical velocity is a linear function of a temperature (i.e. Eqs. (15) and (16) are valid). This solution, as well for the passive scalar case, coincides with the Sinai–Yakhot [3] solution. Measurements done by Thoroddsen and van Atta [13] show that temperature pdf's collapse quite well and they are nearly Gaussian. Thus one can conclude that above measured pdf's are described by Eq. (14) and $n \approx 1$.

4. Conclusion

The present work gives the derivation and the analysis of the self–similar closed–form solution for the equation for the one–point pdf of a passive scalar, in decaying grid–generated turbulence with a uniform mean transverse scalar gradient. The new self–similar solution is a function of the conditional expectations of scalar dissipation and transverse velocity. Under the assumption of a linear dependence of conditional expectation of velocity on scalar (experimental data [9] support this assumption) it follows that the self–similar solution depends only on the conditional expectation of scalar dissipation. Moreover, it looks exactly the same as the asymptotic solution for the pdf equation in decaying turbulence with no mean gradient found by Sinai and Yakhot [3]. This explains why the pdf that was calculated by Jayesh and Warhaft [1] from Sinai and Yakhot's asymptotic solution using the measured conditional expectation of scalar dissipation is remarkably close to the measured pdf. Finally, the exact expression that relates the conditional expectation of molecular diffusion and the conditional expectation of transverse velocity is found. It is concluded, using this expression, that, when the conditional expectation of velocity is a linear function of scalar, the same dependence will exist for the conditional expectation of molecular diffusion. The case of the temperature fluctuations in grid–generated turbulence under the action of a stable (negatively buoyant), linear, mean temperature profile is briefly studied. A quasi–self–similar solution is obtained for this case. When the conditional expectation of vertical velocity is a linear function of temperature the latter solution is the strictly self–similar one and coincides with Sinai and Yakhot's [3] solution.

Acknowledgments

This work was carried out while the author was Visiting Professor at the École Centrale de Lyon, Laboratoire de Mécanique des Fluides et d'Acoustique URA CNRS 263 in November 1994 - July 1995. I gratefully acknowledge discussions with Prof. J.Mathieu and Prof. J.N.Gence. I would also like to thank Prof. D.Jeandel, Drs S.Simoëns and F.Nicolleau for their hospitality.

References

1. Jayesh and Warhaft, Z. (1992) Probability distribution, conditional dissipation, and transport of passive temperature fluctuations in grid–generated turbulence, *Phys. Fluids* **A 4**, 2292–2307.

2. Sahay, A. and O'Brien, E.E. (1993) Uniform mean scalar gradient in grid turbulence: Conditioned dissipation and production, *Phys. Fluids* **A 5**, 1076–1078.

3. Sinai, Y.G. and Yakhot, V. (1989) Limiting probability distributions of a passive scalar in a random velocity field, *Phys. Rev. Lett.* **63**, 1962–1979.

4. Kuznetsov, V.R. and Sabelnikov, V.A. (1980) *Turbulence and Combustion*, Hemisphere, New York.

5. Li, J.D. and Bilger, R.W. (1994) A simple theory of conditional mean velocity in turbulent scalar–mixing layer, *Phys. Fluids* **A 6**, 605–610.

6. Valiño, L., Dopazo, C., and Ros, J. (1994) Quasistationary probability density functions in the turbulent mixing of a scalar field, *Phys. Rev. Lett.* **72**, 3518–3521.

7. O'Brien, E.E. (1980) The probability density function (PDF) approach to reacting turbulent flows, in P.A.Libby and F.A.Williams (eds.), *Turbulent Reacting Flows*, Springer–Verlag, New York.

8. Sirivat, A. and Warhaft, Z. (1983) The effect of a passive cross-stream temperature gradient on the evolution of temperature variance and heat flux in grid turbulence, *J. Fluid Mech.* **128**, 323–346.

9. Venkataramany, K.S. and Chevray, R. (1978) Statistical features of heat transfer in grid generated turbulence: constant gradient case, *J. Fluid Mech.* **86**, 513–543.

10. Kailasnath, P., Sreenivasan, K.R., and Saylor, J.R. (1993) The conditional scalar dissipation rates in turbulent wakes, jets, and boundary layers, *Phys. Fluids* **A 5**, 3207–3215.

11. Pope, S.B. and Ching, E.S.C. (1993) The stationary probability density functions: An exact result, *Phys. Fluids* **A 5**, 1529–1531.

12. Yoon, K. and Warhaft, Z. (1990) The evolution of grid–generated turbulence under conditions of stable thermal stratification, *J. Fluid Mech.* **215**, 601–638.

13. Thorõddsen, S.T. and van Atta, C.W. (1992) Exponential tails and skewness of density–gradient probability density functions in stably stratified turbulence, *J. Fluid Mech.* **244**, 547–566.

PDF AND HIGHEST CONCENTRATION EVOLUTION IN TURBULENT JETS

V.A. SABELNIKOV*, S. SIMOËNS** and M. AYRAULT**

*Central Aerohydrodynamic Institute (TsAGI), Zhukovsky, Moscow region, 140160 Russia.
**Laboratoire de Mécanique des Fluides et d'Acoustique, ECL,UCB,UMR CNRS 5509, 36 Av. G. de Collongues, BP 163, 69131 Ecully Cédex, France

ABSTRACT.-The streamwise evolution of the highest concentration C_{max} of a passive scalar in a turbulent round jet is measured in the range ($0 \leq x/d \leq 20$). The Schmidt number was $Sc \cong 5.10^4$ and the jet Reynolds number $Re_d \cong 13340$. Beyond the near field of the jet nozzle ($x/d \leq 4$) the C_{max} decreases. This finding needs to be taken into account in scalar probability density function (PDF) modeling.

1. Introduction

Correct description of the highest value (C_{max}) and the smallest value (C_{min}) of concentration is very important when considering the tails of the concentration probability density function (PDF). It is usually assumed that C_{min} and C_{max} do not change with turbulent mixing and are equal to their initial values, which we normalize to 0 and 1, respectively. The concentration PDF describes the likelihood of different concentrations within the range $C_{min} \leq C \leq C_{max}$ and evolves downstream. Whether C_{min} and C_{max} also change with mixing is the point at issue. Direct numerical simulation and stochastic mixing models (Miller et al., 1993; Girimaji, 1992; O'Brien and Sahay, 1992) show that for isotropic velocity and concentration fields C_{min} and C_{max} do indeed change with time. There have been remarkably few experimental studies of the above problem. Measurements of centerline and radial profiles of highest and smallest temperatures in plane buoyant jets at different Richardson numbers were done by Kotsovinos (1977). He shown that the non-dimensional centerline highest temperature excess above the uniform ambient temperature (non-dimensionalization was made using the mean centerline temperature excess) and the non-dimensional centerline smallest temperature excess depend on the local Richardson number. The values of non-dimensional highest temperature excess were in the range(1.2-1.4) at $x/d \leq 50$ for the smallest initial Richardson number and increase with this number. Papantoniou and List (1989) had measured radial highest

159

L. Fulachier et al. (eds.), IUTAM Symposium on Variable Density Low-Speed Turbulent Flows, 159–166.
© 1997 *Kluwer Academic Publishers.*

concentration profiles in the far field (x/d = 105) of buoyant round jets using laser induced fluorescence. Centerline non-dimensional highest concentration value was around 2.3. Highest-to-mean concentration ratios, $C_{max}/<C>_{axis}$, were also measured by Wanta (in Gifford (1959)) at various distances downwind from a point source installed on a micrometeorological tower in the atmospheric boundary layer. Measured ratios are in the range of 1.5 to 2.5 for large distances (from 1km to 6 km) from source. The experiment reported here is aimed to study streamwise evolution of the highest concentration C_{max} of a passive scalar in a turbulent axisymetric jet.

2. Experimental set-up

The experiment was carried out in an axisymmetric turbulent air jet. The jet discharges from a d = 2 cm internal diameter nozzle into quiescent ambient air. Initial velocity of the jet was equal to U_0 = 10 m/s. Thus the corresponding Reynolds number $Re_d = U_0 d/\nu$ was equal to 13340. Preliminary hot-wire anemometry measurements of mean and fluctuating velocities were performed and compared with those of Wygnansky and Fiedler (1969). The results were in good agreement with their data. The passive scalar consisted of incense particles (diameter $\cong$ 1 μm). The value of the Schmidt number was $\cong$ 5×10^4. Basis of the measurement is Mie scattering which is proportional to the particle concentration (see e.g. details in Balint (1983)). The flow was illuminated with a vertical light sheet passing through the jet axis. Light source was a 300 mJ Yag laser and the plane was 1 mm thick. The thickness of the light plane was reduced to δ = 0.7 mm using a spherical lens. Concentration measurements were made using a video CCD camera used in linear mode and having an array of 512x512 pixels. Initial concentration was maintained constant and corresponded to maximum intensity level accepted by the CCD camera. It was checked on each instantaneous digitised image that the grey-level value Z was maintained constant in the potential core zone. The dynamic range, D, was 155. Absolute error, E_1, produced by this technique is about 1 giving a relative error, E_2 (= E_1/D) of about 0.7 %. Corrections to account for light absorptions through the measuring volume have been described in more detail by Ayrault et al. (1993). The acquisition frequency was 50Hz (video frequency). This is longer than the time taken for the large scale eddies of the jet to pass at the convection velocity. This convection time can be estimated by the relation (Wygnansky and Fiedler, 1969) : $T_c = 0.066 (d/U_0)(x/d)^2$ (0.01s in our case). Lowest frequency in the flow is of the order of $1/T_c$, thus the sample time separation, T_f (=0.1s), is greater than T_c ($T_f \geq 10.T_c$) allowing statistical independent samples. Spatial resolution for these measurements is limited by the dimension of the sampling volume, i.e. the pixel distance size, δ_p, and δ. For our measurements $\delta \cong \delta_p$. An area of 40cmx40cm was recorded, yielding δ_p to be about 750μm. Estimation of the Kolmogoroff micro-scale, η, is obtained from the relation $\eta = L \, Re_L{}^{-3/4}$, where L is the integral turbulence length scale, $Re_L = u'L/\nu$ is the turbulent Reynolds number, u' is the r.m.s. of longitudinal velocity. On centerline of the jet at location x/d ≈12, L ≈ d, $u'/U_0 \approx 0.2$ so $Re_L \approx 2670$, giving a value of η which was about $\eta \approx 54$ μm ($\delta_p \approx 14\eta$). Furthermore the Batchelor scale of the scalar field, η_B, is smaller than η by a factor of $Sc^{1/2} \approx 2. \, 10^2$. Thus spatial resolution of our measurements was not sufficient to resolve finest scales of turbulent fluctuations. In spite of this limited spatial resolution, we believe, that our measurements of highest concentration are not greatly affected.

3. Image processing

Digitisation allows integer local grey-level values Z from 0 to 255. Number of images used for the statistics was 1000. As the optical signal due to particle scattering propagates through the flow, some distortions and absorptions accumulate and corrections of each instantaneous record I(x,y,t) ($(x,y) \in [1,512]^2$, fixed t) were performed. These corrections result from geometrical distortion, non-uniformity of the light sheet, background scattering and attenuation due to absorption along the optical path. More details can be found in Ayrault et al. (1993). The normalised concentration, C ($0 \leq C \leq 1$, is determined by the relation : $C = (Z - Z_{min}) /(Z_{max} - Z_{min})$ where Z_{min} is the grey-level value of the background and Z_{max} is the grey-level value at the nozzle exit.

4. Results

4.1. C_{max} MEASUREMENTS

For every instantaneous image k (k = 1, ..., 1000) we determined the maximum of concentration C_{max}^k (x) corresponding to each radial position x (300 axial locations covering the axial range from 0 to 20 jet exit diameters). Finally the highest concentration was determined as the maximum of the instantaneous maxima, i.e.

$$(1) \qquad C_{max} (x) = \max_{k \in [1,1000]} [C_{max}^k (x)]$$

Streamwise evolutions of $C_{max}(x)$ and $<C>_{axis}$, non-dimensional by the initial concentration value C_0 were compared in fig. 1a and b.. Furthermore it may be seen that the highest concentration, $C_{max}(x)$, is equal to initial nozzle concentration (C = 1) only if X/d ≤ 4. Beyond this location, the C_{max} decreases with axial distance from the nozzle.

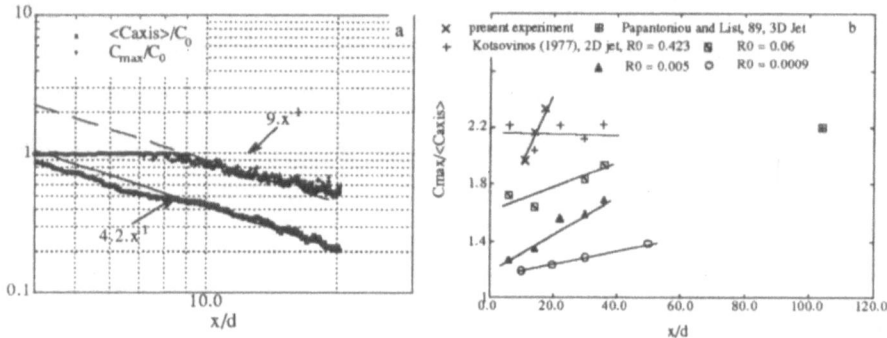

Fig.1. - Streamwise evolution of (a) highest concentration C_{max}/C_0 and $<C>_{axis}/C_0$, (b) highest-to-mean concentration ratios $C_{max}/<C>_{axis}$ for this experiment, Kotsovinos (77) (2D jet) with various Richardson numbers and Papantoniou and List (89) (round Jet).

It is seen from figure 1b that the highest-to-mean concentration ratios $C_{max}/<C>_{axis}$ are within the range of 2 to 3. It is quite difficult to give direct comparison of our results for $C_{max}/<C>_{axis}$ ratio with a few other published ones (Kotsovinos, 1977; Papantoniou and List 1989; Gifford, 1959), since experimental conditions are different. Kotsovinos's (1977) measurements of $C_{max}/<C>_{axis}$ (where C was proportional to temperature excess) were performed in plane buoyant jets at x/d ≤ 50 for different Richardson numbers, and Papantoniou and List (1989) have studied concentration field in the far field (x/d = 105) of round jet. Ratios $C_{max}/<C>_{axis}$

measured by Wanta and presented in Gifford (1959) were obtained for a point source installed on a tower in the atmospheric boundary layer.

Ratios of highest-to-mean excess temperature measured by Kotsovinos (1977) depend on the local Richardson number and increase with this number. They are within the range of 1.2 to 1.4 for the smallest initial Richardson number. Measurements of $C_{max}/<C>_{axis}$ reported in Papantoniou and List (1989) and Gifford (1959) gave respectively ratios around 2.3 and 1.5 to 2.5. Our results show a better agreement with the measurements of Papantoniou and List (1989) and Wanta (see in Gifford, 1959).

Kotsovinos (1977) and Papantoniou and List (1989) presented also transverse profiles of highest-to-mean concentration ratios. They noted that C_{max} is not constant but profiles of C_{max} are more flat than mean concentration profiles and fall down sharply at the boundaries of jet. This conclusion raises some doubts. Indeed, measured values of C_{max} clearly depend on the records time length of realisation. We can only hypothesize here that with increasing the length of the record time of realisation the C_{max} profile will be flatter, and at infinitely large times, $t \to \infty$ (or k tends to ∞), C_{max} will lose dependance from transverse coordinates.

4.2. CONCENTRATION PDF DETERMINATION

Concentration PDF's were determined at the locations $x/d = 12$ and 20 for various radial positions (one point in 12 have its correspondant on the same ray in 20). Two normalisations of the concentration values were used. In the first normalisation, instantaneous concentration C was divided by the local mean concentration $<C>$ and the PDF of the ratio $\eta = C/<C>$ was considered (Figure 2.a and b). In the second normalisation, the PDF of the ratio $s=(C-<C>)/\sigma$ (where $\sigma = < (C- < C >)^2 >^{1/2}$) was determined (Figure 3.a and b). The PDF's were normalised by the requirement that their integrals be unity : i.e.

(2)
$$\int_0^{\eta max} PDF(\eta)\, d\eta = 1 \,,\, \eta_{max} = (C_{max}/<C>), \text{ and}$$

(3)
$$\int_{s min}^{s max} PDF(s)\, ds = 1 \,,\, s_{min} = -<C>/\sigma \text{ and } s_{max} = (C_{max} - <C>)/\sigma.$$

The PDF's of normalised concentrations η and s are presented in Figures 2.a-b and 3.a-b for a number of distances from the jet axis.

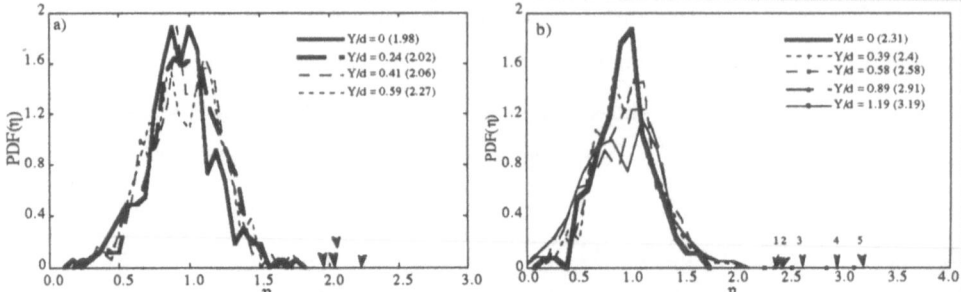

Fig. 2. - PDF of normalised concentration η = C/<C> for different radial positions in axisymetric jet at locations x/d = 12 (a) y/d = 0, 0.24, 0.41, 0.59, and x/d = 20 (b) y/d = 0,0.39,0.58,0.89,1.19..

In papers of Schefer et al. (1994) and Dowling and Dimotakis (1991) it was concluded that the PDF are closely symmetric and nearly Gaussian on jet axis. Away from the axis the peak in the PDF shifts to lower concentrations than the mean and tails of the distribution are skewed towards higher concentration to compensate. Making allowance for the scatter in the data due to finite number of samples, these conclusions are bore out by a comparison in figures 2 and 3, which give results close to the jet axis, and far from the axis. It may be observed that at the centerline, the PDF's are nearly symmetric and Gaussian. PDF at outer radial locations are skewed toward higher concentrations. Our results are in good correspondance with the data of (Schefer et al. 1994; Dowling and Dimotakis (1991).

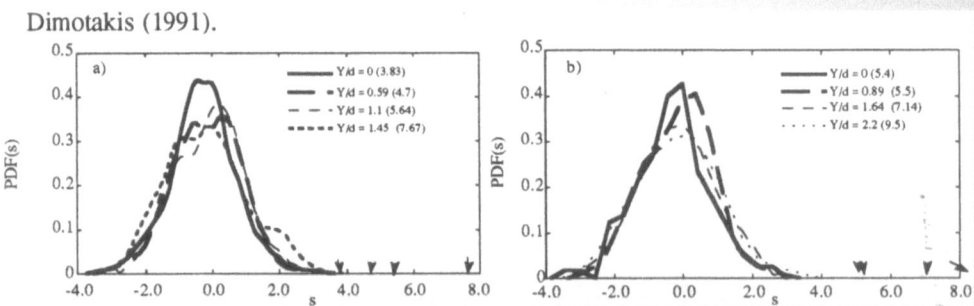

Fig. 3. - PDF of normalised concentration s = (C - <C>)/σ for different radial positions in axisymetric jet at locations x/d = 12 (a) y/d = 0, 0.24, 0.41, 0.59, and x/d = 20 (b) y/d = 0, 0.89, 1.64, 2.2..

In figures 2 and 3 the solid arrows show values of $\eta_{max} = C_{max}/<C>$ and $s_{max} = (C_{max} - <C>)/\sigma$ when C_{max} is defined by (1) . The table 1 lists the η_{max} and s_{max} calculated using equations (2) and (3) at different points shown in the figures at location $x/d = 12$ and 20. Table 1 contains also values $\eta'_{max} = 1/<C>$ and $s'_{max} = (1-<C>)/\sigma$, i.e. it is assumed that $C_{max} = 1$ is equal to concentration values at the nozzle exit. It can be concluded that C_{max} determined by (1) provides a better upper bound on the PDF than does $C_{max} = 1$, reinforcing our conclusion that it is important to recognise that the highest concentration is reduced significantly by mixing.

It is seen that using for the right (maximum) bound of concentration field value of C_{max} determined by (1) allows us to localise PDF with more accuracy than in the frame of the usual assumption that right bound is equal to unity.

Y/d (x/d=12)	η'_{max}	η_{max}	s'_{max}	s_{max}	Y/d (x/d=20)	η'_{max}	η_{max}	s'_{max}	s_{max}
0	2.86	1.98	7.22	3.83	0	4.54	2.31	15.6	5.4
0.24	2.86	2.02	8.13	4.27	0.39	4.76	2.4	13.6	4.88
0.41	2.94	2.06	8.25	4.8	0.58	5	2.58	13.3	5.13
0.59	3.23	2.27	8.63	4.7	0.89	5.9	2.91	13.8	5.5
0.76	3.70	2.58	9.13	5.7	1.19	6.25	3.19	14	5.8
1.10	4.35	3	9.63	5.64	1.64	7.7	3.94	17.4	7.14
1.45	5.56	3.88	11.71	7.67	2.2	12.5	6.7	20.4	9.5

Table 1: η_{max}, s_{max}, η'_{max} and s'_{max} for different radial positions at locations $x/d=12$ and 20.

5. Conclusion

We have measured the streamwise evolution of the highest concentration in an axisymmetric turbulent air jet at a Reynolds number of $Re_d = 13340$ and Schmidt number Sc of about 5. 10^4 in the axial range from 0 to 20 jet exit diameters. Our experiments indicate that the highest concentration is equal to the initial nozzle concentration $C = 1$ only in the near field of the jet $X/d \leq 4$. Beyond this location the highest concentration begins to fall. It should be noted that the highest concentration at every axial location in our experiment was determined from analysis of instantaneous radial concentration profiles, not from entire planar cross-section images. A better understanding of evolution of the maximum concentration would, of course, require that entire planar images to be analysed.

Perhaps one of the more interesting questions concerning PDF and streamwise evolution of C_{max} is how this evolution depends on Re_d, Sc and R_0 numbers. The importance of this problem also indicates the need for further experiments with higher spatial resolution to determine the influence of Re, Sc and R_0 on concentration (Dowling and Dimotakis (1991) reported first very interesting results in this subject). The consequences of changes of concentration bounds observed in the experiment also imply that existing PDF modeling, based on consideration of the PDF within a fixed concentration range (see e.g. Kutznetsov and Sabelnikov, 1990), needs to be refined to take into account the evolution of the scalar concentration bounds.

ACKNOWLEDGEMENT
V.A.S. is grateful to Prof. Denis JEANDEL for his kind hospitality in his laboratory (LMFA, ECL, UCB, UMR CNRS 5509) where this work was done.

REFERENCES

AYRAULT M. , SIMOENS S., MEJEAN P.,1993, Effects of two-dimensionnal obstacle on the dispersion of dense gas releases, 3rd Symp. on Experimental and numerical flow visualisation, New-Orleans, USA, Nov. 28-Dec. 3.

BALINT J.L., 1983, Application d'une méthode de visualisation laser et de traitement d'images à l'étude de la dispersion dans des écoulements turbulents, *Thèse de 3.° cycle*, Ecole Centrale de Lyon.

DOWLING D.R., DIMOTAKIS D.F., 1991, Similarity of the concentration field of gas-phase turbulent jets, *J. Fluid Mech.*, **218**, 109-142.

GIFFORD F.G., 1959, Statistical properties of a fluctuating plume, *Adv. in Geophys.* **6**, 117-137.

GIRIMAJI S.S.,1992, Towards understanding turbulent scalar mixing, *NASA* Contract Rep. C.R. 4446.

KOTSOVINOS N., 1977, Plane turbulent jets; Part. 2 : Turbulence structure, *J. Fluid Mech.*, **81**, 1-45.

KUZNETSOV V.R., SABELNIKOV V.A., 1990, Turbulence and combustion, Hemisphere publishing corporation, New-York.

MILLER R.S., FRANKEL S.H., MADNIA C.K., GIVI P., 1993, Johnson-Edgeworth translation for probability modeling of binary scalar mixing in turbulent flows, *Combus. Science and Tech.*, **91**, 21-52.

O'BRIEN E.E., SAHAY A., 1992, Asymptotic behaviour of the amplitude mapping closure, *Phys. Fluids*, A, **4** , 1773-1775.

PAPANTONIOU D., LIST E.J., 1989, Large-scale structure in the far field of buoyant jets, *J. Fluid Mech.*, **209**, 151-190.

WYGNANSKI E., FIELDLER H., 1969, Some measurements in the self-preserving jet, *J. Fluid Mech.* , **38**, 577-612.

III. Modelling and Experiments. Buoyancy Effects

TURBULENT TRANSFER MODELLING IN TURBULENT FLOWS WITH DENSITY VARIATIONS AND BUOYANCY FORCES

V. A. FROST
Institute for Problems in Mechanics, Russian Academy of Sciences, pr Vernadskogo, 101, Moscow, 117526, Russia

The effect of stable stratification is of profound importance in turbulent flows occurring in the environment and chemical engineering. To be able to predict these flows, it is necessary to know the intensity of turbulent micromixing. The modelling of the turbulent micromixing is very difficult problem because the effect of buoyancy forces on the turbulent micromixing is not well understood. On the other hand, the experimental data on effect of buoyancy forces on the micromixing can be useful only when we have adequate mathematical models for simulating the main features of the mixing process.

Turbulence in stratified fluid has much in common with turbulence in combustion. In both cases we deal with effect of bulk forces (gravity, centrifugal forces, inertia) on fluid particles of various densities resulting in considerable variation in motion of particles. The method of the joint probability density function (PDF) for velocity and concentration proposed formerly for the simulation of chemically reacting flows [1, 2] is considered. The usage of the joint-pdf method for turbulence prediction in the presence of dynamically active buoyant forces allows us to take into account the differences of particles motion due to difference of their density.

In the case of turbulent combustion, the single-point velocity-concentration PDF equation is used for calculation of averaged rate of chemical reaction as well as all product moments of velocity and concentration. To adapt the method for prediction of stratified turbulent flows is sufficient to suppose that the density is a function of concentration or temperature only.

This approach allows us to calculate so-called conditionally averaged velocity (CAV). The CAV is calculated for the fixed value of density and makes us possible to take into account the difference in motion of fluid particles of different densities under the action of bulk forces. For example, if buoyancy forces act, the light-weight particles are to lift, whereas the heavy particles are to move down.

So-called Langevin set of equations is used instead of Navier-Stokes and molecular diffusion equations to obtain the equation for single-point joint PDF [3]. This approach allows us to concentrate difficulties of physical modelling in the initial

L. Fulachier et al. (eds.), IUTAM Symposium on Variable Density Low-Speed Turbulent Flows, 169–172.
© *1997 Kluwer Academic Publishers.*

stage of analysis and then to pay attention only to the calculations. The equation obtained is unclosed and contains turbulence scales and micromixing terms which have to be obtained from experiments.

The joint PDF is a function of a number of independent variables. Therefore the calculation of the joint PDF is very difficult problem. Meanwhile, the problem can be reduced to the set of equations for the moments. But in this case we face obstacles similar to the closure problem for Reynolds equations.

The method proposed is tested with the experimental data on decaying turbulence downstream turbulizing grids for uniform and stratified cases and for uniform shear flows [4-7]. Our experimental data are also used for testing [8].

The calculation has been performed [9] for the set of five differential equations for the first and the second product moments of the velocity components and concentration (C is averaged concentration, $S_1 = <u'^2> = <v'^2>$ and $S_3 = <w'^2>$ are turbulent intensities, $S_4 = <c'^2>$ is intensity of concentration fluctuation, $S_{34} = <cw'^2>$ is vertical turbulent mass flux):

$$\frac{\partial C}{\partial t} = -\frac{\partial S_{34}}{\partial z},$$

$$\frac{\partial S_1}{\partial t} = -2.\alpha.S_1 + \varepsilon - \frac{\partial a_{113}}{\partial z},$$

$$\frac{\partial S_3}{\partial t} = -2.\alpha.S_3 + \varepsilon - 2.S_{34}.Fr - 4\frac{S_4}{\rho}.\frac{\partial S_1}{\partial z} - 3.\frac{\partial a_{333}}{\partial z},$$

$$\frac{\partial S_4}{\partial t} = -2.\beta.S_4 - S_{34}.\frac{\partial a_{344}}{\partial z},$$

$$\frac{\partial S_{34}}{\partial t} = -(\alpha + \beta).S_{34} - 2.S_4.Fr - S_4.\frac{\partial S_1}{\partial z} - S_3.\frac{\partial C}{\partial z} - 2.\frac{\partial a_{334}}{\partial z},$$

where u, v, w are the velocity components, z is downward vertical axis, α and β are intensities of velocity and concentration dissipation, ε is intensity of pressure and friction fluctuations. Factor Fr (Froude number) represents effect of bulk and pressure forces:

$$Fr = \frac{-g - 2.\frac{\partial S_1}{\partial z}}{(a_\rho + b_\rho.C).b_\rho},$$

where g is acceleration of gravity, a_ρ and b_ρ are the coefficients in relationship between concentration and density which can be accepted in this case as the simplest linear relationship between inverse ρ and c.

The third moments $a_{344} = <w'c'^2>$, $a_{113} = <u'^2w'>$, $a_{333} = <w'^3>$, and $a_{334} = <w'^2c'>$ are determined by equilibrium consideration.

In the whole, the calculations and experiment are coincident. However, in this case it is impossible to take into account a plethora of initial PDF and the fact that the range of concentration variations is limited.

Two ways of improving the method are examined:

i) the equations for the product moments are numerically solved simultaneously with the equation for concentration PDF;
ii) the set of equations for the second moments of velocity and equations for concentration PDF and CAV are simultaneously solved.

The equation for CAV is derived on the base of joint PDF equation for concentration and velocity with Langevin model of micromixing [1] and is rewritten for the case considered in a form of equation for vertical component $<w>_c$ (here and below the dimensional form of equations are used):

$$\frac{\partial <w>_c P(c;z,t)}{\partial t} + <w'^2>_c \frac{\partial P}{\partial z} = \frac{\alpha}{\rho} <w>_c P(c;z,t) + g.(1 - \frac{<\rho>}{\rho})P(c;z,t) + \beta.M_w[P].$$

Here $P(c;z,t)$ is PDF of concentration c, $\rho = \rho(c)$ is density, $\beta.M_w[P]$ and $\beta.M_p[P]$ below are the terms describing micromixing influence on CAV and PDF respectively. PDF equation is used in form:

$$\frac{\partial P(c;z,t)}{\partial t} + \frac{\partial <w>_c P(c;z,t)}{\partial z} = \beta.M_p[P].$$

In the first case, we have to use the CAV to allow for buoyancy effects and to set boundary conditions. To simplify the calculations, the CAV and turbulent diffusivity can be obtained from the equilibrium condition. In the second way, we need minimum number of assumptions and can describe motions of particles of various densities correctly.

Turbulence decaying in a vessel after strong turbulization of stratified fluid is considered. The calculations are performed for the set of equations for PDF and CAV. The nonpermiable condition $<w>_c = 0$ is put at the top and bottom boundaries. Some kinds of this condition for different kinds of interaction between buoyancy forces and turbulent diffusion are considered. The initial conditions for PDF are varied from δ-functions (black-and-white distribution) to the uniform distribution (case of constant density).

References

1. Frost, V. A.: Proceedings of III Allunion Meeting on Combustion Theory, Moscow, Nauka, vol. 1, pp121-125, 1960.

2. Chung, P. M.: AIAA J., vol. 7, pp1982-1991, 1969.

3. Frost, V. A.: the preprint of IPMech RAS, No.483, pp28, 1990.

4. Stillinger, D. C., Helland, K. N., Van Atta, C. W.: J. Fluid Mech., 131, pp91-122, 1983.

5. Itswier, E. C., Helland, K. N., Van Atta, C. W.: J. Fluid Mech., 162, pp299-338, 1986.

6. Lienhard, J. H., Van Atta, C. W.: J. Fluid Mech., 210, pp57-112, 1990.

7. Tavoularis S., Karnik U.: J. Fluid Mech., 204, pp457-478, 1989.

8. Emelianov, V. M., Frost, V. A.: the preprints IV Int. Symp. on stratified flows, Grenoble, France, vol. 2, June 29- July 2, 1994.

BUOYANCY-GENERATED VARIAELE-DENSITY TURBULENCE

D. L. SANDOVAL AND T. T. CLARK
Los Alamos National Laboratory
Los Alamos, New Mexico USA 87545

AND

J. J. RILEY
University of Washington
Seattle, Washington USA 98145

1. Introduction

Because of the importance of turbulence mixing in many applications, a number of turbulence mixing models have been proposed for variable-density flows. These engineering models (one-point statistical models) typically include the transport of the turbulent kinetic energy and the turbulent energy dissipation rate (i.e., $k - \epsilon$ models). The model presented by Besnard, Harlow, Rauenzahn and Zemach (1992) (herein referred to as BHRZ) is a one-point model intended to describe variable-density turbulent flows. Transport equations for the Reynolds stress tensor, R_{ij}, and the turbulent energy dissipation rate, the density-velocity correlation, a_i, and the density-specific volume correlation, b are derived. This model employs techniques and concepts from incompressible, constant-density turbulence modeling and incorporates ideas from two-phase flow models.

Clark and Spitz (1994) present a two-point model for variable-density turbulence. Their derivation is based on transport equations that are based on two-point generalizations of R_{ij}, a_i, and b. These equations are Fourier transformed with respect to the separation distance between the two points. Transport equations are derived for R_{ij}, a_i, b. As in the one-point model, this model contains many *ad-hoc* assumptions and unknown model coefficients that must be determined by comparison with experimental and numerical data. However, the two-point formalism requires fewer equillibrium assumptions then does a single-point model.

L. Fulachier et al. (eds.), IUTAM Symposium on Variable Density Low-Speed Turbulent Flows, 173–180.
© *1997 Kluwer Academic Publishers.*

Our primary concern in this paper lies in the nonlinear processes of turbulence and the influence of large density variations (not within the Boussinesq limit) on these processes. To isolate the effects of variable-density on the turbulence we restrict our flow to be incompressible, statistically homogeneous buoyancy-generated turbulence. To our knowledge there have not been any simulations reported for this problem.

2. Equations of Motion

We shall consider the turbulent mixing of two miscible, incompressible fluids of different densities. The Mach number is zero. The velocity field for the mixing of two miscible, incompressible fluids is not in general divergence free, i.e., $\nabla \cdot \vec{u} \neq 0$ [Joseph (1990)].

The equations of motion used in this study [Sandoval (1995)] are the conservation of mass, the conservation of momentum for a Newtonian fluid and the conservation of species equation. Fick's law [see, e.g., Bird, Stewart and Lightfoot (1960)] for the diffusion of two species of different densities gives

$$\frac{\partial \rho}{\partial t} + u_j \frac{\partial \rho}{\partial x_j} = \rho \frac{\partial}{\partial x_n} \left\{ \frac{\mathcal{D}}{\rho} \frac{\partial \rho}{\partial x_n} \right\} \tag{1}$$

Comparison of the conservation of mass equation with (1) leads to the following result for incompressible mixing flows:

$$\frac{\partial u_n}{\partial x_n} = -\frac{\partial}{\partial x_n} \left\{ \frac{\mathcal{D}}{\rho} \frac{\partial \rho}{\partial x_n} \right\} . \tag{2}$$

Thus, the incompressible velocity field is divergent.

These equations will be solved with periodic boundry conditions and "random" initial conditions for density and a nearly zero initial velocity.

3. Averaged Equations

We examine homogeneous turbulence subjected to an acceleration. Letting a tilde denote a mass-weighted average and an overbar denote a volume-weighted average, we have $u_i = \overline{U}_i + u_i' = \tilde{U}_i + u_i''$. Where, $\tilde{u}_i = \overline{\rho u_i}/\overline{\rho}$. Note that $u_i'' = a_i + u_i'$ where $a_i = \overline{u_i''}$. The frame of motion is chosen so that the mean volume-weighted velocity is zero; $\overline{U}_i = 0 = \tilde{U}_i + a_i$. Using the exact averaged equations for the subsequent correlations and exploiting homogeneity [for more complete detail see Sandoval (1995)];

$$\underbrace{\frac{\partial \overline{P}}{\partial x_i}}_{A} = \left(\frac{\overline{\rho}}{1 - \overline{\rho}b} \right) \left\{ \underbrace{g_i}_{B} + \underbrace{\overline{v' \frac{\partial \tau_{ni}'}{\partial x_n}}}_{C} - \underbrace{\overline{v' \frac{\partial p'}{\partial x_i}}}_{D} + \underbrace{\overline{u_i' \frac{\partial u_n'}{\partial x_n}}}_{E} \right\} , \tag{3}$$

where $b = -\overline{\rho'(1/\rho)'}$. Substituting (3) into the mass flux equation, and in the limit of small time we neglect the velocities;

$$\frac{\partial a_i}{\partial t} = \left(\frac{\overline{\rho} b}{1 - \overline{\rho} b}\right) g_i - \left(\frac{1}{1 - \overline{\rho} b}\right) \overline{v' \frac{\partial p'}{\partial x_i}}. \tag{4}$$

The second term on the right side is modeled (e.g., in BHRZ) as a simple "drag" (i.e., a destruction of a_i). If this modeling is accurate, then this term is proportional to a_i, and hence asymptotically tends to zero in the limit of small time. The variations in this correlation can be studied via direct numerical simulation of the equations of motion using (1) and (3).

4. Numerical Simulations

A numerical algorithm solving the equations of motion with (1) and (3) was developed from the algorithm of McMurtry (1987). It is modified for the incompressible mixing of miscible fluids subjected to a constant acceleration. This algorithm is related to the projection method but takes into account the fact that the velocity field is divergent. A pseudo-spectral method is used so that the spatial derivatives are computed in wavenumber space whereas the nonlinear terms are computed in physical space via the use of fast Fourier transforms. The aliasing errors are ameliorated by truncation of the Fourier fields. The temporal discretization is Adams-Bashforth and the fields are time-advanced in wavenumber space. As in the work of Batchelor, Canuto and Chasnov (1992), the fluctuations are assumed to be periodic in all three spatial directions. However, the Boussinesq approximation has not been made. The equations of motion are solved in nondimensional form following Batchelor, et al. (1992). Simulations used a grid size of 128^3.

The density field is initialized using the method of Eswaran and Pope (1988). This method creates an initial density field that approximately conforms to a double-delta function probability density function (pdf), where the density values of the initial field corresponds closely to either the high density, ρ_{max}, or low density, ρ_{min}, value.

The initial velocity field is set to zero, then slightly modified to account for the divergent velocity [eq. (2)] condition. Because the initial velocity field is nearly zero, density-velocity correlations are initially nearly zero. An acceleration is applied in the z-direction and the developing flow is statistically axisymmetric about the z-axis.

Table 1 lists the initial density statistics for the simulations. Case NCD is a nearly constant-density case performed to provide comparisons with the fully variable-density case, FVD.

Run	R_o	ν	σ	g	ρ_{max}	ρ_{min}	ρ_{max}/ρ_{min}	θ_o
NCD	256	7.800e-3	1.0	10.	1.05	0.95	1.105	0.0434
FVD	256	8.543e-3	1.0	1.0	1.60	0.40	4.000	0.5206

TABLE 1. List of initial statistics for buoyancy-driven cases

5. Numerical Results

In the two cases considered, the initial conditions are such that R_o is the same in both cases but the initial density ratios are different. The initial velocity is nearly zero. Through the action of an acceleration, the velocity increases rapidly and the fluid is set into motion. The energy reaches a maximum and begins to decrease as the density field mixes and diffuses towards its constant mean value, and as viscous dissipation becomes appreciable. The velocity is initially highly correlated with the density field. The density field corresponds to a source of potential energy which is converted to kinetic energy ("turbulence") and finally into "heat" (viscous work). As the density field diffuses and mixes towards its mean value the source of potential energy decays and the remaining kinetic energy decays away. The kinetic energy will tend towards zero when the density field is uniformly mixed (there is no available potential energy) and dissipation removes the remaining kinetic energy.

The mean presure gradient is obtained from eq. (3) and plotted, normalized by the hydrostatic pressure, in Fig. 1a for both cases. For the NCD case, the mean pressure is close to the hydrostatic balance. As time evolves, the mean pressure remains at this value. For the FVD case, the mean pressure gradient is initially lower than the hydrostatic balance. This plot shows that the mean pressure gradient for case FVD varies more in time. Again, as the density fluctuations decay, the mean pressure gradient approaches the hydrostatic balance.

Figure 1b plots the values of each term in the equation for the mean pressure gradient [eq. (3)] for case FVD. In this case, in the limit as the density variations tend to zero, the mean pressure gradient tends to the value $\bar{\rho}g = 1$. Initially, the mean pressure gradient is approximately 0.75 and the acceleration term, $g/\bar{v}$, is approximately 0.67. The correlation $\overline{v'(\partial p'/\partial x_3)}$ (labeled "D") is approximately 0.08 and represents 11 percent of the initial mean pressure gradient. All other terms are initially nearly zero. As the flow develops the mean pressure gradient increases to a value slightly larger than 1.0. At the last time shown it has very nearly a constant value of 1.0, which is consistant with the Boussinesq approximation. The second

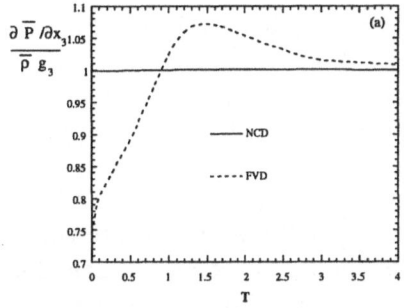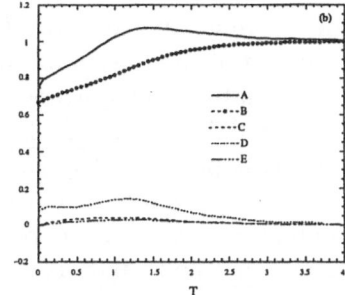

Figure 1. (a) Evolution of the mean pressure gradient for both cases and (b) Evolution of terms in the mean pressure gradient [eq. (3)] for FVD case

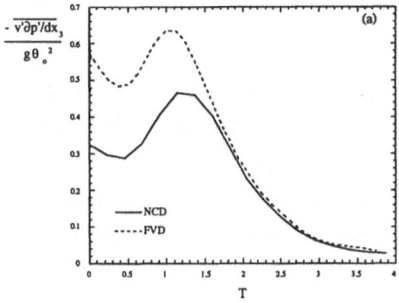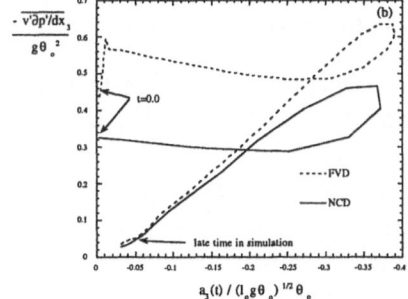

Figure 2. (a) $-\overline{v'\,\partial p'/\partial x_3}/g\theta_o^2$ vs. T and (b) $-\overline{v'\,\partial p'/\partial x_3}$ vs. a_3 for both cases

and third terms in the equation (3) for the mean pressure gradient grow slightly from their nonzero values and contribute only a few percent to the total mean pressure gradient. The correlation $\overline{v'(\partial p'/\partial x_3)}$ remains a large contribution to the mean pressure gradient up to late times.

Figure 2a shows $\overline{v'(\partial p'/\partial x_3)}/g\theta_0^2$ as a function of time for both cases. The magnitude of this correlation increases with initial density fluctuations. The nonzero value at the initial time represents a "rapid" part of this correlation. This correlation increases in time due to the "slow" part which is analogous to a drag for the mass flux. At late times this correlation is decaying and represents slow part decrease in the drag for the turbulent mass flux (see below in this section).

The correlation, $\overline{v'(\partial p'/\partial x_i)}/g\theta_o^2$, has been postulated by some modelers to behave as a "drag" term which impedes the growth of the turbulent mass flux [see, e.g., Besnard, et al., (1992)]. We have shown that the presence of this correlation impedes the growth of the turbulent mass flux. These results, however, suggest that this correlation behaves as both a "drag" or "slow" term and as a "rapid" term. The instant that the fluid is accelerated, this correlation immediately ("rapidly") takes a nonzero value, even though the mass flux is zero. To understand how this correlation behaves as a function of the mass flux, it is plotted as a function of the mass flux in Fig. 2b. Initially, $-\overline{v'(\partial p'/\partial x_3)}$ is approximately 0.32 for case NCD and approximately 0.56 for case FVD. At $T = 0$, when the acceleration is applied, the dominant part of $\overline{v'(\partial p'/\partial x_3)}$ is the "rapid" part. As the flow evolves, the "rapid" part vanishes as the density fluctuations decay and the "drag" part increases causing an increase in $\overline{v'(\partial p'/\partial x_3)}$. Ultimately, at late times, the "drag" or "slow" part is dominant as the "rapid" part has vanished so that $\overline{v'(\partial p'/\partial x_3)}$ decays nearly linearly with the mass flux. At late times in this buoyancy-driven turbulent flow this correlation represents a "drag" on turbulent mass flux.

6. Model Comparisons

In this section we address the efficacy of the extensions of traditional k-ϵ model formulation and a heuristic two-point closure to variable-density turbulence.

The model of Besnard, Harlow, Rauensahn and Zemach (herein refered to as the "one-point model") is an extension of one-point k-ϵ closures to variable-density turbulence. For the incompressible case, this model includes equations for R_{ij}, the turbulent energy dissipation rate, species concentrations, and correlations for the density-velocity and specific volume-density fluctuations. The higher-order unknowns in these equations are closed using typical constant density assumptions [e.g., Launder, Reece and Rodi (1975)] suitably extended to account for a variety of variable-density effects. The details of the model can be found in Besnard et al. (1992).

The model equations are derived for general inhomogeneous flows at high Reynolds number in BHRZ. It should be noted that the time scales for dissipation of a_n and b are "constructed" from a time scale associated with the energy cascade, $\tau \sim R_{nn}/(\rho \epsilon)$. This assumption is based on (1) turbulence characterized by a "large" inertial range where the dissipation rate is independent of viscosity (or diffusivity) and (2) that this cascade also dominates the dissipation of "a_n" and "b". This assumption is questionable for the case of initially quiescent, turbulence subjected to an acceleration. Thus the DNS provides a stern test of the single-point model.

The model of Clark and Spitz (herein referred to as the "two-point model") (1994) is a two-point (spectral) phenomenological model. The advantage of a two-point (spectral) formulation is that it eliminates the need for length-scale/dissipation equations and corollary assumptions which are employed in one-point modeling. The two-point model is an attempt to relieve some of the limitations inherent in the one-point k-ϵ or R_{ij}-ϵ formalism. Details of the development may be found in Clark and Spitz (1994).

6.1. RESULTS

The comparison between the one-point model, the two-point model and the DNS is made by examining one-point statistics. For details of the initial conditions see Sandoval (1995).

Figure 3a shows the evolution of b for the DNS and the model results. The modified one-point model for the decay of b is somewhat different then the DNS result. The initial value for ϵ is artificially small, leading to an inadequate dissipation of b at the early times. This may represent a deficiency in the methodology of one-point closures and also suggest an inclusion of molecular diffusive and viscous effects in the one-point formulation. The flows studied in the DNS simulations are dominated by viscous diffusion, at early and late times and this is not accounted for in the one-point model. Thus, we can choose ϵ to give either the correct energy dissipation, or the correct turbulent time-scale, or the correct length scale of b, but not all three. Therefore, b does not dissipate fast enough and, as a result, the subsequent behavior of b is incorrect although the initial growth rates of a and R_{nn} are adequate (see Figure 3b and 5c). At late stages, the values of a and R_{nn} are overpredicted, probably due to an overprediction of b.

The two-point model which does include molecular viscous and diffusive effects does substantially better at predicting the evolution of b. As a consequence, it does better at late times for both a and R_{nn} than the one-point model. This model does slightly overpredict the values of a and underpredicts the value of R_{nn} at their extrema; the one-point model does slightly better here.

Figure 3d shows the evolution of $\partial \bar{p}/\partial x_n$ for the DNS and the models. The model comparisons show relatively good agreement with the DNS.

7. Conclusions

The results of the direct numerical simulations indicate that the correlation $\overline{v'\partial p'/\partial x_i}$ can be considered the sum of a "slow part" and a "fast part", in analogy with the well-known decomposition of pressure-strain correlations of constant density turbulence. This modification, among others was incorporated into a single-point model and a two-point model of variable density turbulence. Predictions of the modified models are compared to the DNS results, and indicate that two-point model performed slightly better than the one-point model. It is suggested that the one-point model should be extended further to account for molecular effects at early times and to permit a length scale for the density fluctuations that is independent of the length scales for the velocity fluctuations.

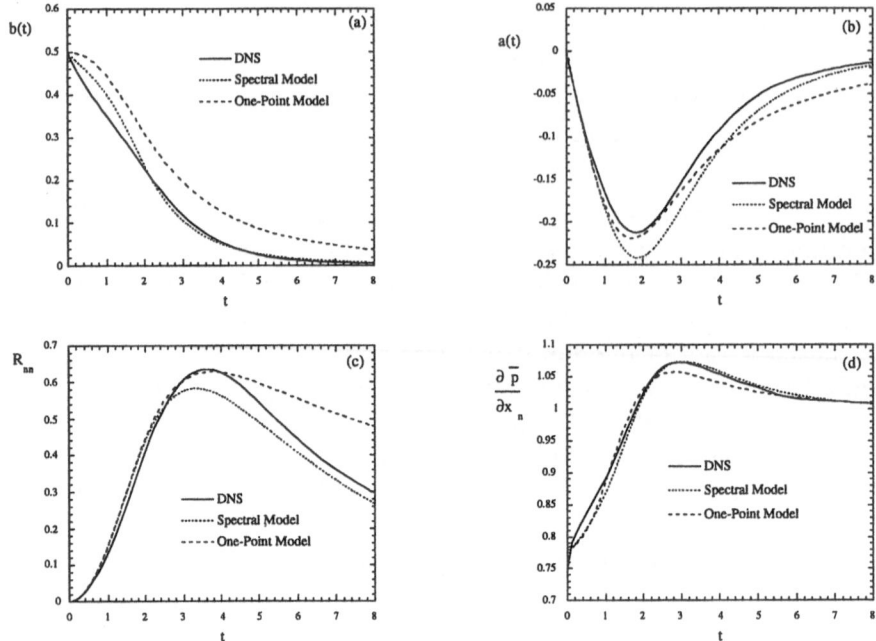

Figure 3. Evolution of (a) b, (b) a, (c) R_{nn} and (d) $\partial \overline{P}/\partial x_3$ for DNS case FVD and the one-point and two-point models

References

Batchelor, G. K., Canuto, V. M. and Chasnov, J. R., 1992, "Homogeneous buoyancy-driven turbulence", J. Fluid Mech, **235**

Besnard, D., Harlow, F. H., Rauenzahn, R. M., and Zemach, A. C., 1992, Turbulence Transport Equations for Variable-Density Turbulence and Their Relationship to Two-Field Models, Report No. LA-12303-MS, Los Alamos National Laboratory

Bird, R. B., Stewart, W. E. and Lightfoot, E. N., 1960, Transport Phenomena, John Wiley & Sons

Clark, T. T., and Spitz, P., 1994, A Spectral Model for Variable-Density Turbulence, Report No. LA-12671-MS , Los Alamos National Laboratory

Eswaran, V. and Pope, S. B., 1988, "Direct numerical simulations of turbulent mixing of a passive scalar", Phys. Fluids, **31**, 3

Joseph, D. D., 1990, "Fluid Dynamics of two miscible liquids with diffusion and gradient stresses", Eur. J. Mech., B/Fluids, **9**, 6,

Launder, B. E., Reece, G. J. and Rodi W, 1975, "Progress in the development of a Reynolds- stress turbulence closure", J. Fluid Mechanics, **68**

Leith, C. E., 1967, "Diffusion Approximation to Internal Energy Transfer in Isotropic Turbulence", Phys. Fluids, **10** (7)

McMurtry, P. A., 1987, PH.D thesis, University of Washington

Sandoval, D. L., 1995, "The dynamics of variable-density turbulence" Dissertation, University of Washington, (also Report No. LA-13037-T, Los Alamos National Laboratory, 1995).

TURBULENCE STRUCTURE AND ITS MODELING OF A NATURAL CONVECTION BOUNDARY LAYER

T. TSUJI, M. SHIMADA AND Y. NAGANO
Department of Mechanical Engineering,
Nagoya Institute of Technology
Gokiso-cho, Showa-ku, Nagoya 466, Japan

Abstract.   In the near wall region of a turbulent natural convection boundary layer in air along a vertical plate, the velocity-pressure gradient correlation plays an important role in turbulent energy production and thermal energy is directly converted into kinetic energy through this correlation. Existing turbulence models can never represent such behavior. In the present study, a new turbulence model of the velocity-pressure gradient correlation for thermally driven flow is proposed by taking account of the occurrence of pressure fluctuations due to temperature fluctuations and its verification is made in comparison with experimental data.

1. Introduction

The turbulent natural convection boundary layer in air along a vertical plate is typical of turbulence for variable density low speed flows. When experimentally investigating the boundary layer, some marked characteristics different from those observed in usual turbulent boundary layers have been found regarding the viscous sublayer, the Reynolds shear stress and the turbulent heat fluxes near the wall (Tsuji and Nagano, 1988a; Tsuji and Nagano, 1988b). Also, coherent structures such as intermittent bursts and low-speed streaks in the near-wall region could not be detected in this boundary layer (Tsuji and Nagano, 1992). To clarify the origin of the structure peculiar to thermally driven turbulence, we examined the budgets of turbulent energy, Reynolds shear stress and turbulent heat fluxes. Consequently, it was found that the velocity-pressure gradient correlation plays an important role in turbulent energy production and that thermal energy is directly converted into kinetic energy through this correlation (Tsuji *et*

181

L. Fulachier et al. (eds.), IUTAM Symposium on Variable Density Low-Speed Turbulent Flows, 181–188.

al., 1991). Existing turbulence models never express such behavior, because most of them have been developed using information on forced convection. In the present study, we have contrived a new turbulence model which reproduces the characteristics of velocity- and temperature-pressure gradient correlations in a turbulent natural convection boundary layer.

2. Governing equations for thermally driven flows

The governing equations based on the Boussinesq approximation (Boussinesq, 1903) have been widely used for analyses of thermally driven flows. With increasing temperature difference, however, these equations cannot represent the correct behavior of thermally driven flows (Chenoweth and Paolucci, 1986; Frölich *et al.*, 1992; Mlaouah *et al.*, 1996). Therefore, we have attempted to construct a turbulence model based on the following modified governing equations applicable to arbitrary fluids under realistic temperature conditions (Mlaouah *et al.*, 1996):

$$\frac{\partial U_i}{\partial x_i} = \beta \frac{DT}{D\tau} = \alpha\beta \frac{\partial^2 T}{\partial x_j \partial x_j} \tag{1}$$

$$\frac{DU_i}{D\tau} = -\frac{1}{\overline{\rho}_\infty}\frac{\partial P}{\partial x_i} + \nu \frac{\partial^2 U_i}{\partial x_j \partial x_j} - g_i\beta(T - \overline{T}_\infty) \tag{2}$$

$$\frac{DT}{D\tau} = \alpha \frac{\partial^2 T}{\partial x_j \partial x_j} \tag{3}$$

where U_i, T and P represent the instantaneous velocity, temperature and pressure, α, ν, β, $\overline{\rho}_\infty$ and g_i are the thermal diffusivity, the kinetic viscosity, the volumetric expansion coefficient, the mean ambient density and the gravity acceleration, respectively, and $D/D\tau$ denotes the substantial derivative. These equations are derived as the first-order approximation of the density variation from the exact governing equations, and only the continuity equation (1) differs from the Boussinesq approximation [similar governing equations have been proposed by Oberbeck (1879)]. The expression appearing in the last term on the right-hand side of Eq. (1) is obtained by employing the thermal energy equation (3). Also, these equations are essentially equivalent to the low-Mach number equations (Rehm and Baum, 1978; Paolucci, 1982) in the case where the temperature difference is relatively small.

By employing the continuity equation (1) without the Boussinesq approximation, the turbulent energy production in the near-wall region may be written in terms of the pressure-strain correlation (i.e., the pressure work due to thermal expansibility). Following the procedure for forced convection (Launder *et al.*, 1975), we approximately integrated the Poisson equation for pressure fluctuation and estimated the order of magnitude of the pressure-strain correlation. However, results consistent with the experiment could not be obtained, because the order of the last term in Eq. (1) was very small. Therefore, it was concluded that the turbulent energy production through the velocity-pressure gradient correlation should be modeled from a different viewpoint.

3. Generation of pressure fluctuation with temperature fluctuation

The variation of fluid density in thermally driven flows is usually assumed to be a function of temperature, independent of pressure (i.e., the low Mach-number flows). For gases, the relation $\rho T = $ const. is widely applied. As far as this relation exactly holds, temperature fluctuation never affects the pressure linked directly with fluid motion. However, there is some doubt whether the relation $\rho T = $ const. is maintained strictly in a turbulent flow with violent temperature fluctuations as observed in the experiment. It is naturally expected that temperature fluctuations may cause pressure fluctuations. Hence, we write pressure fluctuation as follows:

$$p = p_h + p_{th} \qquad (4)$$

where p_h is the pressure fluctuation related to fluid motion and p_{th} is the pressure fluctuation originating directly from temperature fluctuations. Thermodynamic pressure variation depends on the thermal condition of the system and is usually treated as a time function independent of location in the low-Mach number approximation. Then, thermodynamic pressure becomes constant for boundary layers under steady ambient conditions. However, pressure fluctuations of very small order are expected to depend on the local characteristics of the thermal field, i.e., the mean absolute temperature $\overline{T}$ and the intensity of temperature fluctuation $\sqrt{\overline{t^2}}$. Thus, we assume the pressure fluctuation p_{th} produced by temperature fluctuation t as follows:

$$p_{th} \propto F\left(\overline{T}, \sqrt{\overline{t^2}}\right) t \qquad (5)$$

We determine the specific form of Eq. (5) under the following conditions, referring to the budget balance of turbulent quantities (Tsuji $et\ al.$, 1991) to be shown later in Figs. 1 - 5: (i) the conversion of thermal energy into kinetic energy through the velocity-pressure gradient correlation $(1/\overline{\rho}_\infty)\overline{u_i(\partial p/\partial x_i)}$ is isotropic as inferred from the balance of turbulent energy components, (ii) the temperature-pressure gradient correlation $(1/\overline{\rho}_\infty)\overline{t(\partial p/\partial y)}$ included in the transport equation for normal turbulent heat flux $\overline{vt}$ has a wall-limiting behavior in accordance with the experiment (gradual approach to

zero from the positive side), and (iii) both the velocity- and temperature-
pressure gradient correlations related to p_h are of very small order in the
near-wall region. Consequently, Eq. (5) becomes

$$p_{th} = -f(\overline{T}/\overline{T}_\infty)\left(t/\sqrt{\overline{t^2}}\right) \tag{6}$$

where $\overline{T}_\infty$ is the mean ambient temperature. The minus sign in Eq. (6)
can be readily understood by considering that the state variation of a gas
and the generation of buoyancy force are due to temperature variation.
Moreover, by assuming an exponential form for the function of mean absolute temperature in Eq. (6) and fitting the peak values of the velocity- and
temperature-pressure gradient correlations for the wall-normal component
linked via Eq. (6) to the near-wall measurements, the following expression
is obtained:

$$p_{th} = -c\overline{p}_\infty R\overline{T}_\infty(\overline{T}/\overline{T}_\infty)^m \left(t/\sqrt{\overline{t^2}}\right) \tag{7}$$

Here, $c = 1.15 \times 10^{-7}$, $m=6$ and R is the gas constant. The pressure fluctuation estimated from Eq. (7) under the relevant experimental conditions is
less than about 0.02 Pa, so there is a very fair possibility that small pressure fluctuations directly caused by temperature fluctuation control the
turbulence structure near the wall. This may be the key to constructing
turbulence models for thermally driven flows.

The results described above are obtained from a discussion on the balances of turbulent energy components and turbulent heat flux, and the
velocity- and temperature-pressure correlations for these balances can be
estimated. However, for the balance of Reynolds shear stress showing peculiar characteristics near the wall, it is difficult to estimate the velocity-
pressure gradient correlation, i.e., $(1/\overline{\rho}_\infty)[\overline{u_i(\partial p/\partial x_j)} + \overline{u_j(\partial p/\partial x_i)}]$, even
though the relation of Eq. (7) is given. Therefore, to model a general form
of the velocity-pressure gradient correlation, we assume that the cross-
correlation coefficients between velocity and pressure gradient fluctuations
may be equivalent as follows (Shikazono and Kasagi, 1993):

$$\overline{u_{(i)}\frac{\partial p_{th}}{\partial x_{(j)}}}\Big/\sqrt{\overline{u_{(i)}^2}}\sqrt{\overline{\left(\frac{\partial p_{th}}{\partial x_{(j)}}\right)^2}} = \overline{u_{(j)}\frac{\partial p_{th}}{\partial x_{(j)}}}\Big/\sqrt{\overline{u_{(j)}^2}}\sqrt{\overline{\left(\frac{\partial p_{th}}{\partial x_{(j)}}\right)^2}} \tag{8}$$

where the summation convention is not applied to the suffix put in paren-
theses. Then, the velocity-pressure gradient correlation Π_{ij} in the Reynolds
stress equation can be expressed by the following equation:

$$\Pi_{ij} = -\frac{1}{\overline{\rho}_\infty}\left(\overline{u_i\frac{\partial p}{\partial x_j}} + \overline{u_j\frac{\partial p}{\partial x_i}}\right)$$

$$= -\frac{1}{\overline{\rho}_\infty}\left(\overline{u_i\frac{\partial p_h}{\partial x_j}}+\overline{u_j\frac{\partial p_h}{\partial x_i}}\right)+\frac{c}{3}R\overline{T}_\infty\left(\frac{\sqrt{\overline{u^2_{(i)}}}}{\sqrt{\overline{u^2_{(j)}}}}+\frac{\sqrt{\overline{u^2_{(j)}}}}{\sqrt{\overline{u^2_{(i)}}}}\right)\left(\frac{\overline{T}}{\overline{T}_\infty}\right)^m$$

$$\times\left(\frac{1}{\sqrt{\overline{t^2}}}\right)\left\{\left[\frac{m}{\overline{T}}\frac{\partial \overline{T}}{\partial x_k}-\frac{1}{\sqrt{\overline{t^2}}}\frac{\partial \sqrt{\overline{t^2}}}{\partial x_k}\right]\overline{u_k t}+\frac{\partial \overline{u_k t}}{\partial x_k}\right\} \qquad (9)$$

On the other hand, the temperature-pressure gradient correlation in the turbulent heat flux equations is expressed as follows:

$$\Pi_{it} = -\frac{1}{\overline{\rho}_\infty}\overline{t\frac{\partial p}{\partial x_i}} = -\frac{1}{\overline{\rho}_\infty}\overline{t\frac{\partial p_h}{\partial x_i}}+mcR\left(\frac{\overline{T}}{\overline{T}_\infty}\right)^{m-1}\sqrt{\overline{t^2}}\frac{\partial \overline{T}}{\partial x_i} \qquad (10)$$

4. Verification of proposed model

We verify Eqs. (9) and (10) by directly substituting experimental values into the terms related to p_{th} and estimating the terms for p_h with a conventional pressure-strain model used in the numerical study of Peeters and Henkes (1992). These results are presented in Figs. 1-5, compared with the balances of turbulent quantities (Tsuji et al., 1991), in which each term is normalized with the molecular viscosity, the friction velocity and the friction temperature. The present model for p_{th} plus the model for p_h used by Peeters and Henkes (1992) gives substantially the profiles of the velocity- and temperature-pressure gradient correlations.

For the balances of turbulent energy components $\overline{u^2}/2$ and $\overline{v^2}/2$ shown in Figs. 1 and 2, Eq. (9) well expresses the turbulent energy production,

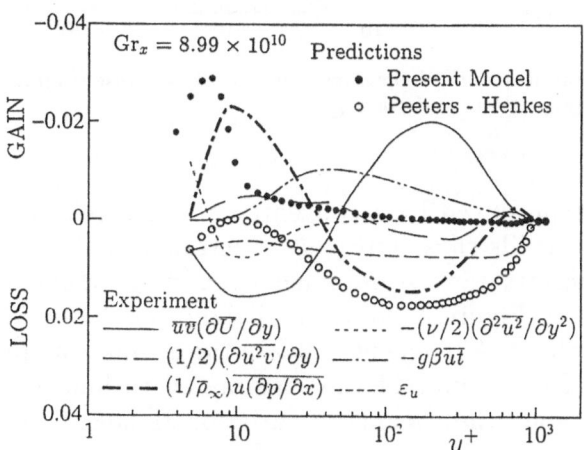

Figure 1.   Balance of turbulent energy component $\overline{u^2}/2$ [Eq. (9) = Present model + Conventional model (e.g., Peeters-Henkes)]

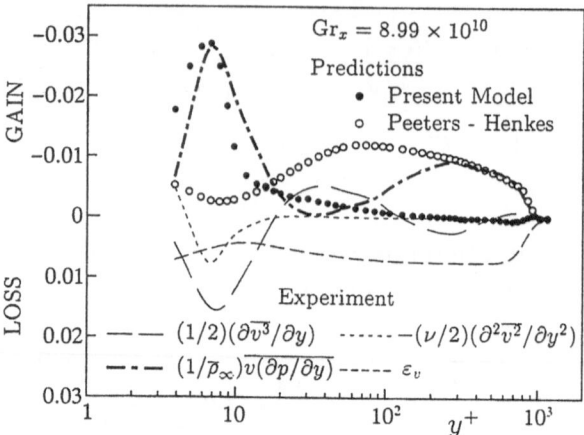

Figure 2.   Balance of turbulent energy component $\overline{v^2}/2$ [Eq. (9) = Present model + Conventional model (e.g., Peeters-Henkes)]

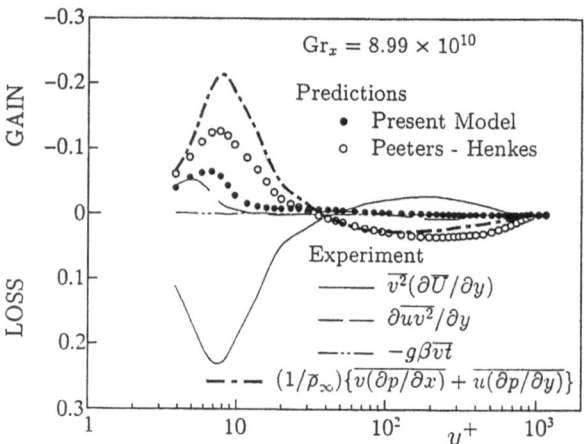

Figure 3.   Balance of Reynolds shear stress $\overline{uv}$ [Eq. (9) = Present model + Conventional model (e.g., Peeters-Henkes)]

which is never reproduced by a conventional pressure-strain model alone. Also, for the Reynolds shear stress $\overline{uv}$ illustrated in Fig. 3, a profile close to the measurements can be obtained by adding values estimated from the present model. Figure 4 shows the balance of the turbulent heat flux normal to the wall, $\overline{vt}$. Agreement between the estimate and the experiment is fairly good. The balance of the streamwise turbulent heat flux $\overline{ut}$ is shown in Fig. 5. Since the streamwise variation of mean temperature is very small, the contribution of the pressure fluctuation p_{th} becomes negligible. Consequently, the profile of the temperature-pressure gradient correlation for $\overline{ut}$ can be well expressed with the model for p_h.

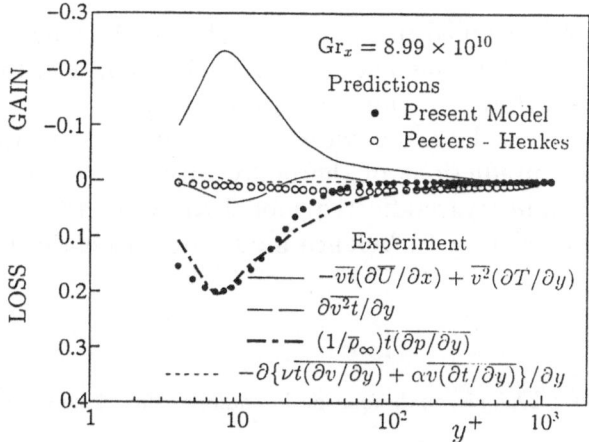

Figure 4.   Balance of normal heat flux $\overline{vt}$ [Eq. (10) = Present model + Conventional model (e.g., Peeters-Henkes)]

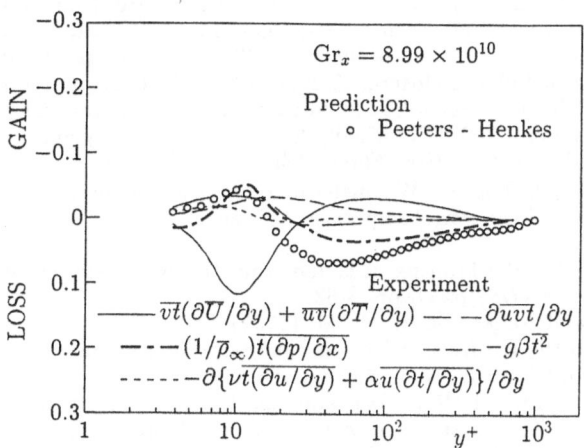

Figure 5.   Balance of streamwise heat flux $\overline{ut}$

5.   Concluding remarks

Natural convection boundary layers are originally formed by thermal energy supplied from the heated surface, so that the turbulence structure near the wall may show peculiar characteristics. As mentioned above, the modeling of velocity-pressure gradient correlation holds the key to the accurate prediction of turbulence behavior. As far as the pressure variation with fluid motion is merely considered, the role of turbulent energy production cannot be given to the velocity-pressure gradient correlation. This also implies that the actual turbulence behavior near the wall cannot be predicted even with direct numerical simulations based on the Boussinesq governing equations and those with a low Mach number approximation. Of course, the model developed in the present study might be insufficient and more advanced investigations may be required to apply it to various thermally driven flows. However, since the non-Boussinesq effect on turbulence characteristics seems to be rather weak, we consider for the present that improved turbulence predictions of thermally driven flows could be realized by including a temperature-affected model for velocity-pressure gradient correlation in the existing turbulence model based on the Boussinesq approximation.

References

Boussinesq, J. (1903) *Théorie analytique de la chaleur*, Vol. **2**, pp. 154-176, Gauthier-Villars, Paris

Chenoweth, D. R. and Paolucci, S. (1986) Natural convection in an enclosed vertical air layer with large horizontal temperature differences, *J. Fluid Mech.* Vol. **169**, pp. 173-210

Fröhlich, J., Laure, P. and Peyret, R. (1992) Large departures from Boussinesq approximation in the Rayleigh-Bénard problem, *Phys. Fluids A* Vol. **4**, pp. 1355-1372

Launder, B. E., Reece, G. J. and Rodi, W. (1975) Progress in the development of a Reynolds-stress turbulence closure, *J. Fluid Mech.* Vol. **68**, pp. 537-566

Mlaouah, H, Tsuji, T. and Nagano, Y. (1996) Behavior of laminar and transitional buoyancy driven flows in a 2-D differentially heated square cavity and its governing equations, submitted to *Int. J. Heat Fluid Flow*

Oberbeck, A. (1879) Ueber die Wärmeleitung der Flüssigkeiten bei Berücksichtigung der Strömungen infolge von Temperaturdifferenzen, *Ann. Phys. Chem.* Vol. **7**, pp. 271-292

Paolucci, S. (1982) On the filtering of sound from the Navier-Stokes equations, *Sandia Nat. Lab. Rep. SAND82-8257*, pp. 3-52

Peeters, T. W. J. and Henkes, R. A. W. M. (1992) The Reynolds-stress model of turbulence applied to the natural-convection boundary layer along a heated vertical plate, *Int. J. Heat Mass Transfer* Vol. **35**, pp. 403-420

Rehm, R. G. and Baum, H. R (1978) The equations of motion for thermally driven, buoyant flows, *J. Res. National Bureau Standards* Vol. **83, no. 3**, pp. 297-308

Sikazono, N. and Kasagi, N. (1993) Modeling Prandtl number influence on scalar transport in isotropic and sheared turbulence, *Proc. 9th Symp. Turbulent Shear Flows*, Kyoto, Vol. **2**, pp.18.3.1-18.3.6

Tsuji, T. and Nagano, Y. (1988a) Characteristics of a turbulent natural convection boundary layer along a vertical flat plate, *Int. J. Heat Mass Transfer* Vol. **31 no. 8**, pp. 1723-1734

Tsuji, T. and Nagano, Y. (1988b) Turbulence measurements in a natural convection boundary layer along a vertical flat plate, *Int. J. Heat Mass Transfer* Vol. **31 no. 10**, pp. 2101-2111

Tsuji, T., Nagano, Y. and Tagawa, M. (1991) Thermally driven turbulent boundary layer, *Proc. 8th Symp. Turbulent Shear Flows*, Munich, Vol. **2**, pp. 24.3.1-24.3.6

Tsuji, T. and Nagano, Y. (1992) Experiment on spatio-temporal turbulent structures of a natural convection boundary layer, *Trans. ASME, J. Heat Transfer* Vol. **144**, pp. 901-908

TURBULENT STRUCTURE IN HORIZONTAL CHANNEL
FLOW UNDER UNSTABLE DENSITY STRATIFICATION

O. IIDA
Department of Mechanical Engineering, Nagoya Institute of Technology
Gokiso-cho, Showa-ku, Nagoya 466, Japan

AND

N. KASAGI
Department of Mechanical Engineering, The University of Tokyo
Hongo 7-3-1, Bunkyo-ku, Tokyo 113, Japan

Abstract. Direct numerical simulations are performed for a fully-developed horizontal channel flow under unstable density stratification. Moreover, to investigate the effects of the penetrative thermal plumes and counter-gradient heat flux on near-wall turbulence, the ad-hoc calculations are carried out, in which the buoyancy force acts only conditionally, depending on the local state of the turbulent heat flux.

Introduction

When the flows are bounded by the wall and the thermal stratification is dynamically unstable, there takes place some large-scale thermal convection. The occurrence of thermal convection then affects significantly the dynamics of turbulence. The classical problem of thermal convection is to determine the motion of a layer of fluid contained between two horizontal planes, which is uniformly heated from below and cooled from above. An interesting feature of this convection is a reversal of the temperature gradient in the central region of the layer. This should be because thermal plume penetrates right across the layer, generating transient stable blobs of fluid close to the opposite boundary. This has been observed in both laboratory experiment [1] and atmosphere [2].

When shear is imposed, convective rolls are organized in the direction of mean velocity while the rolls are suppressed in the horizontal direction. The another important role of shear is its generation mechanism of turbulence. Over the last three decades, there has been continuous interest in the near-

L. Fulachier et al. (eds.), IUTAM Symposium on Variable Density Low-Speed Turbulent Flows, 189–196.
© 1997 *Kluwer Academic Publishers.*

wall turbulence driven by shear. From the studies of wall-bounded turbulent shear flows, the organized motion, *i.e.* the so-called bursting phenomenon, was identified as a primary turbulence mechanism for the production of the Reynolds stress and turbulence energy in the near-wall region. From this view, some experimental studies were made to reveal buoyancy effects on the turbulent wall shear flows [3, 4, 5, 6]. However, because of the limited experimental measurements, it is still unknown how thermal plumes affect the streamwise vortices in unstable stratification.

Thus, we carried out direct numerical simulation of a fully-developed turbulent flow between horizontal planes, uniformly heated from below and cooled from above. In a previous study [7], it was found that, as the Grashof number increased, the large-scale thermal convection associated with buoyant thermal plumes diminished the quasi-coherent streamwise vortices which are characteristic of the near-wall turbulence.

The suppression of bursting phenomenon is also observed in the channel flow under system rotation. Bech & Andersson (1995) [8] showed that the large scale roll cells aligned with the streamwise direction suppressed the bursting phenomena because the secondary flow extracted a comparable amount of kinetic energy from the mean flow as does turbulence. However, in both studies [7, 8], the destruction mechanisms of the bursting is not clearly explained. Especially, it remains unknown why the energy cascade from the secondary flow to turbulence is suppressed even if turbulence can not extract kinetic energy directly from the mean flow.

Our objective is to clarify the suppression mechanisms of the near-wall turbulence under unstable density stratification. As a result, it was found that the suppression of the near-wall turbulence was closely associated with the penetrative thermal plumes. When the penetrative thermal plumes generate stable fluid blob, the buoyancy could suppress near-wall turbulence through the counter-gradient heat flux [9]. To verify this supposition and evaluate the effect of the counter-gradient heat flux, two ad-hoc calculations were carried out, in which the buoyancy force acted only conditionally, depending on the local state of the turbulent heat flux.

Nomenclature

C_f friction coefficient, $2\tau_w/\rho U_b^2$

Gr Grashof number, $g\beta\Delta T(2\delta)^3/\nu^2$

Gr_a Grashof number of ad-hoc calculations

g gravitational acceleration

Nu Nusselt number, $2\delta q_w/\alpha$

p static pressure

Pr Prandtl number, ν/α

q_w total flux between two walls

Re_τ Reynolds number, $u_\tau\delta/\nu$

t	time
U	mean velocity in x-direction
U_b	bulk mean velocity
u, v, w	velocity fluctuations in x-, y- and z-directions
u_τ	friction velocity, $\sqrt{\tau_w/\rho}$
x, y, z	streamwise, wall-normal and spanwise directions
α	thermal diffusivity
β	volumetric expansion coefficient
ΔT	temperature difference between two walls
δ	channel half width
Θ, θ	mean and fluctuating temperatures
θ_τ	friction temperature, q_w/u_τ
ν	kinematic viscosity
π	modified pressure, $p + \int \bar{\rho} g dy$
ρ	density of fluid
τ_w	wall shear stress

Subscripts and superscripts

$(\ )_0$	at $Gr = 0$ and $Pr = 0.71$
$\overline{(\ )}$	ensemble average
$(\ )^+$	non-dimensionalized by wall variables, u_τ and ν
$(\ )^*$	non-dimensionalized by u_τ and δ

Numerical Procedure

Consider a plane Poiseuille flow of incompressible fluid, which is driven by a constant mean pressure gradient imposed in the x-direction. The two walls are assumed at different, but constant temperatures without fluctuations. As a result, there is a constant positive temperature difference ΔT between the lower and upper plates. The gravitational acceleration g is assumed in the y-direction to impose an unstable buoyancy effect. The ordinary no-slip boundary condition is imposed on all velocity components at the wall. The periodic boundary conditions are assumed at the periods of $5\pi\delta$ and $2\pi\delta$ in the x- and z-directions, respectively.

The equations used in this study are the standard set of hydrodynamic equations with the Boussinesq-Oberbeck approximation. The Reynolds number is given as $Re_\tau = 150$. The Prandtl number is fixed at $Pr = 0.71$. A spectral method is adopted with Fourier series in the x- and z-directions and a Chebyshev polynomial expansion in the y-direction.

In this study, five different simulations were carried out as shown in Table 1. At first, three cases with different Grashof numbers are simulated by using relatively coarse grid points, i.e., $64 \times 47 \times 64$ in the x-, y- and z-directions, respectively. The validity of the calculation were assured by carrying out the computation with 128×128 Fourier modes and Chebyshev polynomials up to the 96th order [7].

TABLE 1. Grashof numbers

	Case U1	Case U2	Case U3		Case UA1	Case UA2
Gr	0.0	1.3×10^6	4.8×10^6	Gr_a	5.0×10^4	1.0×10^5

Moreover, two ad-hoc calculations are carried out, in which the buoyancy force acts only conditionally, depending on the local state of the instantaneous turbulent heat flux $v\theta$. The equation of the wall-normal velocity should be written as follows:

$$\frac{Dv}{Dt} = -\nabla\pi + \frac{1}{8}\frac{Gr}{Re_\tau^2}\theta + \frac{1}{Re_\tau}\nabla^2 v, \tag{1}$$

where the variables are nondimensionalized by u_τ, δ and ΔT. In Cases UA1 and UA2, the temperature fluctuation θ is assumed to become active only when the instantaneous turbulent heat flux $v\theta$ takes a positive value and thus contributes to the acceleration of the wall normal velocity.

All computations were started from the same initial condition, $i.e.$ the fully developed turbulent channel flow without density stratification. Each of the computations was continued until the flow field was judged to reach a fully developed state after the Grashof number was prescribed.

Results

Figure 1 shows the time evolution of the normalized skin friction coefficient C_f/C_{f0} and Nusselt number Nu/Nu_0 at $Gr = 1.3 \times 10^6$. The friction coefficient first increases up to 1.1 at $t^* = 480$, then decreases to 0.97 at the fully developed state. On the other hand, Nu/Nu_0 simply increases and reaches 1.4. Thus, at the fully developed state, heat transfer is enhanced between the two walls without the drag increase. As discussed in Tennekes and Lumley (1972), the buoyancy generated eddies cause relatively little momentum transport, although they are quite effective in transporting heat. Thus, the

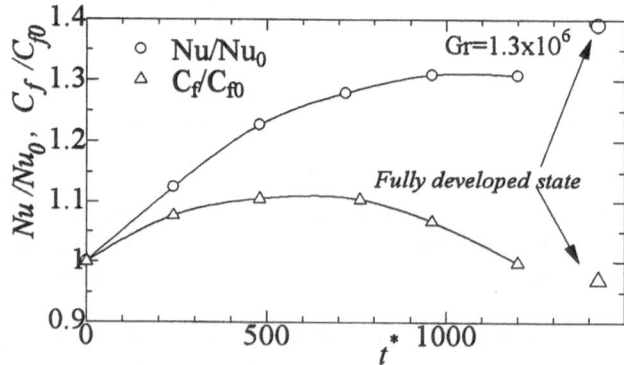

Figure 1.   Time evolution of skin friction coefficient and Nusselt number in Case U2

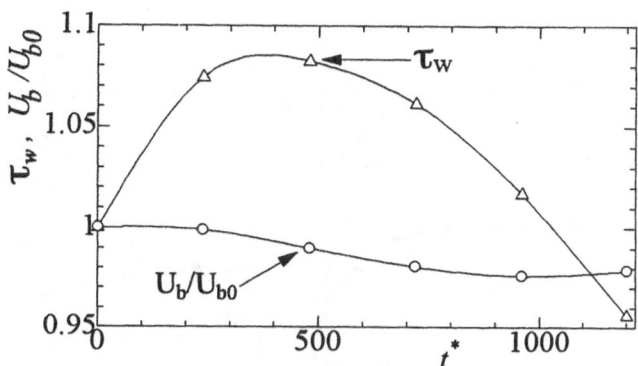

Figure 2.   Time evolution of wall stress and bulk mean velocity in Case U2

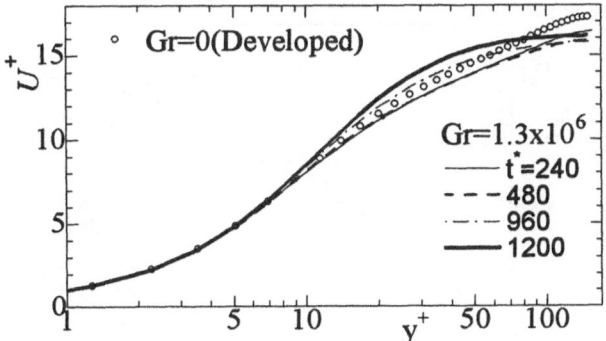

Figure 3.   Time evolution of mean velocity profile in Case U2

heat transfer could be preferentially enhanced without the increase of skin friction. However, the relatively small change in skin friction should be attributed not only to the weak unstable buoyancy effect on the momentum transfer but also to the suppression of the near-wall turbulence as discussed later.

The time evolution of τ_w and U_b is shown in Figure 2. It is clear that the decrease in C_f is due to the decrease in τ_w. This has resulted from the streamwise vortical structures in the near-wall region, which is substantially affected by buoyancy.

Figure 3 shows the time evolution of mean velocity profile U^+. It first decreases especially in the center of channel. Then, it begins to increase in the near-wall region after $t^* > 960$. From Figs 2, 3 and 4, it is evident that there are two stages before turbulence becomes fully developed under unstable stratification. At first, the buoyancy activates turbulence especially in the channel central region, and secondly, the near-wall turbulence is suppressed.

Figure 4 shows the mean temperature and the root-mean-square streamwise vorticity at several spanwise positions when t^* becomes 1200. Note that these quantities have been averaged in the x-direction at a fixed y- and z-position. The temperature profile is shifted toward the hot wall at

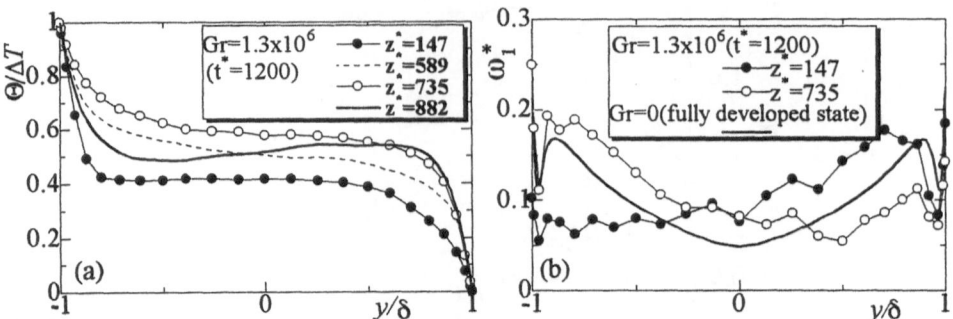

Figure 4. Mean temperature and rms streamwise vorticity at different spanwise positions in Case U2. (a)Mean temperature, (b)RMS streamwise vorticity

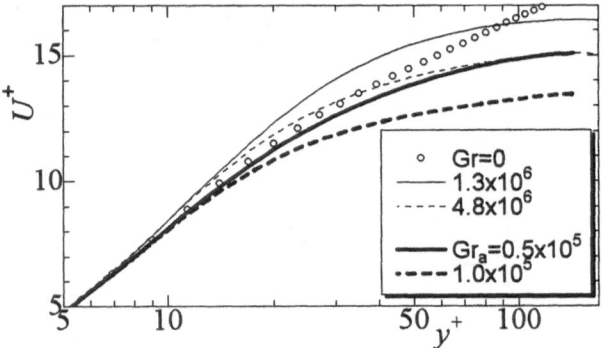

Figure 5. Mean velocity profile

$z^* = 735$, while toward the cold wall at $z^* = 147$. Both temperature profiles show large temperature gradients close to the walls and a nearly isothermal region in the channel central region. This indicates the occurrence of large scale hot and cold plumes between two walls. It should also be noted that a reversal of the temperature gradient can be observed at $z^* = 882$, indicating a stable blob of fluid generated by the thermal plumes. At the wall where the thermal plumes impinge, the streamwise vorticity decreases remarkably while ω_x is enhanced at the near-wall region where the thermal plumes arise. Thus, the suppression of the near-wall turbulence is consider to be caused by the penetrative thermal plumes on the opposite boundary. The stable fluid blob generated by thermal plumes is dynamically important because it would damp the wall normal velocity through the counter-gradient heat flux. On the other hand, the large-scale thermal convection extracts a considerable amount of energy from the mean flow and also could suppress smaller scale turbulence.

Figure 5 shows the mean velocity profiles at the fully developed state in all cases. In cases of the preferential buoyancy, the mean velocity profile decreases at the center of the channel without an increase in the buffer region. The decrease of the mean velocity profile indicates that the buoyancy

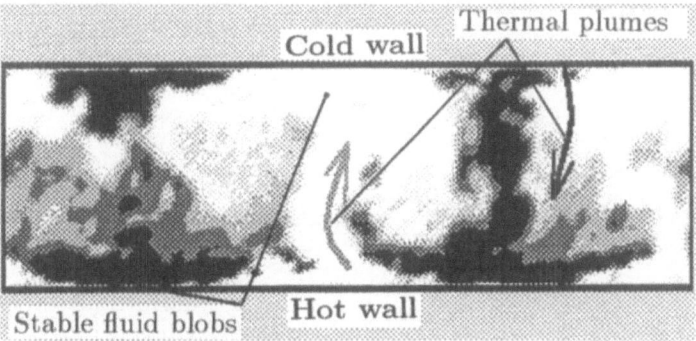

Figure 6. Instantaneous distribution of $\theta/\Delta T$ in the vertical $(y\text{-}z)$ plane at $Gr_a = 1.0 \times 10^5$: $\Delta y^+ \times \Delta z^+ = 300 \times 943$ (Black to white; $\theta/\Delta T = -0.05$ to 0.05).

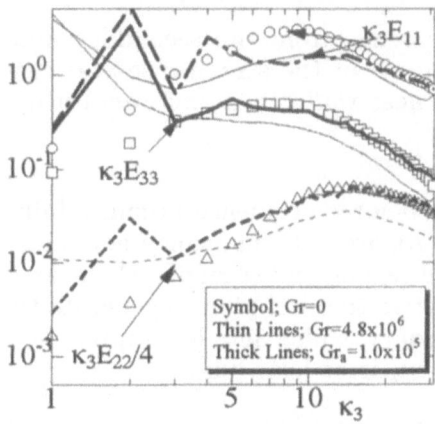

Figure 7. One-dimensional spanwise energy spectrum at $y^+ = 20$

generated eddies extract kinetic energy from the mean flow. Thus, in Cases UA1 and UA2, the larger kinetic energy is removed from the mean flow by thermal convection than in other cases.

Figure 6 shows an instantaneous distribution of temperature fluctuation $\theta/\Delta T$ in Case UA2. The plume-like structure in the temperature field extends throughout the entire depth of the convective layer. The stable fluid blob is also found to be generated in the near-wall regions. There is a great deal of similarity in large-scale structures between Cases UA2 and U3 (not shown here).

Figure 7 shows the spanwise energy-spectrum at $y^+ = 20$. It is seen that in Case U3 the energy cascade of $\overline{v^{+2}}$ and $\overline{w^{+2}}$ is suppressed and more energy is piled up at low wave-numbers in comparison to Case UA2. Figure 8 shows the streamwise vorticity ω_1. In Case U3, ω_1 decreases remarkably in the near-wall region. On the other hand, in Cases UA1 and UA2, the decrease of ω_1 is very small if compared with Case U3 although larger kinetic energy is removed from the mean velocity field to the large-scale thermal plumes. Thus, there is no doubt that the counter-gradient heat flux sup-

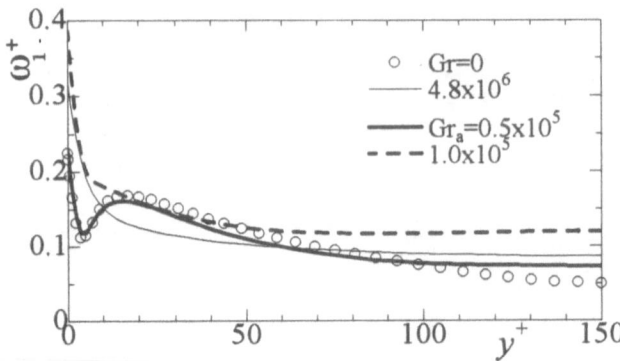

Figure 8.   RMS streamwise vorticity fluctuations

presses the energy cascade from the secondary flow to the smaller-scale turbulence. The decrease of the streamwise vorticity should result in the increase of U^+ in the near-wall region and even the drag reduction observed in Case U2.

Conclusions

There are two stages before turbulence becomes fully developed under unstable stratification. At first, the buoyancy activates turbulence especially in the channel central region, and secondly, the near-wall turbulence is suppressed. The penetrative thermal plumes generate the stable blob of fluids at the opposite boundary, which should suppress the near-wall turbulence through the counter gradient heat flux.

The authors are grateful to Professors Y. Nagano and T. Tsuji for their useful discussions during the course of this work.

References

1. Chu, T. Y. and Goldstein, R. J., 1973, *J. Fluid Mech.*, **60**, pp. 141-159.
2. Warner, J. and Telford, J. W., 1967, *J. Atmos. Sci.*, **24**, pp. 374-382.
3. Kasagi, N. and Hirata, M., 1977, *Heat Transfer and Turbulent Buoyant convection*, D. B. Spalding and N. Afgan, Eds., 1, Hemishere, Washington, D. C., pp. 27-38.
4. Mizushina, T., Ogino, F., and Katada, N., 1982, *Int. J. Heat Mass Transfer*, **25**, pp. 1419-1425.
5. Fukui, K. and Nakajima, M., 1985, *Int. J. Heat Mass Transfer*, **28**, pp. 2343-2352.
6. Fukui, K. and Nakajima, M., 1991, *Int. J. Heat Mass Transfer*, **34**, pp. 2373-2385.
7. Iida, O. and Kasagi, N., 1995, *Proc. of 4th ASME-JSME Thermal Eng. Joint Conf.*, 1, Hawaii, March, pp. 417-424.
8. Bech, K. H. and Andersson, H. I., 1995, *Proc. of 10th Turbulent Shear Flows*, Pennstate, August, 8−1-8−6.
9. Baines, W. D. and Turner, J. S., 1969, *J. FLuid Mech.*, **37**, pp. 51-80.
10. Tennekes, H. and Lumley, J. L., 1972, *A First Course in Turbulence*, M.I.T. Press, Cambridge, Mass.

SECOND ORDER MODELLING OF INCOMPRESSIBLE TURBULENT BUOYANT FLOWS : SOME COMMENTS ON THE MAXIMUM PRINCIPLE

J.M. HERARD
EDF- DER. Département Laboratoire National d'Hydraulique
6, quai Watier. 78400. Chatou. France.

1. Introduction

When focusing on incompressible turbulent flows with buoyant effects, the problem of the Maximum Principle for the mean Temperature (MPT afterwards) simply arises from the fact that : $T_{min} \leq T(x,t) \leq T_{max}$ implies : $T_{min} \leq <T> (x,t) \leq T_{max}$, where $<.>$ denotes some statistical average. This is a standard feature for laminar flows, and it was early pointed out by Du Vachat [3], Lumley [13] and Schumann [15] within the frame of turbulent flows. Actually the latter MPT means that two new positivity constraints should hold, namely : ($T_{max} -<T> (x,t)) \geq 0$, and : $(<T> (x,t) - T_{min}) \geq 0$. Hence, the problem is : do second order closures enable to preserve the MPT ? The present contribution aims at providing some enlightenments on this particular topic. In order to investigate the suitability of some one-point second-moment closures, we assume that the basic set of equations reads :

$$U_{k,k} = 0 \tag{1.a}$$
$$U_{i,t} + U_k(U_i)_{,k} + (R_{ik})_{,k} = (\Sigma_{ik})_{,k} + \beta_i(T-T_0) \tag{1.b}$$
$$(R_{ij})_{,t} + U_k \ (R_{ij})_{,k} + R_{ik}U_{j,k} + R_{jk}U_{i,k} = \phi_{ij} - (d_{ijk})_{,k} + (\beta_iX_j+\beta_jX_i) \tag{1.c}$$
$$T_{,t} + U_k \ (T)_{,k} + (X_k)_{,k} = (\lambda \ T_{,k})_{,k} \tag{1.d}$$

From now on, we omit brackets except for the variance $<\theta^2>$. U is the mean velocity, T stands for the mean temperature, R is the Reynolds stress tensor, X is the turbulent heat flux. The vector β is zero when gravity effects are neglected. We also set : $\rho_0 \Sigma_{ik} = - P \ \delta_{ik} + v \ \rho_0 \ (U_{i,k} +U_{k,i})$ where P denotes the mean pressure and : $d_{ijk} = <u'_iu'_ju'_k>$. We even more assume that ϕ_{ij} and d_{ijk} are chosen in such a way that the over-realizability and the strong-realizability hold (see [5], [7], [8], [13], [14], [15], [16], [17], [18], [19], [22], [23] for basic definitions and various proposals), when : $T = T_0$ and : $X_i = 0$. There are at least two distinct strategies to close the system : one may either use a gradient type law to account for the turbulent heat flux : $X_i = - D_{ik} T_{,k}$; otherwise, one may as usual derive an equation of the form :

$$X_{i,t} + U_kX_{i,k} + X_kU_{i,k} + R_{ik}T_{,k} = \phi_i - (d_{ik})_{,k} + \beta_i <\theta^2> \tag{1.e}$$

L. Fulachier et al. (eds.), IUTAM Symposium on Variable Density Low-Speed Turbulent Flows, 197–204.
© *1997 Kluwer Academic Publishers.*

and add the following :

$$<\theta^2>_{,t} + U_k<\theta^2>_{,k} + 2X_kT_{,k} = \phi - (d_k)_{,k} \qquad (1.f)$$

2. Full second moment closure.

It was quite recently suggested in [22] that the realizability requirement is not necessary to compute complex turbulent flows with second moment closures ; this actually seems to be a rather surprising suggestion and, though it is not intended to present new closures for standard contributions such as ϕ_{ij} it will be assumed that one of the following closures is chosen (see [5], [7b], [16] also) :

$$\phi_{ij} = \alpha_1 R_{ij} + \alpha_2 R_{ij}^2 + \alpha_3 R_{ij}^3 + \alpha_4(R_{ik}U_{j,k}+R_{jk}U_{i,k}) + \alpha_5(R_{ik}U_{k,j}+R_{jk}U_{k,i})$$

$$(2.1)$$

$$\phi_i = \frac{(\alpha_1+\eta_1)}{2} X_i + \alpha_2 (R_{ik}X_k - \frac{X_lX_l}{2<\theta^2>}X_i) + \alpha_3 (R_{ik}^2X_k - \frac{R_{lk}X_lX_kX}{2<\theta^2>}X_i)\dots$$

$$+ \alpha_4X_kU_{i,k} + \alpha_5 X_kU_{k,i} \qquad (2.2)$$

$$\phi = \eta_1 <\theta^2> \qquad (2.3)$$

also setting :

$$d_{ijk} = d_{ij} = d_i = 0 \qquad (2.4)$$

(odd joint pdf assumption), and :

$$\alpha_1 = \alpha'_{10} (I, II, III) \left(\frac{\varepsilon}{I}\right) + \alpha'_{11} (I, II, III) \left(\frac{R_{lk}U_{1,k}}{I}\right)$$

$$\alpha_2 = \alpha'_2 (I, II, III) \left(\frac{\varepsilon}{II}\right) \quad ; \quad \alpha_3 = \alpha'_3 (I, II, III) \left(\frac{\varepsilon}{III}\right)$$

$$\alpha_i = \alpha'_i (I, II, III) \ (i=4,5)$$

$$\eta_1 = \eta'_1 (I, II, III) \left(\frac{\varepsilon_\theta}{<\theta^2>}\right) \quad ; \qquad (2.5)$$

(Notations : $I = R_{ll}$; $II = R_{lk}R_{kl}$; $III = R_{lk}R_{km}R_{ml}$). All primed functions are non dimensional bounded functions which depend on the three invariants I, II, III. We introduce : $Z_{ij} = <\theta^2> R_{ij} - X_i X_j$ and derive the governing equation (for regular enough, e.g. C^1 solutions) for the second order tensor Z and the variance $<\theta^2>$:

$$(Z_{ij})_{,t} + U_k \ (Z_{ij})_{,k} = Z_{ik}H_{jk} + Z_{jk}H_{ik} \qquad (2.6)$$

$$<\theta^2>_{,t} + U_k<\theta^2>_{,k} = (\eta_1-2C_{ll}) <\theta^2> \qquad (2.7)$$

setting :

$$H_{jk} = \left(\frac{\alpha_1 + \eta_1}{2} - C_{ll}\right)\delta_{jk} + C_{jk} + \alpha_2\frac{Z_{kj}}{2<\theta^2>} + \alpha_3\frac{R_{km}\ Z_{mj}}{2<\theta^2>} + (\alpha_4 - 1)\ U_{j,k} + \alpha_5\ U_{k,j}$$

(2.8)

and noting : $<\theta^2>C_{ij} = X_i\ T_{,j}$. Coefficient functions should agree with :

$$\alpha_{10}'\ (I, II, III) + \alpha_2'\ (I, II, III) + \alpha_3'\ (I, II, III) = -2$$

(2.9)

Note that, for instance, Lumley 's slow model [13] belongs to the class (2.1-2.9). The simplest model is of course obtained setting :

$$\alpha_4'\ (I, II, III)\ =\ \alpha_5'\ (I, II, III) = \alpha_{11}'\ (I, II, III) = 0$$

(2.10a)

Any choice among QLOR (Quasi Linear Objective Realisable) or RIP (Realisable Isotropisation of Production) closures, namely:

$$\alpha_4'\ (I, II, III)\ =\ \alpha_5'\ (I, II, III) = 3/10\quad ;\quad \alpha_{11}'\ (I, II, III) = -6/5$$

(2.10b)

$$\alpha_4'\ (I, II, III) = 3/5\quad ;\quad \alpha_5'\ (I, II, III) = 0\quad ;\quad \alpha_{11}'\ (I, II, III) = -6/5$$

(2.10c)

is in agreement with the following requests : $\phi_{ij}^{rapid}\ (\ R_{kl} = \frac{1}{3}\ \delta_{lk}\ ) = \frac{1}{5}\ (U_{i,j} + U_{j,i})$ and $\phi_{ll}^{rapid} = 0$. An advantage of QLOR (2.10b) is that it represents an objective contribution (see [21]) ; obviously, more complex closures to account for rapid terms, such as those detailed in [5] and [16], might be considered either, since associated rapid ϕ_{ij} contributions may be written as :

$$\phi_{ij}^{rapid} = R_{ik}\ H_{kj} + R_{jk}\ H_{ki}$$

(2.11)

Solutions of system (2.6, 2.7) fulfil the strong-realizability concept, provided that both the mean velocity gradients, the C_{ij}'s and the inverse of the turbulent mechanical and scalar time scales remain bounded. Moreover (see [9] for a similar result) :

Proposition II.1 : The turbulent field issuing from system (2.6, 2.7) cannot degenerate provided that the following quantity

$$\frac{3}{2}(\alpha_1 + \eta_1) - 2C_{ll} + \left(\frac{\varepsilon}{I}\right)\left|\frac{\alpha_2'}{2}\ (1 - \frac{X_lX_l}{<\theta^2>\ I}\ )\left(\frac{I^2}{II}\right) + \frac{\alpha_3'}{2}(\frac{R_{lm}}{I})\ (\frac{R_{ml}}{I} - \frac{X_mX_l}{<\theta^2>\ I}\ )\left(\frac{I^3}{III}\right)\right|$$

remains bounded over time and space.

This is simply due to the fact that, owing to (2.6) and (2.11), the governing equation for the determinant of **Z** reads :

$$(\delta_3)_{,t} + U_k\ (\delta_3)_{,k} = 2\ \delta_3\ H_{kk}$$

(2.12)

Consequently, (2.12) may integrated along the characteristic curves, which ensures the strict positivity of δ_3, provided that the initial conditions for the latter quantity are strictly positive. Anyway, it is not clear whether the previous conditions hold ; more precisely, we know that :

$$0 \le \frac{X_l X_l}{<\theta^2> I} \le 3 \quad ; \quad 1 \le \frac{I^2}{II} \le 3 \quad ; \quad 1 \le \frac{I^3}{III} \le 9$$

$$0 \le n_l \left\{\frac{R_{lm}}{I}\right\} n_m \le n_l \left\{\frac{I\,\delta_{lm}}{I}\right\} n_m = 1 \qquad \text{(for given value of } \mathbf{n} \text{ such that : } n_l n_l = 1)$$

$$0 \le n_l \left\{\frac{X_l X_m}{I <\theta^2>}\right\} n_m \le \left\{\frac{X_p X_p}{I <\theta^2>}\right\} \le 3$$

However, both the boundedness of the C_{ij} components and the mean velocity gradients do not clearly arise. Even more, no theoretical argument enables to conclude in a simple manner that the MPT is ensured, even if the initial data and the boundary conditions for the temperature are chosen in the range (T_{min}, T_{max}). This results from the fact that the turbulent heat flux $\mathbf{X}$ does not vanish when T reaches the bound T_{min} (or T_{max} either). Surprisingly, we may get more information by investigating the convective subset associated with (1.a-1.f), at least when removing so-called rapid terms. It was noticed in [7b] that the convective subset is hyperbolic in that case, as soon as the over-realizability of the second-rank tensor R holds (the inter-realizability is not compulsory to insure hyperbolicity). Even more, the one dimensional Riemann problem associated with the latter non conservative convective system may be solved, considering approximate jump conditions (see [1], [11] and [12]), and the classical theory of hyperbolic systems ([20]) ; thus, it occurs that the analytical solution is over realisable (including the inter-realizability) ; *however, the MPT does not hold.* The construction of the solution is straightforward but tedious, and the reader is referred to reference [7b] for details. We may even use a numerical argumentation which confirms this annoying result. For that purpose, we proceed as in [9] and introduce an extension of the fractional step method introduced in [24]. The whole consists in solving successively within each time step, the following "laminar" system :

$$U_{k,k} = 0 \tag{2.13}$$
$$U_{i,t} + U_k(U_i)_{,k} = (\Sigma_{ik})_{,k} + \beta_i(T-T_0) \tag{2.14}$$
$$T_{,t} + U_k (T)_{,k} = (\lambda\, T_{,k})_{,k} \tag{2.15}$$

together with the "turbulent" system :

$$(R_{ij})_{,t} + U_k\,(R_{ij})_{,k} = \phi_{ij}^{slow} - (d_{ijk})_{,k} + (\beta_i X_j + \beta_j X_i) \tag{2.16}$$
$$X_{i,t} + U_k X_{i,k} = \phi_i^{slow} - (d_{ik})_{,k} + \beta_i <\theta^2> \tag{2.17}$$
$$<\theta^2>_{,t} + U_k<\theta^2>_{,k} = \phi^{slow} - (d_k)_{,k} \tag{2.18}$$

Then the following initial boundary value problem is solved (the initial data is provided by the solution of (2.13 to 2.18)) :

$$U_{i,t} + (R_{ik})_{,k} = 0 \tag{2.19}$$
$$(R_{ij})_{,t} + R_{ik}U_{j,k} + R_{jk}U_{i,k} - \phi_{ij}^{rapid} = 0 \tag{2.20}$$

$T_{,t} + (X_k)_{,k} = 0$ (2.21)

$X_{i,t} + X_k U_{i,k} + R_{ik} T_{,k} - \phi_i^{rapid} = 0$ (2.22)

$<\theta^2>_{,t} + 2X_k T_{,k} = 0$ (2.23)

The regular (C^1) solutions provided by (2.16 - 2.18) fulfil the inter-realizability concept, and a similar result holds for those of (2.19 - 2.23), provided that the closures are in agreement with (2.1 - 2.5), or belong to the class [7b]. Obviously, the MPT holds through (2.13 - 2.15). Moreover :

Proposition II.2 : . If the rapid part is neglected, the system (2.19 - 2.23) is hyperbolic ; eigenvalues are in the statistically two-dimensional case :

$$\lambda_1 = -\left(2\,\mathbf{n^t R n}\right)^{1/2} \quad ; \lambda_9 = \left(2\,\mathbf{n^t R n}\right)^{1/2}$$
$$\lambda_2 = -\left(\mathbf{n^t R n}\right)^{1/2} \quad ; \lambda_8 = \left(\mathbf{n^t R n}\right)^{1/2}$$
$$\lambda_3 = \lambda_4 = \lambda_5 = \lambda_6 = \lambda_7 = 0$$

All fields are Linearly Degenerate except the 1-wave and the 9-wave. The one dimensional Riemann problem admits a unique over-realisable solution (n stands for some normal unit vector). However, solutions of (2.19 to 2.23) are not in agreement with the MPT.

The main features which are necessary to derive this result are recalled in appendix A. The behaviour of variables U and R_{11} is displayed on figures 1 and 2, when the initial data of (2.19 - 2.23) is such that a one dimensional "double-rarefaction" wave develops in the (x,t) plane. We may turn now to the second strategy.

3. Mixed first order-second order closure

We now focus on (1.a-1.d) and still assume (2.1). However, we introduce :

$X_i = -\tau R_{ik} T_{,k}$ (3.1)

where τ represents some positive bounded function. Then :

Proposition III.1 : The MPT holds through (1.d) , provided that (3.1) is considered and that the Reynolds stress tensor is realisable.

A suitable algorithm to compute the whole set while preserving the over-realizability and the MPT, consists in solving :

$U_{k,k} = 0$ (3.2)

$U_{i,t} + U_k(U_i)_{,k} = (\Sigma_{ik})_{,k} + \beta_i(T-T_0)$ (3.3)

$T_{,t} + U_k(T)_{,k} - ((\lambda\delta_{ik}+\tau R_{ik})\,T_{,i})_{,k} = 0$ (3.4)

$(R_{ij})_{,t} + U_k(R_{ij})_{,k}) - \phi_{ij}^{slow} + (d_{ijk})_{,k} + \tau(\beta_i R_{jk}T_{,k}+\beta_j R_{ik}T_{,k}) = 0$ (3.5) and then :

$$U_{i,t} + (R_{ik})_{,k} = 0 \tag{3.6}$$
$$(R_{ij})_{,t} + R_{ik}U_{j,k} + R_{jk}U_{i,k} - \phi_{ij}^{rapid} = 0 \tag{3.7}$$
$$(T)_{,t} = 0 \tag{3.8}$$

When focusing on Finite Volume techniques, details of some spatial discretizations have been provided in reference [9], which enable to preserve the discrete realizability concept. Actually, the latter require that some CFL-like conditions are accounted for. Another point which seems worth being underlined, concerns numerical techniques. Due to the fact that all these systems are complex non-linear sets of partial differential equations, it has been noticed that "rough schemes" such as those suggested in [6] behave fairly well, since these inherit the capabilities of up-winding techniques such as Roe's linearizations, without generating a tremendous effort of implementation. Eventually, and though not discussed herein, it is also emphasised that the non-conservative convective effects associated with the so-called production tensor $(- R_{ik}U_{j,k} - R_{jk}U_{i,k})$ must be accounted for in a suitable way.

4. Conclusion.

It has been underlined in this contribution that the MPT holds when the turbulent heat flux is accounted for using a gradient-type law, whereas this result is false when applying for a full second-moment closure, though the inter-realizability may hold. Some suitable algorithms to preserve the over-realizability (and the MPT, if valid) have been presented. Similar results of positivity have been obtained in the single phase compressible turbulent framework (see [10]), using the Favre's averaging process ([4]), but also when focusing on some two-fluid compressible models (see [2]).

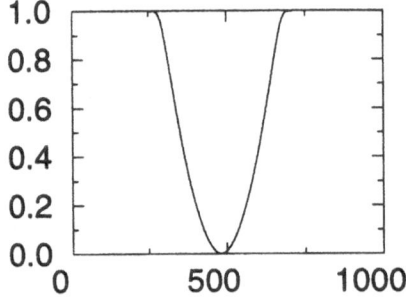

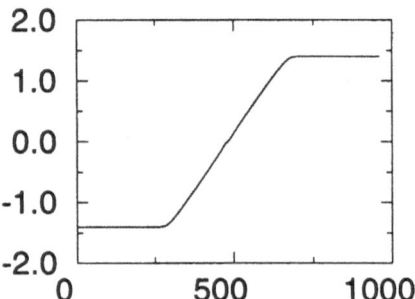

Figure 1   Figure 2
Behaviour of the component R_{11} Behaviour of the mean axial velocity U

References

/1/ Colombeau J.F. "Multiplication of distributions" *Springer-Verlag*, 1992.
/2/ Combe L., Fabre J., Hérard J.M., Simonin O. "On the numerical preservation of the realisability principle for gas-solid flows" submitted for presentation to *ASME FED* Summer Meeting, San Diego, CA , 1996.
/3/ Du Vachat, R. "Realizability inequalities in turbulent flows" *Phys. of Fluids* 20(4), 1977.
/4/ Favre A. "Equations des gaz turbulents compressibles. Parties I et II" *Journal de Mecanique*, 4, pp. 361-390, 391-421, 1965.
/5/ Fu S., Launder B.E, Tselepidakis D.P. "Accomodating the effects of high strain rates in modelling the pressure strain correlations" *UMIST Mech. Eng. Dept. report* TFD 87/5, 1987.
/6/ Gallouet T., Masella J.M. "A rough Godunov scheme" *CRAS Paris*, to appear., 1996.
/7/ Hérard J.M. "Basic analysis of some second moment closures". a) Part I : incompressible isothermal turbulent flows" *TCFD*, Vol.6, n°4, pp.213-233, 1994.b) Part II : incompressible turbulent flows including buoyant effects" *Collection Notes Internes DER* 94NB00053, 1994.c) Part III : compressible turbulent flows with shocks, *in preparation.*
/8/ Hérard J.M."Realisable second moment closures and a priori suitable algorithms" in *Turbulence, Heat and Mass Transfer,* Begell House editors, New-York, NY, 1996.
/9/ Hérard J.M."Suitable algorithms to preserve the realisability of Reynolds stress closures" *ASME FED* Vol 215, pp. 73-80, 1995.
/10/ Hérard J.M., Forestier A., Louis X. "A non strictly hyperbolic system to describe compressible turbulence" *Collection Notes Internes DER* 95NB00003 , 1995.
/11/ Le Floch P. "Entropy weak solutions to some non-linear hyperbolic systems in non-conservative form" *Comm. in Part. Diff. Eq.* 13(6), pp. 669-727, 1988.
/12/ Le Floch P, Liu T.P.. "Existence theory for non linear hyperbolic systems in non conservative form" *CMAP report* n°254, Ecole Polytechnique, 1992.
/13/ Lumley J.L. "Computational modelling of turbulent flows" *Advances in Applied Mechanics*, vol. 18, pp. 123-176, 1978.
/14/ Pope, S.B. "On the relationship between stochastic Lagrangian models of turbulence and second moment closures" *Phys. of Fluids*, vol. 6, n° 2, pp. 973-985, 1994.
/15/ Schumann U. "Realisability of Reynolds stress turbulence models" *Phys. of Fluids*, vol. 20, n°5, 1977.
/16/ Shih T., Lumley J.L. "Modeling of pressure corelation terms in Reynolds stress and scalar flux equations" *Cornell University report* FDA-85-3, 1985.
/17/ Shih T., Lumley J.L. "Critical comparison of second moment closures with direct numerical simulation of homogeneous turbulence" *AIAA Journal*, Vol;.31, n°4, pp. 663-670, 1993.
/18/ Shih T., Lumley J.L. "Second order modelling of a passive scalar in a turbulent shear flow" *AIAA Journal*, Vol;.28, n°4, pp. 610-617, 1989.
/19/ Shih T., Shabbir A., Lumley J.L. "Realisability in second moment turbulence closures revisited" *NASA Technical Memorandum* 106469, 1994.
/20/ Smoller J. "Shock waves and reaction diffusion equations" *Springer Verlag*, 1983.
/21/ Speziale C.G. "Invariance of turbulent closure models" *Phys. of Fluids* 22(6), pp. 1033-1037, 1979.
/22/ Speziale C.G., Abid R., Durbin P.A. "New results on the realisability of Reynolds stress turbulence closures" *ICASE report 93-76*, 1993.
/23/ Speziale C.G., Gatski, T.B. "Assessment of second order closure models in turbulent shear flows" *AIAA Journal*, vol.32, n° 10, pp. 2113-2115, 1995.
/24/ Temam R. "Une méthode d'approximation des équations de Navier-Stokes" *Bull. Soc. Math. France*, vol. 96, pp. 115-152, 1968.

Appendix A

The convective subset (2.19-2.23) is :

$$U_{,t} + (R_{11})_{,x} = 0 \tag{A1}$$
$$V_{,t} + (R_{12})_{,x} = 0 \tag{A2}$$
$$R_{11,t} + 2 R_{11} U_{,x} = 0 \tag{A3}$$
$$R_{22,t} + 2 R_{12} V_{,x} = 0 \tag{A4}$$
$$R_{12,t} + R_{11} V_{,x} + R_{12} U_{,x} = 0 \tag{A5}$$
$$X_{1,t} + X_1 U_{,x} + R_{11} T_{,x} = 0 \tag{A6}$$
$$X_{2,t} + X_1 V_{,x} + R_{12} T_{,x} = 0 \tag{A7}$$
$$<\theta^2>_{,t} + 2 X_1 T_{,x} = 0 \tag{A8}$$
$$T_{,t} + X_{1,x} = 0 \tag{A9}$$

It may be rewritten in the form : $W_{,t} + A (W) W_{,x} = 0$, setting :

$$W^t = (U , V , R_{11} , R_{22} , R_{12} , X_1 , X_2 , <\theta^2> , T). \tag{A10}$$

Then, the Riemann problem associated to system (A1 - A10) and initial conditions :

$$W (x, t = 0) = W_L \quad \text{for} : x < 0 \quad ; \quad W (x, t = 0) = W_R \quad \text{for} : x > 0$$

has a unique solution if initial values are as :

$$U_R - U_L < (2R_{11})_R^{1/2} + (2R_{11})_L^{1/2}. \tag{A11}$$

The solution expending in the (x,t) plane is over-realisable. We choose initial conditions in agreement with (A11) and such that :

$$U_R = - U_L > 0 \; ; \; (R_{11})_R = (R_{11})_L > 0 \; ; \; T_R = T_L \; ; \; (X_1)_R = -(X_1)_L > 0$$

The solution of the Riemann problem does not depend on the initial values of T_L and $(X_1)_L$. Thus, if the 1-characteristic field is a rarefaction wave, we get :

$$T_1 = T_L + \sqrt{2} \; \frac{(X_1)_L}{(R_{11})_L^{1/2}} \; (1 - \left(\frac{(R_{11})_1}{(R_{11})_L}\right)^{1/2}) \tag{A12}$$

with : $(X_1)_1 = (R_{11})_1 \dfrac{(X_1)_L}{(R_{11})_L}$ and : $(R_{11})_1 = \dfrac{1}{2}(U_L + (2 R_{11})_L^{1/2})^2$

The value of the intermediate state of the mean temperature on the right handside of the 1-wave, which is given by (A12), obviously violates the MPT.

VARIABLE DENSITY EFFECTS IN AXISYMMETRIC TURBULENT JETS AND DIFFUSION FLAMES

J.P.H. SANDERS, B. SARH AND I. GÖKALP
Laboratoire de Combustion et Systèmes Réactifs
Centre National de la Recherche Scientifique
45071 Orléans, cedex 2, France

1. Introduction

Turbulent jets and jet diffusion (non-premixed) flames are the basic flow configurations for many applications of reacting flows. In these flows, mixing of the reactants is important to achieve efficient combustion. A varying density can have large effects on the characteristic parameters in the flow, such as turbulence intensities, velocity and mixture fraction decay rates and spreading rates. This paper aims at characterising the variable density effects in axisymmetric jets and diffusion flames using a second order turbulence model, including buoyancy effects. Predictions are compared with experimental data from the literature, when available.

2. Non-reacting turbulent jets

In turbulent non-buoyant jets, where the density variations are due to an initial density ratio $\omega = \rho_a/\rho_j$ (where a denotes 'ambient' and j denotes 'jet') unequal to 1, there are no influences of ω on far field characteristic parameters (Sanders *et al.*, 1996). These numerical findings, with first and second order turbulence closures, are in good agreement with available experimental data for the scalar field (Richards and Pitts, 1993).

In turbulent vertical jets in which buoyancy plays a role, all characteristic parameters depend on ω and on the axial distance (Sanders *et al.*, 1996). They do no longer attain asymptotic (far field) values. For light jets ($\omega > 1$), the mechanical turbulence intensity decreases with axial distance while the scalar intensity increases with x. This is due to a slower veloc-

L. Fulachier et al. (eds.), IUTAM Symposium on Variable Density Low-Speed Turbulent Flows, 205–208.
© *1997 Kluwer Academic Publishers.*

ity decay (caused by the upward buoyant forces) and to a faster mixture
fraction decay (like in a plume) due to conservation of nozzle mass flux,
respectively. Both spreading rates for $\omega > 1$ are smaller than those for
$\omega = 1$. These numerical findings agree qualititavely with experimental data
for the scalar spreading rate. However, the experimental velocity spreading
rate increases with ω according to Panchapakesan and Lumley (1993). Up
to now, no explanation for this discrepancy has been found.

The effect of buoyancy can be split into the mean upward force in the
axial momentum equation and the buoyancy induced turbulence production terms in the transport equations for the Reynolds stresses and scalar
fluxes. These turbulence production terms only have a small influence on
the turbulent fluxes, except for the axial scalar flux $\widetilde{u''f''}$, which is very
much influenced, in agreement with experimental data (Panchapakesan and
Lumley, 1993). This effect is only captured by the second order model.

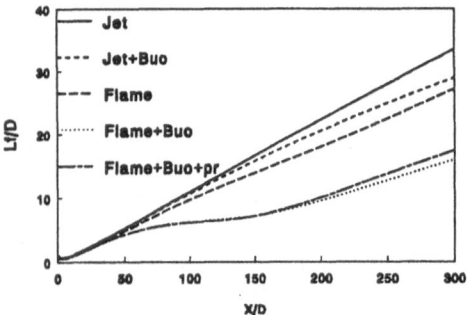

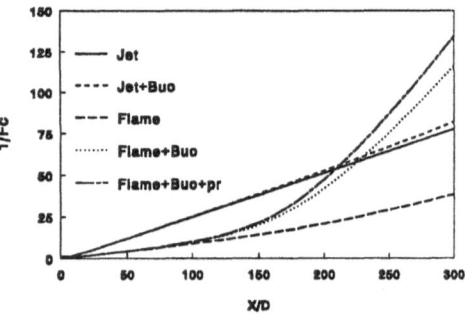

Figure 1. Halfwidths for methane
jets and flames ($\omega = 1.8$). 'Buo':
with upward force; 'pr': with buoyant turbulence production.

Figure 2. Mixture fraction decay
for methane jets and flames. See
also legend of Fig. 1.

3. Turbulent diffusion flames

In turbulent diffusion flames, the density variations are more sustained than
in non-reacting jets; they are important until well beyond the flame tip. The
location of the flame tip is independent of the nozzle diameter D and of the
fuel jet exit velocity U_j. However, it depends on the fuel, its dilution and
the stoichiometry; it is typically located at about $100D$ (pure hydrogen or
hydrocarbon flames).

The non-premixed flames in this study are numerically simulated using the laminar flamelet concept (Peters, 1984; Sanders and Lamers, 1994;

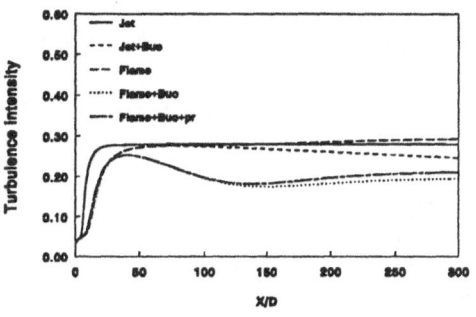

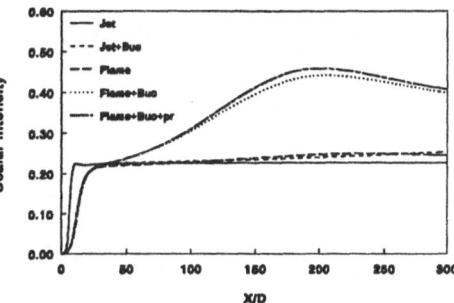

Figure 3.   Centreline $\sqrt{k}/U_c$. See also legend of Fig. 1.

Figure 4.   Centreline $\sqrt{f''^2}/F_c$. See also legend of Fig. 1.

Sanders and Gökalp, 1996). First, in a non-buoyant flame it is found that the sustained density variations have a large influence on the characteristic parameters, in contrast with the absence of any effect in the non-reacting non-buoyant jets. Spreading rates are smaller than in the non-reacting case (Fig.1). The decay of velocity and mixture fraction is approximately linear ($\sim x^{-1}$) while they decrease faster in the far field (like x^{-2}) (Fig.2). Experimental data on flames without buoyancy are hard to find, but the experiments of Drake et al. (1986) show indeed a faster decay in the far field than in the near field, for both variables. Centreline turbulence and scalar intensities ($\sqrt{k}/U_c$ and $\sqrt{f''^2}/F_c$) are only slightly larger in the far field, see Fig. 3 and 4, respectively.

Second, in diffusion flames where buoyancy does play a role, even larger effects on all characteristic parameters are observed. Spreading rates are much smaller than in non-buoyant flames but the influence of the buoyancy induced turbulence production is small (Fig.1). This is also observed in Fig. 5, where predicted halfwidths for the 75 % H_2/ 25 % N_2 flame of Meier et al. (1996) are compared with the experimental halfwidth in the reacting case. The spreading rate in the flame is clearly much smaller than in the jet. Fig. 2 shows that buoyancy has an extremely large influence on scalar decay in a flame.

Mechanical turbulence intensities $\sqrt{k}/U_c$ are smaller than without buoyancy, also in accordance with experiments (not shown) by Wittmer (1980). Due to buoyancy, the scalar intensity $\sqrt{f''^2}/F_c$ increases very strongly with axial distance, and this behaviour is amplified when buoyancy effects become more important, see Fig. 4. In Fig. 6, the scalar intensity is higher for flames with lower fuel exit velocities U_j (lower Froude number). Both experiments and predictions show this behaviour. Buoyancy also has a large influence on decay rates, with effects similar to turbulent plumes: the velocity decays more slowly than without buoyancy while the mixture fraction decays faster (Fig.2).

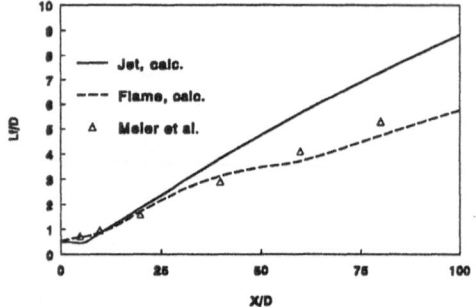

Figure 5. Scalar halfwidths in a 75 % H_2/ 25 % N_2 jet and flame. Symbols of (Meier et al., 1996).

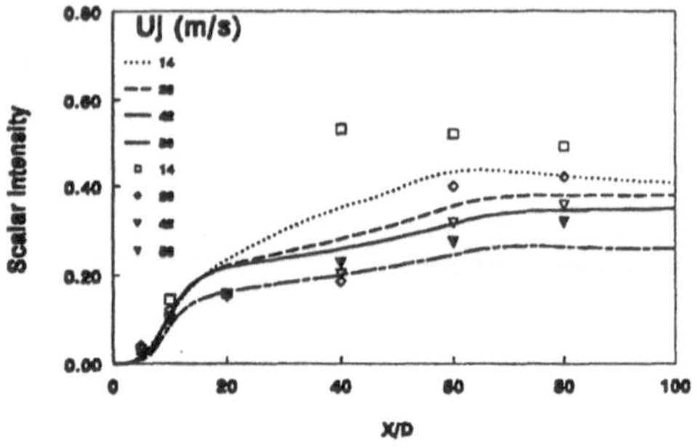

Figure 6. Centreline $\sqrt{f''^2}/F_c$ in the flames at various jet exit velocities (Meier et al., 1996).

References

Drake, M.C., Pitz, R.W. and Lapp, M. (1986) Laser measurements on nonpremixed H_2-air flames for assessment of turbulent combustion models, *AIAA J.* **24**, pp. 905–917.

Meier, W., Vyrodov, A.O., Bergmann, V. and Stricker, W. (1996) Simultaneous Raman/LIF measurements of major species and NO in turbulent H_2/air diffusion flames, *Appl. Phys. B* **62**, to appear.

Panchapakesan, N.R. and Lumley, J.L. (1993) Turbulence measurements in axisymmetric jets of air and helium. Part 2. Helium jet, *J. Fluid Mech.* **246**, pp. 225–247.

Peters, N. (1984) Laminar diffusion flamelet models in non-premixed turbulent combustion, *Progress En. Combust. Sci.* **10**, pp. 319–339.

Richards, C.D. and Pitts, W.M. (1993) Global density effects on the self-preservation behaviour of turbulent free jets, *J. Fluid Mech.* **254**, pp. 417–435.

Sanders, J.P.H. and Lamers, A.P.G.G. (1994) Modeling and calculation of turbulent lifted diffusion flames, *Combust. Flame* **96**, pp. 22–33.

Sanders, J.P.H. and Gökalp, I. (1996) Flamelet based analysis of non equilibrium effects and NO-formation in turbulent hydrogen diffusion flames, *32nd AIAA/ASME/SAE/ASEE Joint Propulsion Conference*, Lake Buena Vista, FL, paper 96-3033.

Sanders, J.P.H., Sarh, B. and Gökalp, I. (1996) Variable density effects in axisymmetric isothermal turbulent jets. A comparison between a first and a second order turbulence model, *Int. J. Heat Mass Transfer*, to appear.

Wittmer, V. (1980) Geschwindigkeit und Temperatur in einer turbulenten Freistrahldiffusionsflamme, PhD. thesis, Karlsruhe, Germany.

NUMERICAL MODELLING OF TURBULENT CONVECTIVE HEAT AND MASS TRANSFER IN TWO-LAYER SYSTEMS

B. I. MYZNIKOVA, I. I. WERTGEIM,
Institute of Continuous Media Mechanics of the Urals Branch of the
Russian Academy of Sciences,
1, Academician Korolyov St., 614061 Perm, Russia

1. Introduction

The results of numerical modelling of turbulent regimes of free convective heat and mass transfer are presented. The model approximately describes the process of liquid oil products storage in the underground reservoir [1,2], where the geothermal gradient causes the strong turbulent convection in a gas-liquid system. The problem under consideration is formulated for a two-layer gas-liquid system filling the vertical cylindrical reservoir formed in solid heat conductive array in which vertical upwards temperature gradient is assigned.

The mathematical simulation of the process is based on the consideration of various physical phenomena having essential influence on it. The existence of the temperature heterogeneities leads to the intensive thermal convection in both media. Gas dissolving over the separation surface and its diffusion into the fluid initiate the concentration convection mechanism. Moreover the fluid evaporation affects the gas convection intensity. The turbulent character of motion in both media must be taken into account as well. As the numerical findings show, the convective flow in the liquid medium slows down the gas propagation velocity that results in the formation of rather thin layer in the region near the interface, where the gas admixture is mainly localised.

2. Statement of the problem

The mathematical statement of the problem includes the equations of momentum, temperature and admixture concentration transfer in the Boussinesq approximation for buoyant terms, and also, depending on the turbulence model used, additional turbulence transport equations. The numerical solution of the problem is based on the two-field approach and is obtained by the finite-difference technique requiring the governing equations writing in terms of the stream function and the vorticity.

Boundary conditions for mean and pulsative velocity fields correspond to non-slip rigid boundaries of the inner enclosure and include conjugation conditions on the liquid-gas interface for velocities, tangential stresses, temperature and heat flux. For admixture concentration external boundaries of the reservoir are considered as impenetrable, on the interface the concentration of gas admixture is equal to its saturation value. The liquid-gas interface is considered as flat and non-deformable. Horizontal boundaries are heat conductive with the fixed temperatures. On the heat

209

L. Fulachier et al. (eds.), IUTAM Symposium on Variable Density Low-Speed Turbulent Flows, 209–212.
© 1997 *Kluwer Academic Publishers.*

conductive lateral surfaces the temperature is linear function of the vertical coordinate, according to geothermal gradient, existing enough far from the reservoir.

The initial conditions are taken usually as motionless pure liquid with equilibrium temperature distribution and admixture concentration having maximum on the interface.

The problem is supposed to have the cylindrical symmetry for the averaged fields.

Some results are obtained not only for the most general two-layer conjugate formulation described above, but for simplified one as well, when convection in gas isn't considered, and boundary conditions on free liquid surface are formulated instead.

There have been used two comparatively simple turbulence models: the zeroth-order model of Smagorinsky type - the effective viscosity model (hereafter - EVM), and the two-parametrical first-order model of the k-ε type, generalized for the case of nonisothermal flows with admixture, using Oberbeck-Boussinesq approximation for buoyant terms - the energy-dissipative model (hereafter - EDM). In the equations for averaged characteristics the approximations have been used of the turbulent stresses tensor $R_{ik}=<u_iu_k>$, of the turbulent heat flux vector $\Gamma_j=<u_j\tau>$ and of the turbulent mass flux vector $F_i=<u_ic>$ in accordance with the Boussinesq gradient hypothesis:

$$R_{ik} - \frac{2}{3}E\delta_{ik} = -\nu_t(\nabla_iU_k + \nabla_kU_i), \qquad \Gamma_i = -\chi_t\nabla_iT, \qquad F_i = -\lambda_t\nabla_iC$$

It has been assumed for the turbulent transfer coefficients that

$$\nu_t = \begin{cases} c_\nu^{3/2}L^2\sqrt{P} & (EVM) \\ c_\nu E^2/D & (EDM) \end{cases}, \qquad \chi_t = \nu_t/Pr_t, \qquad \lambda_t = \nu_t/Sc_t.$$

Here P is the value proportional to turbulence generation due to gradients of averaged fields of velocity, temperature and concentration; L is the macroscopic turbulent scale, E and D are the turbulent energy and the dissipation rate. The last two values have been determined in EDM from their own governing equations. Their form one can find, for instance, in [2-4]; ibid. the results of solving the problem without admixture accounting [2,3], and some preliminary results with it [4] have been presented.

3. Numerical results

The main computations were carried out for "liquid butane - natural gas" system in the stone-salt array. The corresponding values of non-dimensional parameters for this case (Prandtl number for fluid, ratio of kinematic viscosity of fluid to array thermal diffusivity, Schmidt number, ratios of liquid and gas parameters: dynamic viscosities, kinematic viscosities, thermal conductivities, thermal diffusivities, heat expansion coefficients, and ratio of thermal conductivities of fluid and rigid array) are the following:

$$Pr=1.8, \qquad P_s=0.146, \quad Sc=170, \qquad \eta=0.079, \quad \nu=0.657, \quad \kappa=0.371,$$
$$\chi=1.28, \qquad \beta=4.19, \qquad \gamma=0.035$$

The ratio of fluid and gas layer heights $l=L_f/L_g$ is varied from 3:1 to 1:3. The calculations are performed for wide range of governing parameters (Grashof numbers

G - up to 10^{10} and its analog for concentration field G_d - up to 10^{13}), which correspond to regimes from the laminar-turbulent transition to the developed turbulence. In few cases value of Sc is varied to check the influence of this parameter.

The resulting gas convection intensity, depending from value of l, was much greater than liquid one, for l≤1, and of the same order or smaller - for l>1. The qualitative features of the flow - formation of near-surface diffusion layer, temperature and concentration boundary layers are common for both full (F) and simplified (S) formulations, but the differences always exists and are strenghtening for enough large l, when intensive convection in gas layer is of importance. As computational results show, vortex values near the interface (and hence the tangential tensions) essentially (about two orders of magnitude) exceed the volume ones. This result corroborates the necessity of conjugate consideration of heat and mass transfer in the two-layer system. The development of flow and admixture propagation in time is shown on Fig. 1.

The appearance of a new factor - the admixture - essentially changes flow structure comparing with results for pure system [1,2]. Near the upper boundary the region appears which is characterised by high concentration of admixture and comparatively weak intensities of averaged and pulsation velocity fields. Gradually "the second storey" with the tier of weak vortices not far from the boundary, is formed. The turbulence intensity in this region decreases sufficiently, as the suppression of pulsations due to the concentration gradient takes place. The flow patterns calculated by both models and for both formulations are qualitatively similar. The averaged concentration distribution to a great extent depends on the Schmidt number values. The dynamics of admixture in the liquid is also of interest. The time dependencies of the averaged concentration at the initial stage of the process have a rather non-uniform character, with sharp transitions between periods with an almost constant value of mean concentration, and more long periods, when it is almost constant (Fig. 2). As qualitative explanation of such behaviour, we could suppose, that the sharp transition periods correspond to the admixture propagation in developing near-surface diffusion region with strong stable stratification, and periods of almost constant <C> - to the gradual suppression of thermally driven instability in the upper part of the bulk of almost pure liquid layer.

4. Conclusion

The present study allows us to ascertain the main features and the nature of development of the double diffusive turbulent convection regimes in the two-layer gas-liquid system surrounded by solid heat conductive array.

Thus the features of a fairly general numerical technique capable of handling the problem of fully developed turbulent heat and mass transfer have been outlined. It has been confirmed that the conjugate formulation is of importance when the temperature and heat flux distributions on the rigid interface link the temperature fields in the heat dissipating solid and convecting two-layer gas-liquid system. It has been recognized that it is desirable to take into account the difference among typical scales of pulsation components of velocity, temperature and concentration origin on turbulent exchange processes.

The computational results show, out of many controlling parameters, the sufficient effect of the ratio of liquid and gas layers thicknesses and Schmidt number. The model is adaptive for practical use. The results obtained permit one to evaluate the time of gas penetration in the bulk of liquid and other important characteristics of the turbulent double diffusive flow in the gas-liquid system filling the reservoir, and to elaborate recommendations to ensure the choice of optimal conditions of the liquid hydrocarbons safekeeping process.

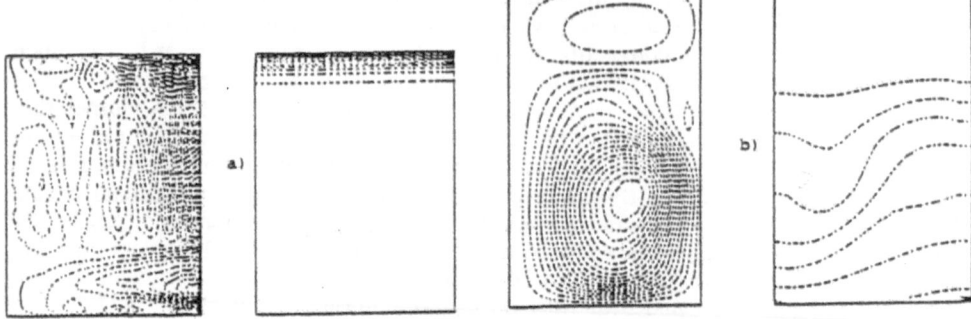

Fig. 1. Stream function and gas concentration patterns obtained on the initial (a) and final (b) stages:
a) - S, Sc=17, G=5.10^8, G$_d$=5.3.10^{12}, b) - F, Sc=170, l=3, G=10^6, G$_d$=6.2.10^9 G=10^7, G$_d$=3.5.10^{10}.

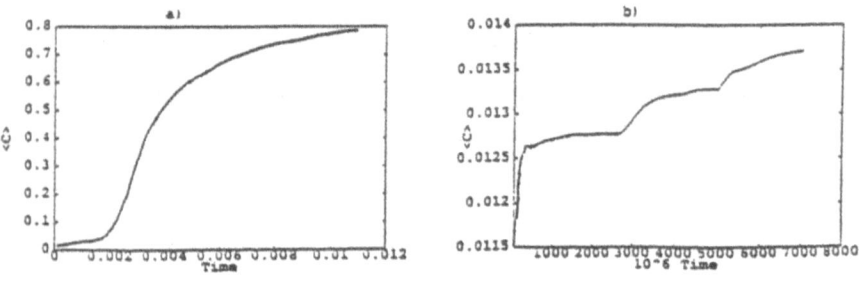

Fig. 2. Time dependencies of mean admixture concentration:
a) - F, Sc=170, l=3, G=10^6, G$_d$=6.2.10^9, b) - S, Sc=17, G=5.10^8, G$_d$=5.3.10^{12}.

References

1. Kazaryan, V. A., Myznikova, B. I., Nepomnyashchy A. A., Tarunin, E. L., Chertkova, E. A.: Numerical investigation of convective heat exchange at storage of liquid hydrocarbons in underground reservoirs, Trans. Acad. Sci. USSR (Izvestiya), Mech. Zhidk. Gaza, 2, pp143-148, 1981.
2. Kazaryan, V. A., Myznikova, B. I., Nepomnyashchy A. A., Polezhaev, A. A., Wertgeim I. I.: Turbulent convective heat and mass exchange in two-layer system of incompressible fluids, Abstracts of the Int. Symp. "Generation of Large-Scale Structures in continuous Media (Perm - Moscow), pp 132-133, 1990.
3. Wertgeim I. I., Chertkova, E. A.: Turbulent convection in cylindrical reservoir, Hydrodynamics and Heat and Mass Transfer Processes, Sverdlovsk: Ural Branch Acad. Sci. USSR, pp55-66, 1989.
4. Wertgeim I. I., Myznikova, B. I.: Numerical analysis of convective heat and mass transfer in two-layer system, Proc. of Heat/Mass Transfer, Minsk Intern. Forum MIF-96, Minsk, Lykov Institute of Heat and Mass Transfer, vol. 1 Convective heat and mass transfer, pp89-93, 1996.

AN EXPERIMENTAL INVESTIGATION OF QUASI-2D

TURBULENCE WITH OR WITHOUT BUOYANCY EFFECTS

T. ALBOUSSIÈRE*, V. USPENSKI**, A. KLJUKIN*** and R. MOREAU*

* *Lab. EPM-MADYLAM, ENSHMG, B.P. 95, F-38402 St-Martin d'Hères*
** *Visitor from Institute of Mechanics, Lomonossov Univ. Moscow, Russia*
*** *Visitor from Institute of Physics, Latvian Acad. of Sciences, Riga, Latvia*

Abstract. An experiment is presented in which a buoyant vertical axisymmetric flow interacts with a 2D horizontal azimutal flow. It appears that the 2D flow becomes predominant upon the buoyant flow as soon as its velocities are of the order of one tenth of the velocity of the buoyant flow or larger. Finally, this turbulent flow is essentially characterised by a small number of large coherent structures which transport the heat as a passive scalar.

1. Introduction.

An experiment has been carried out in a horizontal mercury layer (depth 1cm, diameter 22 cm) in the presence of a uniform vertical magnetic field (0.17 tesla), such that the fluid flow is two-dimensional. Because of well known properties of liquid metal MHD, when a DC electric current is passing from a circular ring of electrodes located into the insulating bottom plate (radius 9.3 cm) to the external vertical wall, an external annulus of fluid is forced to rotate at a given velocity, proportional to the current, whereas the central part of the fluid stays at rest. The sheared region between these two domains becomes unstable when the electric current is above 0.2 Amp and 2D eddies aligned in the direction of the vertical applied magnetic field are generated (streamlines are located within horizontal planes). These eddies interact and the 2D turbulent regime is well developped when the electric current is in the range 1-30 Amp [Fig. 1]. The intensity of this electric current is the control parameter of this sheared flow whose mean velocity may reach values of the order of 30 cm/s. Such MHD 2D sheared flows, which have been qualitatively observed since Lehnert (1955), are now investigated more quantitatively (Kljukin and Kolesnikov, 1989) and modelled (Jüttner and Thess, 1996).

A heated copper piece is located in the middle of the insulating bottom plate, which is separated from the mercury by a thin insulating coating (radius 37.5 mm). And the external vertical wall, which is thick and made of copper, is water-cooled. Therefore, a

L. Fulachier et al. (eds.), IUTAM Symposium on Variable Density Low-Speed Turbulent Flows, 213-219.
© 1997 *Kluwer Academic Publishers.*

horizontal radial temperature gradient may be applied, which is a second control parameter. With or without the magnetic field, the buoyancy due to this temperature gradient generates a toroidal convective flow, whose streamlines are in the vertical meridian planes [Fig. 2]. When the magnetic field is applied (0.17 Tesla, which corresponds to a Hartmann number of 42 in mercury) with no current, this flow is significantly stabilised and braked (velocities are of the order of 10 cm/s without any magnetic field, and 5 to 10 times less with 0.17 tesla).

The purpose of the experimental study presented in this paper is to examine how these two flows may coexist. Typical results on the 2D sheared flow without any heating are presented, just as a reference to be compared with the results in the presence of heating. In particular, a video is shown, where the image of a grid reflected by the free surface of mercury allows to observe the 2D turbulent structures, their merging and their motion around the cell. But in this paper, the emphasis is placed on the results obtained when the buoyant flow has to compete with the 2D sheared flow.

2. Diagnostic techniques.

Beside the flow visualisation, the basic properties of the Hartmann layer which is present along the bottom insulating plate yield a quite interesting diagnostic of the velocity. Indeed, since the electric potential does not vary accross that layer, its value at the wall is exactly proportional to the value of the stream function within the core flow, and, when the two-dimensionality is well achieved, its value does not vary either along the direction of the magnetic field. Therefore, measuring the electric potential at many points along the bottom wall may yield the local velocity (by making differences between values at two neighbouring points) and the local vorticity (by making differences at five neighbouring points). It is noticeable that this diagnostic technique has the unique advantage to give in the same time and at many points the stream function, the velocity and the vorticity. We have developped special sensors which are in the same time electrodes allowing to get the electric potential and thermocouples allowing to measure the local temperature. Each sensor (samples are shown on the poster) includes two metallic circular spots of 0.15 mm located within an insulating cylinder whose diameter is 0.9 mm. One of the spots is made of platinum to give the electric potential (platinum has the same thermoelectric power as mercury), the other is made of constantan, so that the couple Pt-Cst (with mercury to establish the electric contact between them) behaves as a very sensitive thermocouple (extremly small inertia due to the absence of coating). Measuring at the same point and in the same time the radial velocity fluctuation and the temperature fluctuation allows to get the convective part of the heat flux, and the mean temperature profile allows to get the contribution of conduction. The bottom plate of our cell is equipped with 88 such sensors located at chosen positions. In some circumstances, a moving probe carrying five sensors may be introduced from above to get vorticity, stream function and temperature at the same location.

3. Main results

The first idea which arises from these measurements is the fact that the typical velocity of the sheared flow necessary to mask the influence of buoyancy is significantly smaller than the typical buoyant velocity (about 10 times less). This shear flow is always turbulent and

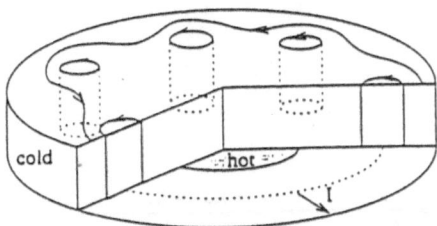

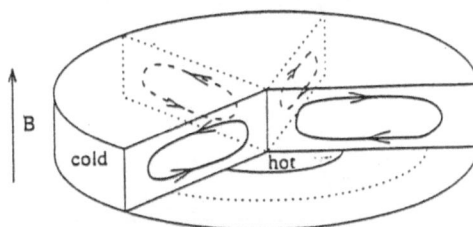

Fig 1: Structure of the 2D turbulent shear flow induced by the electromagnetic force generated by the DC current between the ring of electrodes and the wall and the vertical magnetic field.

Fig 2: Structure of the buoyant flow induced by temperature difference between the heated central region and the cold lateral wall.

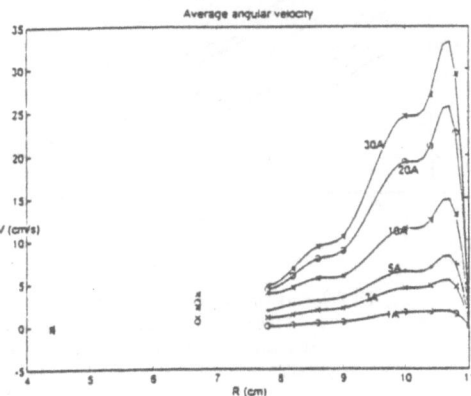

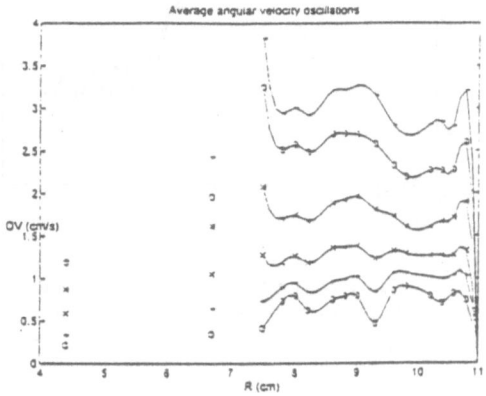

Fig 3: Angular velocity profiles (2D flow) for B=0.17 T and various DC current intensities.

Fig 4: Profiles of the RMS angular velocity fluctuations for B=0.17 T and the same current values as in fig 3.

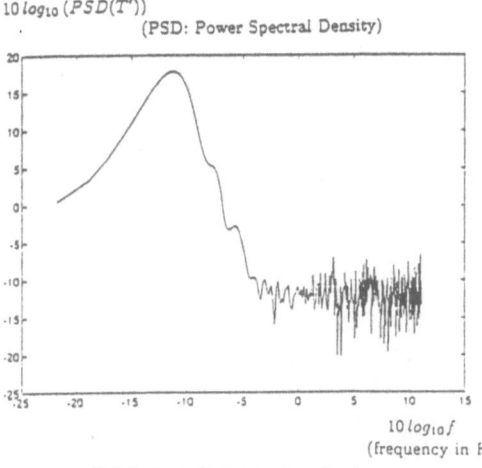

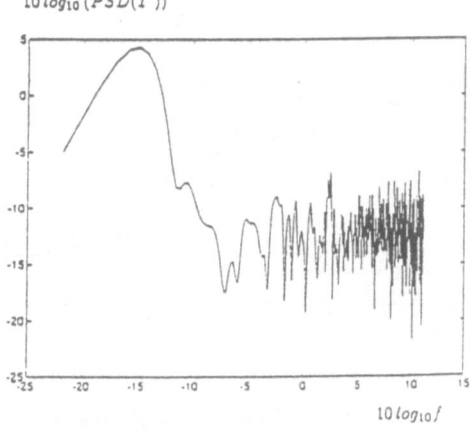

Fig 5: Spectrum of temperature fluctuations in the presence of the heating (8 W) and in the absence of magnetic field and current intensity.

Fig 6: Spectrum of temperature fluctuations with heating (8 W) and magnetic field (B=0.17 T), in the absence of current intensity.

$10 \, log_{10} \, (PSD(T'))$

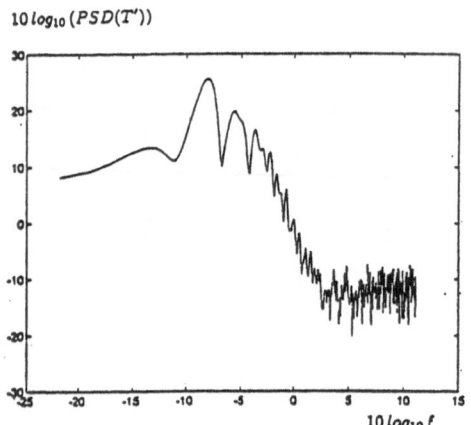

$10 \, log_{10} f$

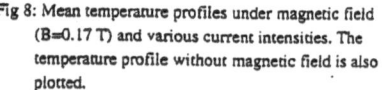

Fig 7: Spectrum of the temperature fluctuations with heating (8 W), magnetic field (B=0.17 T) and DC electric current (1 A).

Fig 8: Mean temperature profiles under magnetic field (B=0.17 T) and various current intensities. The temperature profile without magnetic field is also plotted.

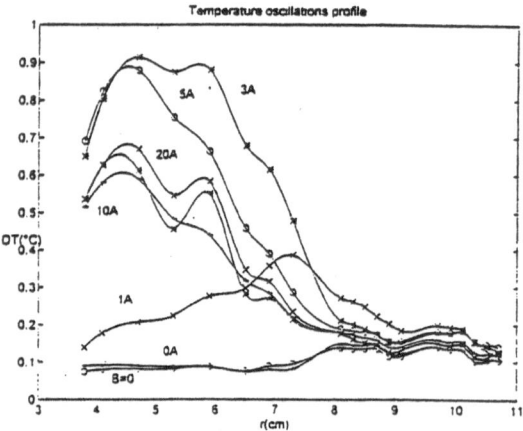

Fig 9: RMS temperature fluctuations profiles in the same conditions as in fig 8.

exhibits large scale coherent structures capable to transport the heat as a passive scalar quantity. Typical profiles are presented (mean angular velocity [Fig. 3], mean temperature [Fig. 8], root mean square of the two velocity fluctuations [Fig. 4] and of the temperature fluctuations [Fig. 9]), as well as characteristic spectra [Fig. 5-7], wich support this idea and illustrate the key properties of this 2D turbulence. It appears that the thickness of the turbulent shear layer is much larger (about 10 times) than in a laminar regime. Within that shear layer, the fluid remains almost isothermal. But in the inner region, where the mean velocity goes to zero, the turbulent kinetic energy remains of the same order as in the shear layer. The temperature difference between the heated part (at a given power: 8 watt) and the cold wall, which is concentrated within the inner region, is reduced from 15 K in the case of the MHD damped convective flow to 5 K in the case of an electric current equal to 10 Amp (this corresponds to a Nusselt number close to 3) and increases again to 9 K in the case of an electric current of 30 Amp (Nusselt number reduced to nearly 2). Such a decrease of the heat transfer when the mean velocity and the velocity fluctuations continue to increase suggests the predominance of particular structural properties of the 2D coherent structures (« cat's eyes » instead of the typical spiral eddies of mixing layers).

These phenomena may also be observed through the spectra of the various quantities. Roughly speaking, with a moderate forcing (I < 3 Amp) the kinetic energy spectra exhibit a slope close to -5/3, which suggests the presence of an inverse cascade of energy. When the forcing is strong enough (I > 10 Amp), we get a regime with a small number (2 or 3) of large coherent structures, capable to feed themselves with the vorticity generated by the forcing. The energy and temperature spectra reveal two main regions [Fig. 7], one with a row of peaks easily explainable and the other with a slope near -4. The frequency of the first peak corresponds to the time necessary for each large structure to transit around the whole cell, and the frequencies of the second and following peaks correspond to the distance between two neighbouring structures and its harmonics. The two dissipations take place essentially within the Hartmann layer at the bottom of each structure and are exactly equal, according to the properties of the Hartmann layer (Somméria and Moreau, 1982).

One noticeable effect is the fact that the turbulent heat flux remains rather small in the shear layer (R ≈ 10 cm), whereas it is quite significant within the inner region (R ≈ 5 to 8 cm). This is namely revealed by the comparison of the signals v_r and T' (radial velocity and temperature fluctuations) versus time. Within the shear layer one is almost always zero when the other is maximum or minimum. On the contrary, within the inner region their extrema coincide remarkably [Fig. 10]. Such a behaviour would be in agreement with an organisation of the turbulent structures as a row of « cat's eyes » separated from the wall by wavy continuous streamlines, and this picture is also in agreement with the video. An other interesting result is shown on Fig. 11. Fourier transforms of the product v_rT' have been computed and cut up into seven bands, as indicated in the caption. Each band has then been re-integrated and divided by the mean value<v_rT'> in order to illustrate the contribution of the corresponding length scale to the turbulent heat flux. It appears that, at low forcing of the 2D flow (I = 1 Amp), or when the buoyant flow is still not completely changed by the 2D flow, 50% of the flux is transported by the small scales (typical length smaller than 0.1 cm). At a significant forcing (I = 3 Amp), the large scales (typical length between 1 and 3 cm) have a

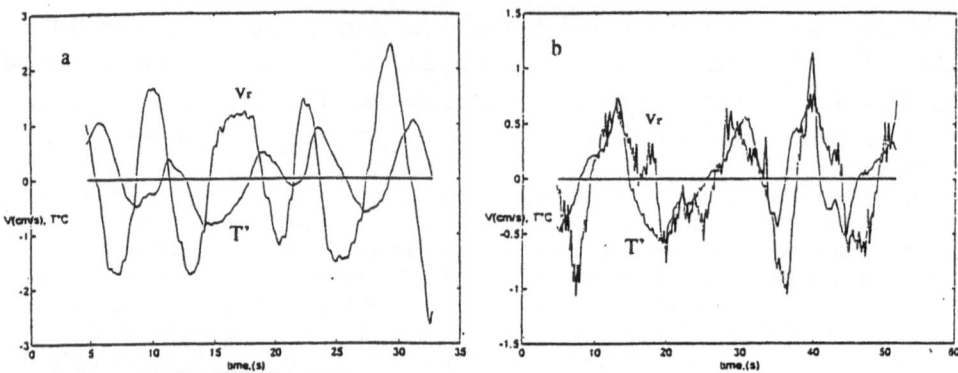

Fig 10 : Variation of the fluctuations v_r and T' versus time at two locations: a) R = 10 cm (sheared zone), b) R = 5 cm (inner zone).

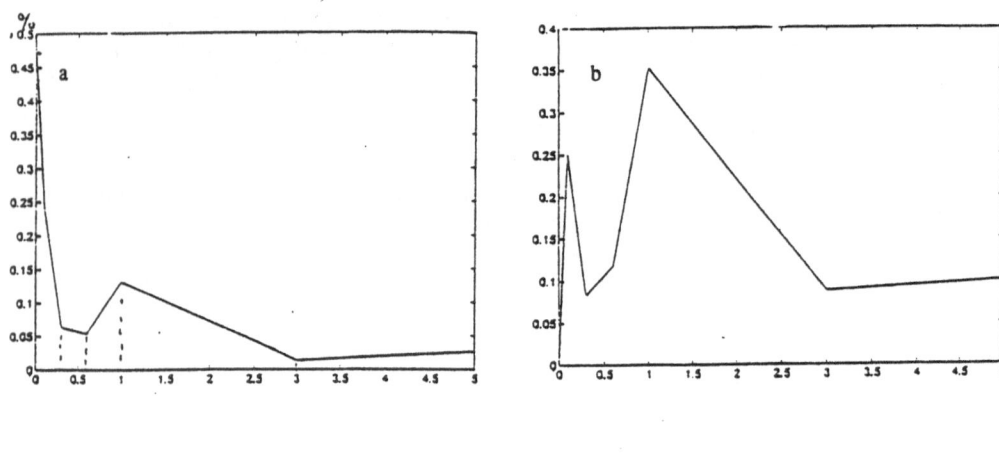

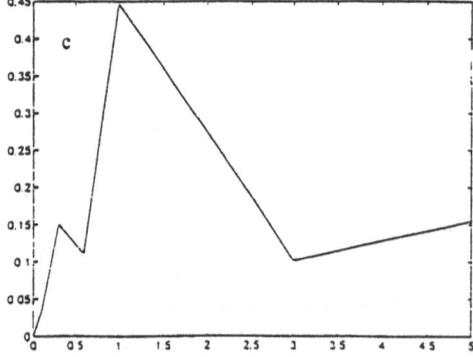

Fig 11 : Contribution of the different length scales to the turbulent heat flux $<v_r T'>$ for three levels of forcing (a: I = 1 Amp, b: I = 3 Amp, c: I = 10 Amp). The Fourier transform of $v_r T'$ is divided into seven bands (l < 0.1 cm, 0.1< l < 0.3 cm, 0.3 < l < 0.5 cm, 0.5 < l < 1 cm, 1< l < 3 cm, 3< l <5 cm, l > 5 cm).

contribution of the same order as that of the small scales. And at high forcing (I = 10 Amp) the larges scales have the most important contribution (50 %).

4. Conclusion.

The main idea suggested by these experimental results is that the 2D horizontal flow is capable to overcome the buoyant axisymmetric flow, as soon as its velocity is of the same order of magnitude, or even less. The authors think that the key to understand this rather surprising result is the fact that both flows are governed by different mechanicisms and almost ignore each other. The buoyant flow is strongly braked by the applied magnetic field and cannot receive any momentum from the 2D turbulence (any 3D turbulent motion capable to transfer some momentum to the buoyant flow is rapidly damped out). On the contrary, the 2D mean flow and its 2D disturbances are supplied in momentum and energy by the external source. It is nevertheless remarkable that this 2D highly turbulent flow is not an efficient way to transfer the heat.

5. References

Lehnert B., 1955, *Instability of Laminar Flow of Mercury caused by an External magnetic Field*, Proc. Roy. Soc. London, A 233, p. 299

Somméria J. and Moreau R., 1982, *Why, how and when MHD turbulence becomes two-dimensional*, J. Fluid Mech., vol. 118, pp. 507

Kljukin A. and Kolesnikov Yu. , *MHD Instabilities and Turbulence in Liquid Metal Shear Flows*, Proc. of the IUTAL Symposium « *Liquid Metal magnetihydrodynamics* », eds. J. Lielpeteris and R. Moreau, Kluwer Acad. Press, 1989, p. 449

Jüttner B. and Thess A., *Inertial Organisation in Forced 2D Turbulence*, 1996, 8th. Beer-Sheva Int. Seminar on MHD-Flows and Turbulence, to appear in Magnetohydrodynamics

PENETRATIVE CONVECTION IN WATER NEAR 4 °C

P. LE GAL

Institut de Recherche sur les Phénomènes Hors Equilibre
UM 138, CNRS - Universités d'Aix-Marseille I & II
12, Avenue Général Leclerc, 13003 Marseille, France

1. Introduction

This article focuses on some experimental aspects of penetrative convection. This type of convection occurs when a layer of fluid, bounded by one or two other stratified layers of the same fluid becomes unstable due to a destabilizing density profile. The convective motions first appear in the unstable layer, and then penetrate into the stably stratified zones. This type of convection can be observed in many natural situations. The atmospheric convective layer above the ground warmed by the sun and penetrating into the upper atmosphere is a typical example. Several experiments have already dealt with this type of convection. There are basically three ways to produce penetrative convection in a laboratory. The first is the application of a sudden change in the thermal boundaries of a stably stratified layer (Krishnamurti (1968), Deardorff et al (1969), Somerscale and Dougherty (1970)). But in this case, penetrative convection is basically non stationary. The main effect of this process is the appearance of hexagonal patterns via transcritical bifurcations (Busse (1967), Krishnamurti (1968)). The second way to create penetrative convection is to place heat sources or sinks within the fluid layer. For example, radiative heat sources can warm a certain region of the fluid and create an inverse density profile that may become unstable. An experimental study of this phenomenon (Whitehead and Chen (1970)) leads to the observation of cold plumes descending into the stably stratified layer in an intermittent manner. In this experiment, no evidence of hysteresis is observed nor a well defined hexagonal convective pattern. The third way to realize penetrative convection, is to use the singular property of water of possessing a parabolic equation of state: $\rho = \rho_0(1 - a(T - T_0)^2)$ where ρ is the density of water and T its temperature in Celsius. T_0 is the temperature where the maximum is reached and its value is about 4 °C. The constant a is equal to 8 10^{-6} °C^{-2}.

2. Penetrative convection

When a layer of water is cooled from below at a temperature lower than 4°C, and warmed from above at a temperature higher than 4 °C, a parabolic density profile is created. The light cold water is initially confined to the bottom of the layer below the

221

L. Fulachier et al. (eds.), IUTAM Symposium on Variable Density Low-Speed Turbulent Flows, 221–228.

4°C isotherm whose position is labelled d. But as soon as this unstable layer is thick enough, convection sets in and the fluid rises above this isotherm. Figure 1 describes this physical situation and also serves to define the various symbols to be used throughout. This hydrodynamic system is controlled by the two temperatures T1 and T2. We construct two dimensionless parameters: a Rayleigh number $Ra = \dfrac{(\rho_0 - \rho(T1))gd^3}{\rho_0 \upsilon \kappa}$ and a penetrative parameter $\lambda = \dfrac{d}{h} = \dfrac{T0-T1}{T2-T1}$ where g is the gravity constant, υ the viscosity of water, and κ its thermal diffusivity. Nevertheless Ra and λ do not in fact describe the actual convective layer whose thickness is unknown. We note that they refer to the convective sublayer initially confined below the 4 °C isotherm.

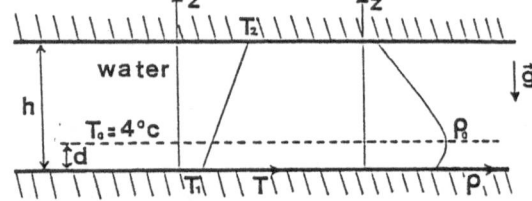

Figure 1: The hydrodynamical system and the symbols used in the text.

This type of physical system using water has been already the subject of several experimental, analytical, and numerical studies. In particular, the position of the criticality curve in the (Ra,λ) plane has been evaluated by a perturbation expansion analysis (Veronis (1963)). Although the expansion does not yield the entire non linear behavior of convection, it does prove that the transition from conductive to convective state, occurs via a subcritical bifurcation. This result has also been confirmed more recently (Roberts (1985), Matthews (1988)). But in these works also, the complete analysis is not obtained: in particular the width of the hysteresis loop is still unknown. Some experiments in water between 0°C and 4°C (convection is not penetrative in this case) confirm the presence of significant hysteresis (Azouni (1983), Azouni and Normand (1983)) contrary to other works (Dubois et al (1978)) which show the occurence of a supercritical hexagonal pattern followed at higher Rayleigh numbers by a transition towards a roll pattern. A very well controlled experiment has also been realized using liquid helium at low temperature (Walden and Ahlers (1981)). Helium like water, possesses a parabolic density law that permits the establishment of penetrative convection. A subcritical loop is confirmed, at least at this low Prandtl number (υ/κ), but visualization of the convective pattern is not possible. These studies show also a region in the parameter space where non stationary convection occurs even at the threshold.

Other experiments performed in water (Townsend (1964 and 1966), Myrup et al (1969), Adrian (1975)) have focused on the temperature profiles rather than in the type of transition. These experiments were usually realized at high Rayleigh numbers in very thick layers of water, and a power law of the type Nu $\propto$ Ra$^{1/3}$ was obtained (Furomoto and Rooth (1961)) as predicted by a theoretical estimate of the heat flux (Malkus (1963)). This law, typically observed in turbulent high Rayleigh number convection (Malkus (1954), Howard (1966)), also occurs in numerical simulations of penetrative convection (Musman (1968), Moore and Weiss (1973)). In these numerical studies an hysteresis was also observed at the convective threshold.

Therefore, we can see that several problems remain to be solved. Our first goal was to visualize some of the convective patterns in water. Aside from some interferograms (Tankin and Farhadieh (1971)), this had not previously been accomplished. One reason for this is the very weak buoyancy forces created by the small density difference between 0 and 4 °C: $\frac{\rho_0 - \rho(0°C)}{\rho_0} = 13 \quad 10^{-5}$, which is two orders of magnitude smaller than in oil for example. Despite this difficulty, we realized some laser tomography, shadowgraphy and schlieren imaging of penetrative convection in water, and we observed the different patterns for several geometrical configurations. Another problem that we addressed is the description of the bifurcation diagram. In measuring the deviation of a laser beam that crosses the upper stable layer of water, we were able to record the Nusselt number as a function of λ, for two different sized narrow rectangular boxes. No hysteresis was observed in contrast with previous experiments (Azouni (1983), Azouni and Normand (1983)) and with the theoretical or numerical calculations. Despite this disagreement, we obtained a 1/3 power law for the Nusselt number even very close to the threshold.

3. Visualization of convection

The experimental set up is a classical system for the study of convection. We use two temperaturecontrolled baths, filled with a mixture of antifreeze and water. These two baths supply water to two circulation loops whose temperatures are monitored by a computer. The stability of this system is +/- 0.05 °C. The cold sources consist of two freon compressors. We can vary the bottom temperature between -5 °C and room temperature and the top temperature between 10 and 40 °C. Different types of containers can be placed between the two water loops but the system always allows visualization through the layer of water. All the temperatures are measured by thermotransistors and the accuracy of the measurement is 0.05 degree. The first visualizations were obtained in a cylindrical geometry. The plexiglass container has an inner diameter of 38 mm, a height of 25 mm and the vertical wall has a thickness of 4mm. This cylinder is sandwiched between two aluminium plates which are cooled or warmed by the water circulations coming from the baths. The fluid is distilled water in which titanium dioxide coated 15 micron flakes were added. We shine a laser light sheet through a vertical plane accross the box. Figure 2 shows the convective layer at the bottom of the cylinder. The approximate values of the parameters are $\lambda=0.2$ and Ra=1500 (the temperatures are measured inside the aluminium plates). We observe that convection takes the form of two concentric tori; the flow is upward at the center and at the wall; secondary cells are also present. They are driven by viscosity at the top of the primitive convective stuctures. This observation is in agreement with theoretical analyses (Veronis (1963), Roberts (1985), Matthews (1988)) and numerical simulations (Musman (1968), Moore and Weiss (1973)). Above these two convective layers, the water is at rest: the flakes have sedimented and light is not deflected.

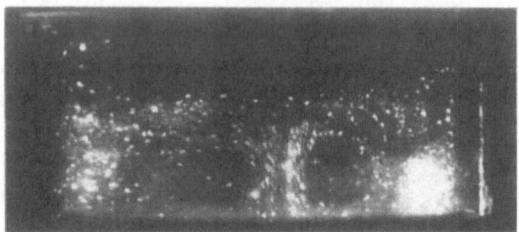

Figure 2: Laser tomography in a small aspect ratio cylindrical container.

The major defect of this kind of visualization is this sedimentation of the flakes: in a few hours the pattern vanishes. It therefore appears necessary to set up optical methods, as they can provide nonintrusive and long-lived visualizations of convective patterns. We set up a standard shadowgraph imaging process. The top and the bottom of the container consist in two 12mm thick glass plates. Their temperatures are controlled by the water circulations, and their diameter is 228 mm. The vertical wall is a 127 mm inner diameter plexiglass cylinder; whose thickness and height are both 25 mm. Figure 3 presents the hexagonal pattern obtained for $l = 1/3$ and Ra = 3500 (temperatures measured on the intern side of the glass plates), values for which the contrast is high enough for visualization. The white dots represent cold rising water. We measure the wavelength of this pattern to be very close to the thickness h of the layer of water and obtain a wavenumber $k = 2\pi/h$ as predicted by the linear theory (Veronis (1963)). This pattern is stationary for parameter values below $\lambda = 0.4$ and Ra = 6000, above which slow transition lasting some hours transforms the hexagonal pattern into a quasi-axisymmetric roll pattern.

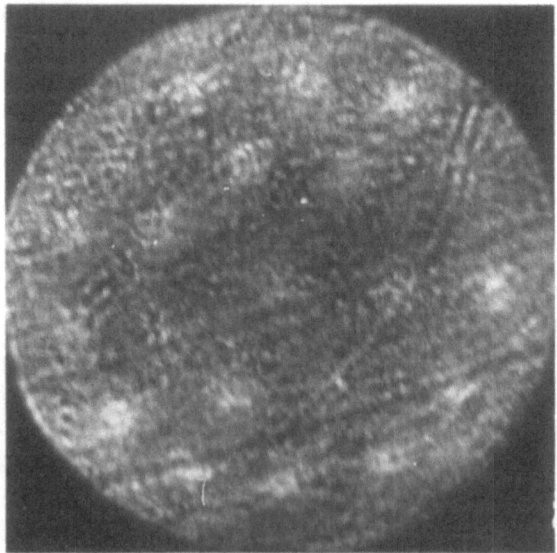

Figure 3: Hexagonal convective pattern at λ=0.33 and Ra= 3500.

To study the vertical structure of convection, schlieren imaging is performed through the vertical sides of two narrow rectangular containers. These containers have a horizontal length of 76 mm, and a height of 25 and 38 mm respectively. The second horizontal direction of 12 mm is rather narrow in order to keep the motion as bidimensional as possible and to allow visualization. The smaller of the two containers has double glass walls with an air gap of 8 mm between the 0.8 mm plates. The top and bottom consist of two copper plates in good thermal contact with temperature-controlled aluminium plates. A classical schlieren technique permits the observation of the convective cells through the vertical sides. We observe the emergence of three structures invading the container to approximatively a third of its height. This imaging technique is then improved by a video acquisition system. The micro-computer is programmed to store the image and to calculate the convective shape every 2 minutes. The schlieren image shows an illuminated region near the bottom of the container. The shape of this region is determined by convection and its upper limit corresponds to the change in the vertical temperature gradient corresponding to the top of the convective layer. By this observation method, the secondary cells do not appear, confirming that they are caused by viscous driving: there is no sharp temperature gradient in this layer allowing optical observation. The computer calculates then the change in brightness of the image. Figure 4-a presents this shape every two minutes. Here it is stationary, at least to the point we studied it (below $\lambda = 0.4$ and Ra = 3500). In contrast, the taller container (height= 38 mm) shows a propagation of these convective rolls (we believe the flow to be nearly bidimensional) as soon as our visualization technique allows any observation. This container has Plexi-glass long vertical sides with a thickness of 12 mm; the smaller vertical sides are made by two 4mm glass plates. It is for technical reasons of construction that we changed the way in which we built the container and we stress here on the possible importance of these vertical walls because they control the boundary thermal conditions.

a)   b)

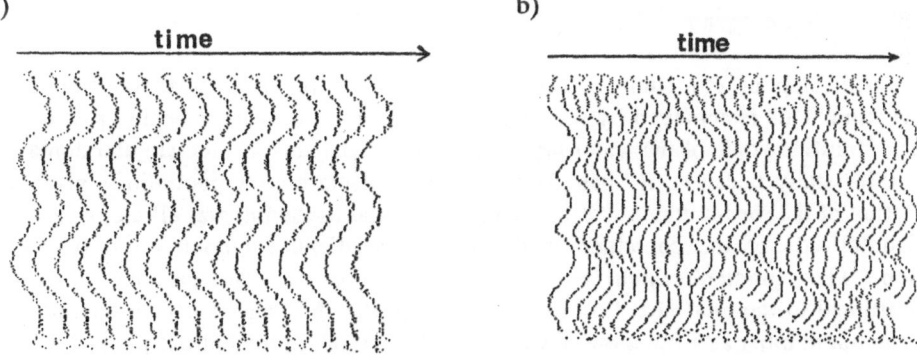

Figure 4: The convection shape computed from the schlieren images
in the 25 mm tall rectangular box (a) and in the 38 mm tall box (b).

The basic dynamics consists of a succession of two states: one state with three wavelengths and another with five. The transition between them occurs as shown in Figure 5 via the emergence of two new structures (i.e. two pairs of convective cells) between those that already exist. These new structures then push the old ones towards the sidewalls where they disappear. This phenomenon is nearly periodic (period around 20 mn) although some chaotic scenarios have also been observed, in which one of the emerging structures is killed before it can push away the previous cells, leaving the convective pattern with four wavelengths for a while.

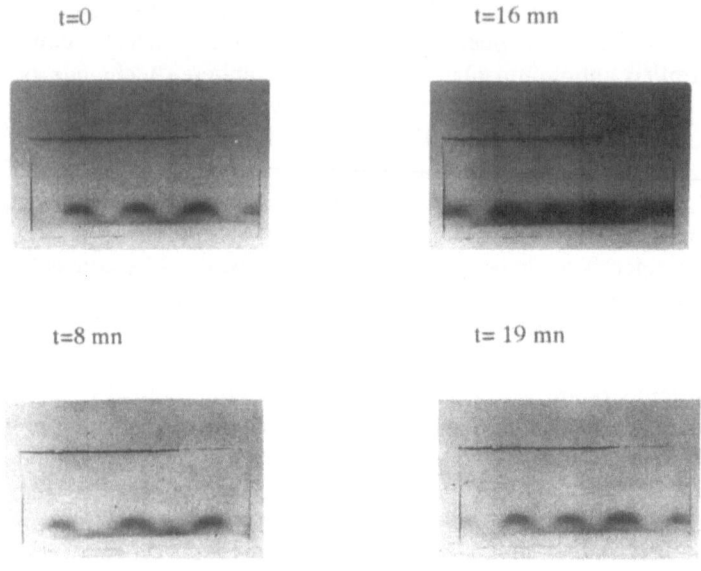

t=0   t=16 mn

t=8 mn   t= 19 mn

Figure 5: Succession of schlieren pictures for different states
of the phase dynamics occuring in the 38 mm tall container.

4. Bifurcation diagram

The next step of our study is the measurement of the heat flux transported by convection. This is achieved by recording the deviation of a laser beam crossing the upper stratified zone of the water layer. We choose the smaller rectangular container for this study because it has better lateral insulation. A photodiode gives the deviation of the beam which is directly proportionnal to the integral all along the long horizontal side of the container of the vertical gradient of temperature in the stable region. This gradient which is constant in the upper conductive zone, is in turn proportional to the total heat flux passing through the box and it is easy to calculate the Nusselt number as a function of λ and Ra. We maintain the bottom plate at a fixed temperature Tl between 0 and 4°C and vary the top plate temperature T2. The starting temperature T2 is approximatively 40 °C; when T2 is decreased at a rate less than 0.2°C per hour, the heat flux decreases and the 4 °C isotherm moves up in the container. Since T1 is fixed in this experiment, λ remains the only parameter. At some values of T2, λ becomes large enough for convection to appear. We performed this scanning for three different values of Tl; first decreasing then increasing T2 and

found no hysteresis on the bifurcation diaqram. Figure 6 presents the bifurcation
diagrams for one of the runs. We have also represented on this figure the width of the
hysteresis loop measured at λ=1 in a previous experiment (Azouni (1983)). This width
is considerably greater than the measurement errors; we therefore believe the lack of
subcriticality in our experiment to be inherent in the physical situation. We note also
the linear behavior of Nu as a function of the penetrative parameter λ. Since T1 is
fixed, the Rayleiqh number is proportional to λ and we recover the 1/3 power law for
the Nusselt number as a function of Ra.

a)   b)

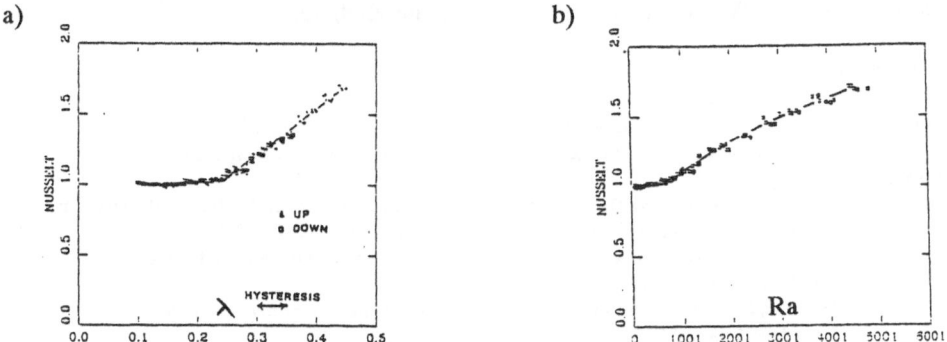

Figure 6: Bifurcation diagram: a) Nusselt number versus penetrative parameter.
b) Nusselt number versus Rayleigh number with the 1/3 power law.

5. Discussion

The visualization of a hexagonal pattern transforming into a roll pattern is in complete
agreement with previous calculations (Busse (1967)) and experiments (Dubois et al
(1978)). The observation of secondary cells above the convective layer is to our
knowledge the first one and it confirms the theoretical predictions (Veronis (1963),
Roberts (1985), Matthews (1988)) and numerical simulations (Musman (1968), Moore
and Weiss (1973)). Although the observation of the phase dynamics is quite surprising,
Normand and Azouni (1992) predict the appearance of a oscillating mode at
convective threshold. But our experimental bifurcation diagram contradicts with the
theoretical predictions of the infinite pattern. The critical values of the Rayleigh
number that we obtained are approximatively double the expected values. We note that
in fact these critical parameters have never been calculated for such low values of λ;
although it seems that for λ less than 0.5, the critical Rayleigh number is constant. The
narrowness of our experimental box can be at the origin of this delay. The comparaison
with the values obtained in Helium (Walden and Ahlers (1981)) is worst: their linear
thresholds are even lower than the predicted values for λ= 0.5. The lack of hysteresis is
also a challenge for which the narrowness of the experimental cell may again provide
an explanation. Theoretical analyses (Veronis (1963), Roberts (1985), Matthews
(1988)) predict a subcritical bifurcation for rolls or hexagons. But Veronis suggested
that the hysteresis may be very small: the solution with higher order terms (which is
not calculated in this paper) displays a very slight subcriticality. Moreover, this
subcriticality of the solution is certainly a function of λ and of the Prandtl number of
the fluid (Walden and Ahlers (1981)), and can be at the origin of this lack in our

228   P. LE GAL

experiment. The observation of the 1/3 power law is usual in turbulent convection when the thermal boundary layers are unstable and emit plumes in a central mixing isothermal zone. The parallel between the two kinds of convection is striking: both of the two physical systems consist in a pattern of convective structures penetrating in an other inactive layer. This law has been observed previously but in turbulent penetrative convection, and its presence near the threshold proves that it is linked to the penetrative nature of convection.

This work was acomplished in the Mathematics Department of M.I.T. in 1986-1988, with the advices of W. Malkus who is deeply acknowledged.

6.-References

-Adrian, R.J. (1975) Turbulent convection in water over ice. *J. Fluid Mech.***69**, 753-781.
-Azouni, M.A. (1983) Hysteresis loop in water between 0 and 4C. *Geophys. Astrophys. Fluid Dynamics* **24**, 137-142.
-Azouni, M.A. and Normand,C. (1983) Thermoconvective instabilities in a vertical cylinder of water with maximum density effects. *Geophys. Astrophys. Fluid Dynamics* **23**, 209-222 and 223-245 .
-Busse, F.H. (1967) The stability of finite amplitude cellular convection and its relation to an extremum principle. *J. Fluid Mech.***30**, 625-649.
-Deardorff, J.W.; Willis, G.E; Lilly, D.K. (1969) Laboratory investigation of non-steady penetrative convection. *J. Fluid Mech.* **35**, 7-31.
-Dubois, M.; Berge, P.; Wesfreid, J. (1978) Non Boussinesq convective structures in water near 4C. *J. Phys.* **39**, 1253-1257.
-Furomoto, T.A.; Rooth, C. (1961) Observations on convection in water cooled from below. *Woods hole Oceanographic Inst. Rep.*, 1-14.
-Howard, L.N. (1966) Convection at high Rayleigh number. *Proc. of the 11th Int. Congr. of Appl. Mech.* Munich, 1109-1115.
-Krishnamurti, R. (1968) Finite amplitude convection with changing mean temperature. Part 1 . Theory. Part 2. An experimental test of the theory. *J. Fluid Mech.***33**, 445-463
-Malkus, W.V.R. (1954) The heat transport and spectrum of thermal turbulence. *Proc. Roy. Soc. A* **225**, 196-212.
-Malkus, W.V.R. (1965) A laboratory example of penetrative convection. *Proc. 3rd Tech. Conf. on Hurricanes and Tropical Met,*. Mexico **5**, 89-95.
-Matthews, P.C. (1988) A model for the onset of penetrative convection. *J. Fluid Mech.***188**, 571-583.
-Moore, D.R. and Weiss, N.O. (1973) Nonlinear penetrative convection. *J. Fluid Mech.* **61**, 553-581.
-Musman, S. (1968) Penetrative convection. *J. Fluid Mech.***31**, 343-360.
-Myrup,L.; Gross,D.; Hoo, L.S.; Goddard,W. (1969) Upside down convection. *Weather* **25**, 150-157.
-Normand, C. and Azouni A. (1992) Penetrative convection in an internally heated layer of water near the maximum density point. *Phys. Fluids A* **4** (2), 243-253.
-Roberts, A.J. (1985) An analysis of near-marginal, midly penetrative convection with heat flux prescribed on the boundaries. *J. Fluid Mech.***158**, 71-93.
-Somerscales, E. and Dougherty, T.S. (1970) Observed flow patterns at the initiation of convection in a horizontal liquid layer heated from below. *J. Fluid Mech.***42**, 755-768.
-Tankin, R.S. and Farhadieh, R. (1971) Effects of thermal convection currents on the formation of ice. *Int. J. Heat Mass Transfer* **14**, 953-961.
-Townsend, A.A. (1964) Natural convection in water over an ice surface. *Quart. J. Roy. Met. Soc.* **90**, 248-259.
-Townsend, A.A. (1966) Internal waves produced by a convective layer. *J. Fluid Mech.* **24**, 307-317.
-Veronis, G. (1963) Penetrative convection. *J. Astrophysical* **137**, 641-663.
-Walden, R.W.; Ahlers,G. 1981: Non-Boussinesq and penetrative convection in a cylindrical cell. J. Fluid Mech.109, 89-114.
-Whitehead, J.A.; Chen, M.M. 1970: Thermal instability and convection of a thin fluid layer bounded by a stably stratified region. *J. Fluid Mech.***40**, 549-576.

IV. Modelling and Experiments. Industrial Applications

STUDYING AND MODELLING VARIABLE DENSITY TURBULENT FLOWS FOR INDUSTRIAL APPLICATIONS

J.P. CHABARD, O. SIMONIN, A. CARUSO, C. DELALONDRE,
S. DALSECCO, N. MÉCHITOUA
EDF- DER. Département Laboratoire National d'Hydraulique
6, Quai Watier. 78400. Chatou. France.

1. Introduction.

Amongst the applications of CFD that EDF has to deal with, a large number are related to high temperature and/or variable density flows : these applications are for example related to the design of new industrial processes using electricity (plasma flows, electric arcs, heating using Joule effect) or to electricity production (pulverised coal burners, gas turbine combustion chambers).

To be able to study these problems, developments were necessary both on the modelization of complex physical phenomena and on their interaction with turbulence. The modelization of turbulence is based on one point closures.

Industrial applications are presented in the various fields of interest for EDF. A first example deals with transferred electric arcs coupling flow and thermal transfer in the arc and in the bath of metal and is related with applications of electricity. The second one is the combustion modelling in burners of fossil power plants. The last one comes from the nuclear power plants and concerns the stratified flows in a nuclear reactor building.

2. Transferred Plasma Arc for Metallic Bath Heating.

Electric arcs are widely used in the metallurgical industry and, for a few years, numerous transferred arc plasma torches have been installed for tundish or ladle heating. Although the global behaviour of these processes is correctly controlled, the knowledge and the understanding of the heat transfer between the electric arc and the steel bath are rather limited, that makes their optimisation difficult. Because of the complex nature of the plasma, experimental work is difficult and only gives global information instead of localised information which may be important to determine the optimum operating conditions in the plasma reactor. But, recent works in modelling [8, 15] demonstrate the interest of using a modelling approach in order to improve a plasma processing.

In the present paper, study is focused on molten metal heating with a transferred arc plasma torch. The interaction between the arc and the metal bath is studied by

231

L. Fulachier et al. (eds.), IUTAM Symposium on Variable Density Low-Speed Turbulent Flows, 231–242.
© 1997 *Kluwer Academic Publishers.*

considering the flow, temperature and electromagnetic fields both in the arc and in the bath which are computed separately using the Mélodie code, a 2D axisymmetrical finite differences code developed at LNH. Moreover, a specific modelling of the anodic layer, corresponding to the boundary layer occurring on the bath surface, is performed by means of a one-dimensional subgrid model.

Numerical model

The computing geometry is approximately a 200 kW pilot furnace supplied by Tetronics to EDF's plasma laboratory (figure 1). It consists of an argon transferred arc between a cathode tip plasma torch and an iron anodic bath under atmospheric pressure. The arc length is 150 mm. A 125 mm radius steel anode is located at the bottom of the furnace. The walls of the furnace are made of high-grade alumina refractory.

Basic assumptions and Modelling in both plasma and bath domains

Steady, two-dimensional axisymmetrical, plasma and molten metal flows are considered. Local thermodynamic equilibrium (LTE) and global electric neutrality are assumed to prevail in the plasma. In consequence, the plasma modelling used the classical fluids mechanics equations coupled with the electromagnetic ones (with Laplace forces contribution in the momentum equation, Joule and radiative source terms in the energy equation). Due to the axisymmetrical geometry, the magnetic field is calculated directly from the current density distribution with the help of the Ampere's theorem [5].

For the bath flow computation, identical equations are solved without, of course, the radiative losses. But, the enthalpy conservation equation is solved using a finite volume method in order to improve the thermal balance. The bath surface (interface) is assumed to remain flat. The Marangoni effect is neglected compared to the surface shear stress and electromagnetic forces. The bottom electrode of the furnace made of iron is also computed. Depending on the furnace working conditions, this electrode, solid at the bottom, can melt close to the bath. The same equations are used both in the solid and liquid part of the metal but, when the temperature decreases under the melting temperature of iron (1809 °K) the velocities are imposed to zero.

In the plasma domain, the iron evaporation is taken into account. In addition to the governing equations of conservation of mass, momentum and enthalpy, a conservation equation of the iron mass fraction is solved. The iron vapour condensation on the walls and in the plasma chamber is taken into account by assuming that the phase transition happens at a constant temperature but its contribution in the energy balance is neglected.

The radiation loss of the plasma is modelled by means of a net emission coefficient function of the local temperature and the iron molar fraction proposed by [7]. For the moment the radiative heat flux to walls (bath surface and refractory) are not taken into account. The complex radiative transfer in the plasma chamber that needs a three dimensional modelling are in progress of development by using a discrete transfer method [13].

In such industrial electric arcs, turbulence must be taken into account. This is performed in the present study by means of a two-equations k-epsilon turbulence model. But for instance, due to the large variations of the molecular viscosity, both laminar and turbulent regimes may be present in the same flow. To account for that, a low-Reynolds k-epsilon turbulence model is used [11, 6]. But the possible effect of the turbulence on the electrical conductivity is neglected. The same turbulence model is used in the anodic metal bath in order to treat laminar as well as turbulent cases.

The thermophysical properties of the plasma are function of the local temperature of the local iron mass fraction. Electrical conductivity is given by simplified formula [9] and thermal conductivity by [1]. The effective binary diffusion coefficient of iron in argon is estimated from the molecular viscosity assuming that the Schmidt number is almost constant and equal to 0.7 [8]. For the liquid iron bath, transport and thermodynamical properties are taken from [2].

Coupling and boundary conditions

As shown in figure 2, the coupling between the two domains is achieved with the help of the condition of transfer of current, heat and momentum at the Arc-Bath interface, according to an iterative scheme. A special treatment of the anodic boundary layer by using a one-dimensional model taking the iron evaporation into account ensures the electrical and thermal coupling between the arc column and the anodic surface.

Boundary conditions for plasma domain

On the walls, global heat transfer coefficients and external temperatures are considered [2, 16]. Those coefficients integrate the wall conductance and the internal convective coefficient. This last one is obtained from the flow computation with the help of the wall-functions.

At the gas inlet, a temperature of 300 °K and a parabolic velocity profile are imposed. The gas outlet is considered as an open boundary. There is of course no electrical current flowing through those boundaries as well as for the refractory walls. On the cathode tip, a zero normal derivative of the temperature and a zero electric potential are specified.

The radial velocity distribution on the bath surface is imposed and taken equal to the one obtained from the metal bath computation. A uniform electric potential, result of the arc computation through the adjustment of the current intensity is imposed at the plasma-metal interface.

Anodic layer treatment of the arc-plasma interface

In the anodic layer, because of the metallic vapour, the LTE is assumed. Furthermore, the partial pressure of the iron vapour at the interface is supposed to be a function of the local temperature according to liquid-vapour transition curve. The radial gradients are negligible compared to the gradients along the axis of the arc. This means that the thickness of the anodic layer is smaller than the radius of the arc. Under these assumptions, the problem is simply one-dimensional in the normal direction to the anode surface and the equations of temperature, mass fraction, continuity and current conservation are solved in order to give the temperature and iron mass fraction profiles in the anodic layer and to ensure electrical and thermal coupling between the arc column and the bath surface [10].

At the upper edge of the 1D anodic layer, the temperature, the iron mass fraction and the current density resulting from the 2D calculation are imposed. At the anodic wall, the metal bath surface temperature is imposed. The iron mass fraction is imposed and calculated as a function of the temperature and the iron vapour pressure.

This anodic model gives also the heat flux transmitted to the bath. According to [10], it can be evaluated from the enthalpy balance in the negative space charge zone near the bath surface and accounts for metal evaporation.

Boundary conditions for iron bath computation

At the bath surface, the heat flux density and the normal current density given by the one-dimensional model are imposed as well as the surface shear stress distribution calculated in the plasma domain. At the bottom of electrode (iron billet), the temperature is set to 773 °K and a zero constant potential is imposed. On the vertical surface of the bottom electrode a zero normal derivative of the temperature and current density are specified. The refractory walls are cooled by high speed forced air [16].

Results and discussion

The computations were carried out for an arc of 2000 A, 150 mm, 50 Nl Ar/min. The convergence was obtained after four plasma-bath iterations. The computed arc voltage is 90 V (not including electrodes falls), that seems rather realistic. The plasma and metal temperature fields are presented on figure 3. For the plasma domain, the computed temperature field is in accordance with those usually found in such argon arcs. The temperature in the bath seems to be probably higher and less homogeneous than the expected one. These results seem to be quite reasonable if one remember that radiative transfer are not included in our model : it would probably modify the plasma temperature field and consequently the bath temperature by changing heat exchange at the interface. On figure 4 we plot the iron vapour mass fraction in the plasma. Due to the very high velocity of the plasma, reaching 1400 m/s near the cathode and flowing from the cathode to the bath surface, the vapour emitted at the interface is transported to the wall and cannot diffuse in the high temperature region of the arc.

There are two main recirculation zones in the metal bath (see figure 5) : a Laplace forces dominated zone where the metal flows towards the bath centre in the vicinity of the arc impinging zone and a surface shear force dominated zone where the metal flows towards the wall in the vicinity of the arc impinging zone. This feature of the metal flow is in accordance with the results obtained by [8] in a transferred arc heated silicon bath. The characteristic velocity of some cm/s seems to be reasonable. The potential drop in the bath is very small due to the high electrical conductivity of the liquid metal.

The present study can be considered as a preliminary step in the modelling of transferred arc on a metal bath. In order to improve the heat transfer from the arc to the bath and also the furnace walls, our simulation has to be completed with a 3D radiative transfer modelling. Measurements will be soon achieved on the pilot furnace of EDF plasma laboratory in order to validate (or not) our computations.

3. Combustion of pulverised coal.

The recent increase in performance of computer hardware makes now possible the 3D simulation of coal combustion in a full-scale power plant. Numerical simulation can be a very efficient tool to study the effects of modifications in boilers (fuels, burner position and orientation, boiler geometry,...) on the general efficiency of the furnace and on the formation of pollutants. A specialised version of the 3D ESTET code [3] has been developed to model pulverised coal flames. In the case of industrial boilers we can assume a no-slip condition between gas and coal particles which is the case for the most part of the furnace, except possibly in the near field of the burners. With such an assumption, the equations for a particle-gas mixture with a mean density can be written. Even with this simplification, the conservation equations constitute a particularly complex system with a strong coupling between non-linear physical phenomena such as two-phase combustion, turbulence and radiation. One of the most difficult and challenging topic remains the turbulence-chemistry interaction for which we have to deal with a large number of species and reactions especially in the field of coal combustion. Furthermore, in an industrial power plant the flow presents some complex features due to burner geometry (swirling air) or injection configuration. Therefore if one wants to reproduce the complex 3D aerodynamics inside the chamber it is necessary to use a sufficiently refined mesh.

The combustion model developed for this specialised version of ESTET takes into account the pyrolysis of the particle and the heterogeneous combustion of the resulting char. A set of two parallel and competitive reactions are used to model the release of volatile matter. It is assumed that heterogeneous combustion cannot occur before pyrolysis is complete because volatiles when released are supposed to prevent the diffusion of gases at the surface of the particle. Char burnout is modelled using a global reaction rate that takes into account the kinetics of the heterogeneous reaction and the diffusion rate of oxygen at the surface of the coal particle. The gaseous combustion of the pyrolysed fuel and of the fuel produced by char burnout is modelled using a fast chemistry assumption [17]. Therefore the composition and the temperature of the gas phase are computed using a transport equation on the mixing rate and on the variance of its fluctuations. Turbulence is modelled assuming a given shape fort the Probability Density Function (a statistical beta function).

Radiation in industrial boiler is the dominant mechanism for thermal exchange. Gases are assumed to be grey and diffusion of radiation is supposed to be negligible. An n-flux model is used for radiation [13], acting on the gas phase and on the coal particles. The absorption coefficient of the gas phase is obtained after Modak [14] in terms of CO_2 and H_2O concentrations given by the gaseous combustion model. The absorption coefficient of the particles is given in terms of the void fraction of the particles and their mean diameter. In industrial applications it is necessary to use a sufficient number of rays along which the radiation transfer equation is integrated. We typically use 32 directions injected from each computational cell with an equal division of the solid angle (the same mesh is used for the flow computation) but sensibility tests have been performed with 128 directions.

Since the flow is very complex and a strict conservation of enthalpy and other scalar variables is required, the balance equations for scalars used for combustion modelling are solved in a strongly conservative form by a finite volume method [12]. The explicit scalar advective term is treated by a Quick/Upwind scheme which allows low numerical diffusion and preserves boundedness. The advection-diffusion source term equation for scalar increment is solved implicitly by an iterative under-relaxed Gauss Seidel algorithm. Large time step can be used as the scalar increment advective term is treated with an upwind scheme.

The Le Havre boiler is a 600 MW tangentially fired utility boiler. Important deformations due to temperature heterogeneity were observed locally on the cooling tubes that cover the walls of the chamber. The distribution of heat fluxes to the walls of the furnace is therefore an important parameter for an efficient running of the plant. Grouping of burners are located at each corner and at three levels of the chamber. Each grouping of burners has two pulverised coal burners with a primary air port for coal dust and secondary air ports for air staging. Burners are directed some degrees off the furnace diagonal in order to obtain a swirling movement intended to improve the gas particle mixture. The computational grid consisted of 80 000 active cells. Because of the small size of the burners, it was not possible to use Dirichlet conditions for inlet velocities with such a grid. Therefore source terms have been prescribed for momentum and also for scalar variables. However in this kind of complex 3D flow, it is important to perform a good prediction of the aerodynamics of the flow and in particular of the jet penetration inside the chamber, because the velocity distribution will actually condition the position of the flame, the mixing efficiency and therefore the heat exchange between the gas and the walls of the furnace. Therefore a mesh sensibility analysis has been performed on a cold flow (only air was injected) using a refined grid with 1 800 000 active cells and this time Dirichlet conditions were used for inlet air velocities. The cold flow has been computed using the two grids, and it was checked that using the coarse grid with source terms would give the same jet penetration when using the dense grid with Dirichlet conditions. Then a computation of the reacting flow in the whole boiler chamber up to the first heat exchangers has been performed. Figure 6 displays the velocity field and the temperature distribution in two horizontal sections of the boiler at full load. The distribution of the net heat fluxes on the walls of the boiler follow from the complex aerodynamic of the flow. From this computation it is obvious that under certain running conditions fairly heterogeneous fluxes can be obtained possibly leading to thermal damages on the cooling tubes.

Another example that shows the importance of a 3D tool is the study of the formation of pollutants. The Vitry power plant is a 250 MW wall-fired plant equipped with Low-NOx burners. Four raws of such burners designed to reduce the formation of NOx using the so-called "air-staging" technique are installed on the front wall (Figure 7). The equipment of primary NOx reduction is completed with two Over Fire Air injection levels located on the front side (burners side) and the rear side. Only one half of the boiler has been represented, the furnace being divided in two equal chambers by a vertical separation. Heat exchangers located just after the arch are not modelled.

A model for predicting the formation of Fuel-NO has been implemented in the ESTET pulverised coal version [4]. We have considered here only the fuel-NO mechanism because it is the major source of NO emissions from pulverised coal flames. It is assumed that HCN is the only intermediate in the formation of fuel-NO. NO is formed by oxidation of HCN and reduced competitively by reaction of NO with HCN. The reaction rates of these two competitive parallel reactions were determined by De Soete. The total mesh uses 40 000 active cells. Each burner is discretized with 25 cells, and each OFA uses 5 cells. The computed results have been compared to temperature, CO_2 O_2 and NO concentration measurements at two levels in the furnace (Figure 7). The measurements show that the actual flame length is somewhat longer than computed. This may be due to the limited knowledge of certain boundary conditions such as the swirling flow fields at inlets. The peak temperature levels are correctly predicted and a good agreement is obtained as far as NO emissions are concerned. The under predicted levels of NO may be attributed to the fact that thermal NO was not modelled, as this mechanism might be not negligible with a low-NOx burners fitted boiler.

A satellite version of ESTET for the study of combustion, heat transfer and nitrogen emissions in pulverised coal power plant has been developed, tested and applied to real cases. The submodels used for combustion have been chosen such that the most essential processes in the boiler can be simulated with reasonable accuracy. The last optimised version of the ESTET software allows to obtain a stationary state for the above computations with less than one hour CPU time on a Cray computer. However a large number of improvements are still needed in the model, in particular the possibility to use a finite number of size particle classes, and a better modelling of the turbulence-combustion interaction in the source terms. A refined mesh could be also useful for predicting the near-burner field, which is essential when dealing with NO emissions.

4. Stratified flow in a nuclear reactor containment building.

Thermal hydraulic numerical predictions are very useful for the detailed analysis of physical mechanisms which should appear in a containment building after a severe accident. In particular, when an accident like loss-of-coolant has occurred, stratification with high temperature gradient and high pressure may appear inducing severe stresses in the structures and local accumulation of hydrogen may take place leading to explosion. The aim of thermal hydraulic studies is to improve the understanding of physical phenomena which can appear into the containment. These phenomena can be various and complex, as flow compressibility, turbulence, two-phase flow aspect, combustion (explosion), possibility to have several species, condensation on walls,....

One objective of the numerical prediction of such flows is the optimisation of the heat exchangers by testing the influence of their position in the building to the global flow. Initially, the two-dimensional research code Mélodie have been applied to study such influence. The geometric configuration used for computations is a schematic geometry describing the upper part of a real building, supposed to be axisymmetric, and neglecting all the equipment such as steam generators, bridges,.... The computational grid is about 5000 nodes. Heat exchangers into the containment building are numerically simulated by sink mass and energy terms. The flow is supposed to be composed by air and saturated steam. Inlet condition near the symmetry axis corresponds to a residual power of 20 MW at 110°C and to 7.33 kg/s of steam.

Figure 8 shows the state flow after about 30 minutes. One can see that a thermal stratification have been created at the top of the building. The flow is mainly generated by the heat exchangers and not a lot by the inlet conditions. This stratification does not disappear. All the vertical positions on the heat exchangers gave the same qualitative result. One have to notify the importance of the behaviour law and the modelization retained for mass and energy

Another objective is to test alternative safety methods to decrease the temperature and homogenise the atmosphere such as spraying water injection. To remove stratified zones, cold water is sprayed from the top of the building, therefore creating a two-phase recirculating flow. The two-phase flow satellite version of Mélodie, named Mélodif, have been applied to such studies. Mélodif solves separate Eulerian conservation equations which are formulated for both phases including interphase transfer terms (mass, momentum and enthalpy). Turbulence fields in the continuous phase are predicted by means of k-epsilon eddy viscosity model. Modelling of the dispersed phase turbulence is achieved by an extension of results obtained in the framework of the Tchen's theory for dispersion of discrete particles transported by homogeneous turbulent fluid flow.

The previous schematic configuration have been used for computations. The initial state corresponds to the stratified state previously obtained. Three annular injections of cold water are performed at the top of the building. The total rate is 280 kg/s. The mean diameter of water droplets is 100 microns and their temperature is 20°C. Walls are supposed to be adiabatic. The total computation represents 70 real seconds. Steam condenses upon the drops as they fall trough gaseous mixture. As shown in figure 9, the stratification is quickly destroyed and the mixture becomes more and more homogenised. This is due mainly by the gravitational effects generated by the temperature gradients. The condensation does not contribute very much to the destruction of the stratification.

More recently, the three-dimensional industrial code ESTET-ASTRID have been applied to compute transient flow configurations in the containment building. A satellite version, named ESTET-IC, is dedicated to such computations. The application presented below mainly concerns the behaviour of hydrogen and the possibility to create local concentration. The flow is single phase and composed by six species which are saturated steam and non condensable gases such as H_2, CO, CO_2, N_2 and O_2. Each specie is numerically treated by solving a transport-diffusion equation. The mixture is supposed to be a perfect gas. The flow is weakly compressible (there are no outlet) and non reactive. Inlet boundary conditions are given by applying the global code MAAP to the lower part of the building. It gives flow rates, concentration of each specie, temperature and pressure. Two main inlets are represented : the annular inlet (70 m^2) and the inlets under the steam generators (about 100 m^2). Due to the small size of the annular inlet, source terms have been applied for momentum and species, instead of Dirichlet conditions. The characteristic of such simulation is the slow transient phenomenon, dominated by natural convection.

As shown in figure 10, the grid used always represents the upper part, but takes now into account the main part of the equipment (bridge, steam generators, pool). This grid consists of 196 000 nodes. The results presented on figures 11 and 12 represent the state of the flow after about 100 minutes. They show the existence of a quasi centred jet which is strong enough to go from the bottom to the top, and generate important convection rolls which homogenise the atmosphere. Concerning hydrogen, the local gradients are weak, but the concentration level increase slowly ; at the last time of the simulation, its maximum level is 0.359 % and its minimum level is 0.355 %. This computation does not take into account physical mechanisms such as steam condensation, heat exchange with the structure, transport and deposition of aerosols, destruction of stratification by spraying water. Further developments in the ESTET code are planed to account these phenomena.

5. Conclusion.

In this paper three applications of variable density turbulent flow are presented. They illustrate the kind of applications EDF has to deal with. The state of the art in the development of physical modelling and numerical techniques enables to apply numerical simulation to these complex and highly non linear situations. More over validation on these complex configurations is a difficult process : even if each physical model implemented in the code is validated individually on representative configurations where detailed data are available, only global validation can be done on the real situation where all the processes are coupled. Nevertheless, results of interest from an engineering point of view are obtained, even if some improvements of physical models are still necessary.

References
/1/ Adachi, Inaba, Amakawa, *10th ISPC Bochum*, August 1991.
/2/ Bouvier, Delalondre, Simonin, Brilhac, *Proceedings of ICHMT Seminar*, July 1994.
/3/ Dalsecco, Méchitoua, Simonin, *EDF report*, HE/44/95-009-A, 1995
/4/ Dalsecco, Méchitoua, Simonin, *EDF report*, HE/44/96-015-A, 1996
/5/ Delalondre, *Ph. D. Thesis, University of Rouen*, France, 1990.
/6/ Delalondre, Zahrai, Simonin, *Proceedings of ICHMT Seminar*, July 1994.
/7/ Essoltani, Proulx, Boulos, Gleizes, *Plasma Chem. and Plasma Proc.*, Vol. 14, n°4, 1994.
/8/ Gu, *Ph. D. Thesis*, University of Trondheim, Norway 1993.
/9/ Ichimaru, Tanaka, *Physical Review A*, pp. 1790-1798, April 1985.
/10/ Kaddani, *Ph. D. Thesis*, University of Orsay, France, 1995.
/11/ Launder, Sharma, *Letters in Heat and Mass Transfer*, Vol. 1, pp 131-138, 1974.
/12/ Mattei, Simonin, *EDF report*, HE/44/92-038-A, 1992
/13/ Méchitoua, EDF report, HE/44/87-015-A, 1987
/14/ Modak, *Fire Research*, Vol. 1 ,pp 339-361, 1978
/15/ Paik, Nguyen, *Int. J. Heat Mass Transfer*, Vol. 38, n°. 7, pp. 1161-1171, 1995.
/16/ Trenty, Bouvier, Delalondre, Simonin, Guillot, *12th ISPC Minneapolis*, August 1995.
/17/ Viollet, *EDF report*, HE-44/85.08, 1985.

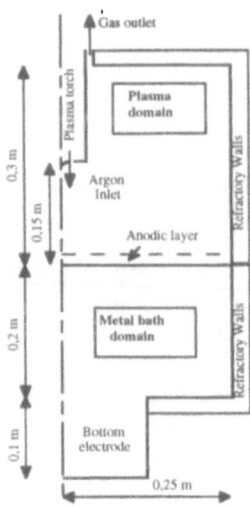

figure 1 : computational domain

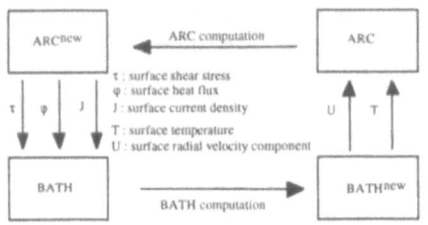

figure 2 : coupling procedure

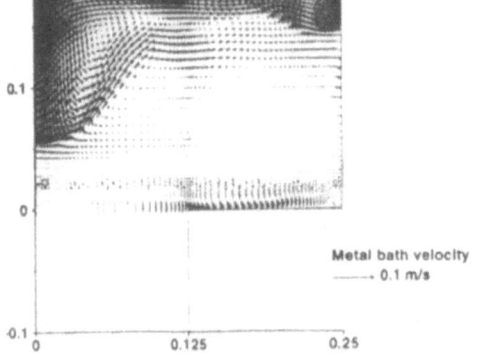

figure 3 : temperature fields both in the plasma and in
the metal bath domains

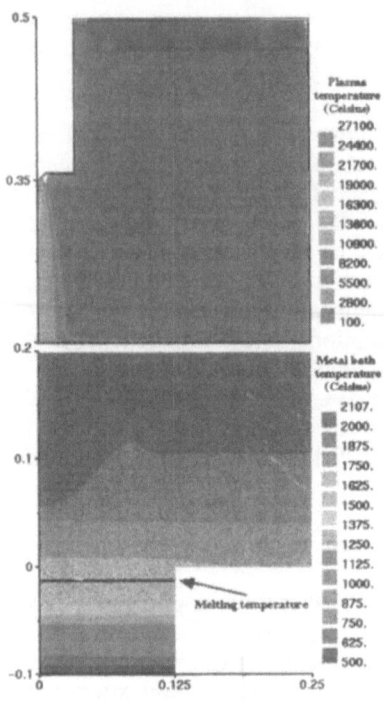

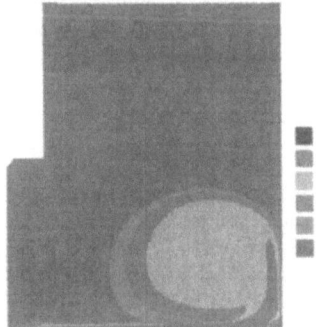

figure 4 : iron vapour mass fraction in the plasma

figure 5 : velocity vectors in the bath domain

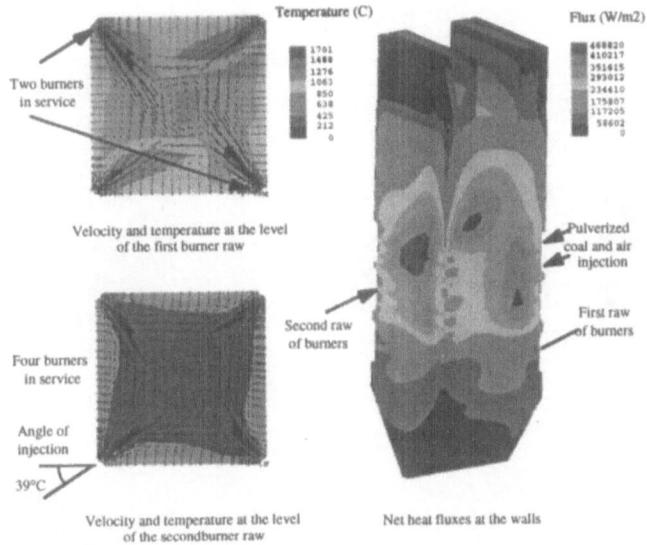

Figure 6 : Simulation of the 600 MW corner-fired Le Havre boiler

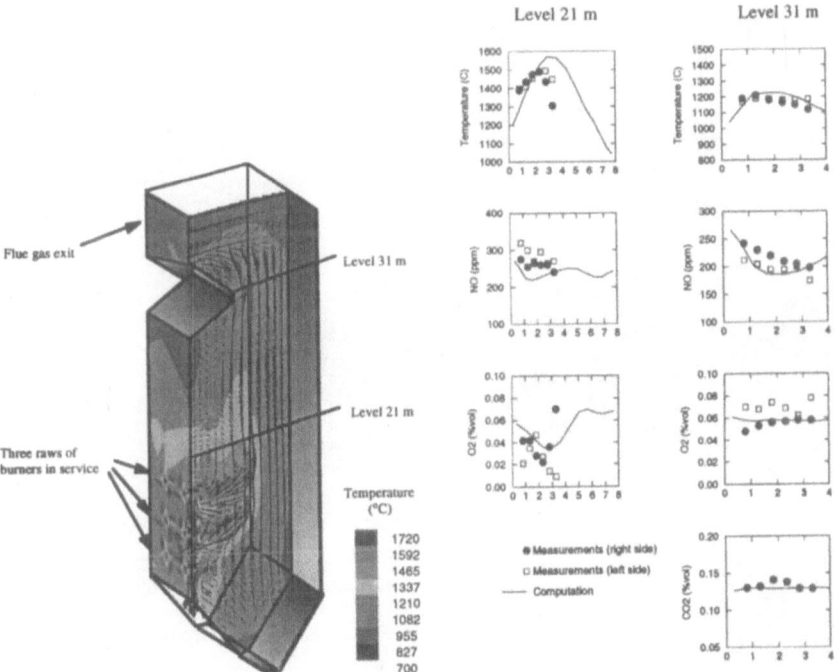

Figure 7 : Simulation of the 250 MW wall-fired Vitry boiler

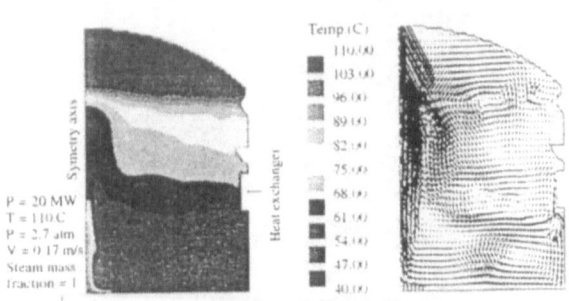

P = 20 MW
T = 110 C
P = 2.7 atm
V = 0.17 m/s
Steam mass
fraction = 1

Figure 8 : Heat exchangers influence on fluid motion
Temperature and velocity fields

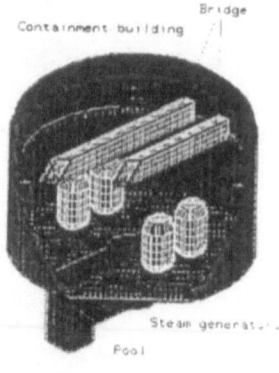

Figure 10 : 3D configuration and mesh
a containment building

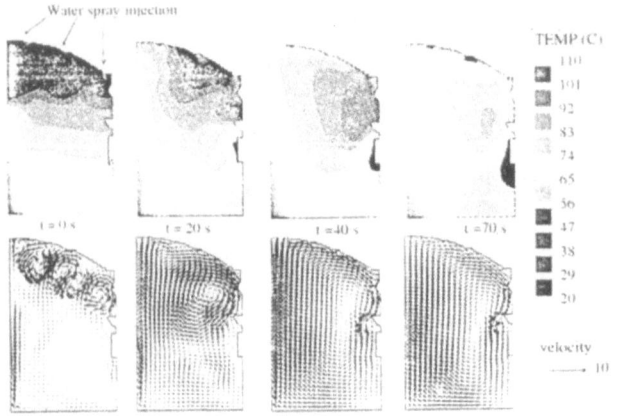

Figure 9 : Two-phase flow computation : Temperature and velocity of the
continuous phase at different times

Figure 11 : hydrogen and velocity
fields on a plane cut of a
containment building

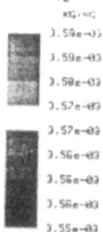

Figure 12 : hydrogen and velocity fields at
different planes cut of a containment
building after 1h 40 mn.

V. Experiments and Measurement Methods

OPTICAL DIAGNOSTICS FOR FLOWS WITH DENSITY VARIATIONS

R.B. MILES, W.R. LEMPERT, J. FORKEY, N.D. FINKELSTEIN,
and S.R. HARRIS
Department of Mechanical & Aerospace Engineering
PRINCETON UNIVERSITY
Princeton, New Jersey 08544 USA

1. Abstract

Two new molecular-based diagnostic methods show great promise for the measurement of turbulence in variable density, low-speed flow fields. The first of these methods is Rayleigh imaging, which is capable of capturing two-dimensional instantaneous cross sections of the flow field density. Through the use of atomic or molecular filters in front of the camera, Rayleigh imaging can be extended to the measurement of temperature and velocity. Flow tagging yields velocity profiles in variable density flow fields and can be used to measure the details of the turbulent structure. These two techniques may be applied simultaneously to capture instantaneous velocity and density fields with high resolution.

2. Introduction

Diagnostic approaches based on molecular scattering and molecular tagging are particularly well suited to the measurement of flows with density fluctuations. This is because the molecules represent a true constituent of the fluid, so molecular scattering and molecular tagging sample the fluid itself. Errors associated with the effects of density variation on a sample probe such as a hot wire or on particle trajectories are not present. Much work has been done using laser-induced fluorescence from molecules such as nitric oxide [1] and OH [2], which are naturally present in hot and combusting flows, but which must be seeded into cold flows. The interpretation of the laser-induced fluorescence signal levels from these molecules is complicated by collisional phenomena, and these effects are particularly troublesome in flows with density fluctuations. Laser-induced fluorescence from oxygen molecules has also been considered for flow diagnostics in air [3]. This approach is attractive since oxygen is strongly predissociative, and, therefore, not subject to quenching. As a result, the oxygen laser-induced fluorescence signal level is a direct measure of the population of oxygen molecules in a particular ground rotational state. Due, in part, to the rapid predissociation, and, in part, to weak absorption, the laser-induced fluorescence signal from oxygen at room temperature is very low. Signals from vibrationally excited oxygen are much stronger because of stronger coupling to the upper electronic level. As a consequence, laser-induced fluorescence from oxygen has been useful in high

L. Fulachier et al. (eds.), IUTAM Symposium on Variable Density Low-Speed Turbulent Flows, 245–256.

temperature flows and as an interrogation step for RELIEF flow tagging, where a line or a pattern of vibrationally excited molecules is created by Raman Excitation (RE) and imaged with oxygen Laser-Induced Electronic Fluorescence (LIEF).

3. Rayleigh Imaging

In order to avoid the problems associated with quenching and seeding, we have explored the application of Rayleigh scattering for quantitative imaging of gaseous flows with density fluctuations. Rayleigh scattering has the advantage of being directly proportional to the density of the gas and lasts only for that period of time during which the gas is illuminated. With high-power, short-pulsed lasers, instantaneous cross-sectional images of the density field can be captured. The drawback of Rayleigh scattering is the low scattering cross section per molecule. For example, oxygen and nitrogen have a scattering cross section of about $3 \times 10^{-27} cm^2$ at 589 nm (orange--the sodium light wavelength) [4]. This cross section is, in effect, the size of the shadow cast by each molecule. At STP (2.69×10^{19} molecules/cc), only one out of every 10^7 photons is scattered per centimeter of path length. Of those photons, only approximately one in 10^4 are usually detected with a camera. This is because of the limited solid angle of the camera collection optics, optical losses in the collection system, and the quantum efficiency of the detector. This scattering can be significantly improved by using ultraviolet light, since the scattering cross section scales as the fourth power of the frequency and as the square of the polarizability. For example, at 193 nm (ArF laser), the total Rayleigh cross section increases to $4 \times 10^{-25} cm^2$ (the differential cross section is $5 \times 10^{-26} cm^2$ /steradian) [5]. High energy laser pulses are also important, since more photons incident produce more photons scattered, and Rayleigh scattering is not subject to saturation. A second drawback of Rayleigh scattering is that the scattered light is at the same frequency or close to the same frequency as the incident source laser. This means that other direct scattering processes, such as scattering from windows, walls, and particles, can obscure the Rayleigh scattering. Thus, the scattering volume must be carefully protected from stray laser light, and the field-of-view of the camera must be masked to minimize the background scattering.

High quality Rayleigh images of compressible flows have been taken in the ultraviolet with argon-fluoride lasers at 193 nm and with frequency-quadrupled Nd:YAG lasers at 266 nm. Figure 1 shows a pair of instantaneous images of the Mach disk region in an underexpanded supersonic air jet taken with an ArF laser. This image clearly show the large increase in density behind the shock, the reflected shock structure, and the turbulent density fluctuations in the free shear layer.

The utility of Rayleigh scattering for density measurements is limited by the low signal level and by interference from background scattering. In the shot noise limit, the signal-to-noise ratio is equal to the square root of the total number of photons collected per resolvable volume element. Thus, for 1% accuracy, at least 10,000 photoelectrons must be collected. This must be done in a single laser shot if the instantaneous density field is to be measured. In essence, this requirement limits the resolution and the field-of-view. For example, a 100 mJ laser at 193 nm, expanded to a 1 cm wide by 100μ thick sheet, gives about 10^4 photoelectrons detected per cubic 100 micron element over a 1 cm by 1 cm field-of-view. Higher resolution will require a smaller field-of-view, a higher energy laser, or a smaller signal-to-noise ratio. Thus flows with large density gradients, such as the one shown in Fig. 1, are easily imaged. Those with smaller density gradients, including low-speed flows with thermally induced density gradients, are more difficult to see.

An alternative method of enhancing the Rayleigh scattering is by using molecules with large Rayleigh scattering cross sections such as Freon [6], or by using a fog of small particles. In supersonic flows there is a fog of condensate particles which occurs naturally in the cold portions of the flow and arises from the condensation of trace species such as water vapor and CO_2 [7]. Since these particles evaporate in the boundary layer and behind strong shocks, they are useful as tools to observe boundary layer and shock structure. For example, Fig. 2 shows two sequential images of a Mach 3 turbulent boundary layer separated in time by 20 μsec [8]. A large boundary layer feature is clearly evident, and its displacement and evolution during this time interval can be observed. Similar images taken at up to a million frames per second have been captured in a small Mach 2.5 supersonic flow using a new pulsed-burst laser and a MHz rate CCD camera [9]. Figure 3 shows a composite of 25 images, taken sequentially at 2 μsec intervals in this flow. The right-hand side of each image is at the wall, flow is from bottom to top, and the scattering is from a dry ice (CO_2) fog created by introducing a small amount of CO_2 into the air far upstream of the nozzle.

Rayleigh scattering can be augmented through the use of optical filters to remove background scattering. This is especially so for high speed flows where the Doppler shift associated with the average motion of the flow is large compared to the Doppler shifts associated with the thermal and acoustic motion of the individual molecules. In this case, a narrow linewidth laser is tuned to overlap the absorption line of a molecular or atomic vapor. A cell filled with that vapor is placed in front of the camera so that light scattered from non-moving objects such as windows and walls is eliminated from the field-of-view. Light scattering from the flow, on the other hand, is shifted in frequency and passes through the cell. This application is shown in diagrammatic form in Fig. 4, and has been used to suppress background scattering for measurements in supersonic inlets [8] and supersonic free jets [10].

For more quantitative measurements, the filter may be used to discriminate the spectral character and spectral shift of the Rayleigh-scattered light [11]. This is accomplished by tuning the laser and observing the scattered light through the atomic or molecular absorption cell as a function of the laser wavelength. This use of the filter is shown diagramatically in Fig. 5. The laser frequency is carefully monitored with respect to a reference laser. The point of maximum absorption occurs when the Rayleigh-scattered light most completely overlaps the absorption feature in the cell (point #3 on the diagram). By measuring the laser frequency and comparing it with the frequency of the filter, the shift is known and the velocity of the flow is measured. This can be done for each resolvable element in the image, and, so, the velocity field can be captured. In a similar fashion, the line broadening of the scattering is indicative of the temperature of the flow. Thus, by observing the rate at which the scattering is cut-off by the absorption filter as the laser is tuned in frequency, the temperature can be determined. Since the total scattering is proportional to the density, these measurements yield velocity, temperature, and pressure. Images taken in a Mach 2 pressure-matched free jet give the velocity, temperature, and pressure profiles shown in Figs. 6, 7, and 8 [12]. For these experiments, the laser was tuned and therefore the flow fluctuations were time averaged. For instantaneous measurements, a single pulse laser can be used together with a multiple set of filters observing the same scattering region.

4. Flow Tagging

Molecular flow tagging is another approach to diagnostics and can be used to complement Rayleigh scattering for more accurate measurements of transport properties. In the RELIEF flow tagging process [13], oxygen molecules are "instantaneously" pumped into their vibrationally excited state by stimulated Raman scattering. This is achieved with a high power, short-pulsed, dual-color laser beam, which is focused to a line through the sample volume. Since oxygen is a homonuclear diatomic molecule, the vibrationally excited state cannot easily radiate its energy, and the molecules remain excited for a relatively long period of time. For example, in pure oxygen they will remain excited for tens of milliseconds [14]. In air, the excitation lifetime is usually determined by the partial pressure of water vapor, and in humid air, the vibrational lifetime may be on the order of 10 μsec or so. Figure 9 shows the lifetime as a function of the water mole fraction for atmospheric pressure air at 300 K [15]. A short time after tagging, the displaced molecules are interrogated by laser-induced electronic fluorescence using an argon-fluoride laser (193 nm). This same laser may be simultaneously used to observe Rayleigh scattering for density measurements. The underexpanded supersonic jet shown in Fig. 1 is reproduced in Fig. 10, this time with a line written into it and interrogated after 5 μsec. The displacement of the line is a quantitative measure of the velocity profile across the jet.

This flow tagging approach has been used to measure velocity profiles in supersonic flows as well as in subsonic flows. The displacement of the line gives an immediate visual appreciation of the velocity profile, and as the time interval between tagging and interrogation goes to zero, the displacement is a true measurement of the velocity profile. It is important to recognize that the line is absolutely straight when it is written, so its position can be found to high accuracy. The displaced line center is found using a Gaussian fitting routine and is typically determined to better than a tenth of a pixel resolution, which usually corresponds to on the order of 2-3 microns. Of course, some displacement must occur in order to measure the velocity, so the time interval must be finite. In general, this finite time interval is chosen to be short compared to the fastest eddy roll-over time in the turbulent flow.

As an example of the utility of this line marking approach, a series of experiments were conducted in a free air jet [16]. Table I shows the conditions for three of these experimental runs (A,B,C). Measurements were taken approximately 40 diameters downstream where the Reynolds number, based on the Taylor microscale, was on the order of 575. The approximate eddy roll-over time at this location was 30 μsec, and the time interval between tagging and interrogation was 5 or 7 μsec, less than one-quarter of the eddy turn-over time. A typical image of the initial tagged line and one after 7 μsec is shown in Fig. 11. In this case, the average velocity was 43 m/sec, and the turbulence intensity was 12 m/sec. Approximately 5,000 lines were analyzed for each run and used to compute the first through sixth order structure functions:

$$S_n(\Delta x) = \overline{\left|u(x) - u(x + \Delta x)\right|^n} \tag{1}$$

Since odd order transverse structure functions are zero by symmetry, the magnitude of the difference has been used. The first through sixth order structure functions vs. Δx are shown in Fig. 12, where the smallest increment is 30μ, corresponding to the resolution of the optical system. These structure functions have been normalized to one at the inertial scale and they show the characteristic logarithmic slope region in the inertial subrange. The portion of the inertial subrange with a linear logarithmic slope extends over approximately one decade before dissipation steepens the slope at small scales. The logarithmic slopes of the first, second, third, fourth, and sixth order structure functions are shown in Fig. 13. Here it is clear that the low order structure functions are reasonably flat over the range between approximately 700 microns and 7 mm. Higher order structure functions are noisy, since they are very susceptible to low probability events. Application of the extended self-similarity method [17], where other structure functions are plotted against the third order structure function, suppresses this noise and extends the region of linearity over four decades. Figure 14 shows the logarithmic derivatives of the ESS plots for the first, second, fourth, fifth, and sixth order structure functions. The extended linear region and reduced noise are apparent when this plot is compared to Fig. 13. Structure function slopes measured in this manner are consistent with those measured using hot-wire probes under similar conditions.

Since the RELIEF technique observes the transverse velocity profile, vortices have a particularly clear signature: a positive and negative excursion from a straight line. Thus, the scale and strength of violent events, even down to scales on the order of the Kolmogorov scale, can be observed with this technique. The probability density distribution function (PDF) of the transverse velocity increment shows that violent events greater than one-half the RMS fluctuations occur across a distance of 1.8 times the Kolmogorov length with a probability of greater than 10^{-4} (Fig. 15). It seems probable that these events correspond to those cases where the line has been tagged through the core of a vortex filament. The frequency of occurrence of these events is consistent with the view that a volume on the order of the integral scale contains one such filament with a length comparable to the integral scale [16].

A similar technology has been developed for observing velocity and vorticity structure in water. This approach is called the PHANTOMM technique (PHoto-Activated Nonintrusive Tracking Of Molecular Motion) and is accomplished by mixing a small amount (mole fractions of several parts per billion) of a caged dye molecule into water [18]. When illuminated with ultraviolet light, the cage is broken and any time after that, the molecule acts as a normal fluorescent dye. Tagging, therefore, is accomplished by using an ultraviolet laser, such as a frequency-tripled Nd:YAG laser, to write a pattern into the fluid, and subsequently using a blue laser, such as an argon-ion laser or pulsed dye laser, to interrogate. Fluorescence is in the yellowish-orange region of the spectrum and is easily imaged with a standard videocamera. An image of a scries of lines written at 1 second intervals into a stationary cylinder filled with water and closed with a rotating top is shown in Fig. 16. The interrogation laser is formed into a sheet containing the initial tagged line. As that line rotates out-of-plane, it appears to become shorter and then longer with a 180° turn. The slowly rotating top draws the fluid up in the center and pushes it out and down on the edges, as is evident from the image. This experimental set-up is being used to study vortex stability and vortex breakdown since the geometry is particularly compatible with cylindrical computational coordinates [19,20,21]. A comparison of the line displacement with the computed result is shown in Fig. 17.

5. Summary

Rayleigh scattering and molecular line tagging are two approaches which may be utilized in a straightforward manner to generate both qualitative and quantitative images of flows with density variations. A wide range of scales can be simultaneously captured, and the images give unambiguous measurements of density and velocity. For density measurements using Rayleigh scattering and velocity measurements using molecular tagging, the character of the flow field is immediately apparent without further data analysis. High repetition rate laser sources allow this data to be captured at rates which are in excess of the highest frequency fluctuations in the flow. Rayleigh scattering can be further augmented with the use of molecular or atomic filter technologies to extract temperature and velocity fields. These same filters are useful for the suppression of background scattering from windows and walls in high-speed flows. Molecular tagging gives a very precise quantitative measure of velocity, so it is particularly useful for the study of flow physics and the validation and verification of computational codes and fluid models.

6. Acknowledgments

The Rayleigh scattering work was supported by the Air Force Office of Scientific Research (Dr. Julian Tishkoff), NASA-Lewis (Dr. Richard Seasholtz), and NASA-Langley (Dr. Richard Antcliff). The pulse-burst laser technology was supported by the Air Force Office of Scientific Research (Dr. Leonidas Sakell) and the New Jersey Photonics and Opto-Electronics Materials Center (POEM) (Dr. Stephen Forrest). Molecular flow tagging work was supported by the National Science Foundation (Dr. Roger Arndt) and a NATO collaborative research grant (CRG-92-0480).

7. References

1.   Lee, M.P., McMillan, B.K., and Hanson R.K., "Temperature Measurements in Gases by use of Planar Laser-Induced Fluorescence Imaging of NO," *Appl. Opt.* **32** (1993) 5370-5396.

2.   Palmer, J.L., Hanson, RK., "Temperature Imaging in a Supersonic Free Jet of Combustion Gases with Two-Line OH Fluorescence," *Appl. Opt.* **35** (1996) 485-499.

3.   Lee, M.P., Paul, P.H., and Hanson, R.K., "Quantitative Imaging of Temperature Fields in Air Using Planar Laser-Induced Fluorescence of Oxygen," *Opt. Lett.* **12** (1987) 75-77.

4.   Shardanand, R., Prasad, and Rao, R.D., "Absolute Rayleigh Scattering Cross Sections of Gases and Freons of Stratospheric Interest in Visible and Ultraviolet Regions," (1977) NASA TN-D-8442.

5.   Miles, R.B., Connors, J.J., Howard, P.J., Markovitz, E.C., and Roth, G.J., "Proposed Single-Pulse Two-Dimensional Temperature and Density Measurements of Oxygen and Air," *Opt. Lett.* **13** (1988) 195-197.

6.   Yip, B., Schmitt, R., and Long, M.B., "Instantaneous Three-Dimensional Concentration Measurements in Turbulent Jets and Flames," *Opt. Lett.* **13** (1988) 96-98.

7. Shirinzadeh, B., Hillard, M.E., and Exton, R.J., "Condensation Effects on Rayleigh Scattering Measurements in a Supersonic Wind Tunnel," *AIAA J.* **29** (1991) 242-246.

8. Forkey, J., Cogne, S., Smits, A., and Bogdonoff, S., "Time-Sequenced and Spectrally Filtered Rayleigh Imaging of Shockwave and Boundary Layer Structure for Inlet Characterization," AIAA-93-2300 (1993).

9. Lempert, W.R., Wu, P.F., Zhang, B., Miles, R.B., Lowrance, J.L., Mastrocola, V., and Kosonocky, W., "Pulse-Burst Laser System for High-Speed Flow Diagnostics," AIAA 96-0179 (1996).

10. Miles, R., Lempert, W., Forkey, J., Finkelstein, N., and Erbland, P., "Quantifying High-Speed Flows by Light Scattering from Air Molecules," AIAA-94-2230 (1994).

11. Forkey, J.N., Finkelstein, N.D., Lempert, W.R., and Miles, R.B., "Demonstration and Characterization of Filtered Rayleigh Scattering for Planar Velocity Measurements," *AIAA J.* **34** (1996) 442-448.

12. Forkey, J., "Development and Demonstration of Filtered Rayleigh Scattering--A Laser Based Flow Diagnostic for Planar Measurement of Velocity, Temperature, and Pressure," Princeton University, MAE Technical Report #2067 (1996).

13. Miles, R.B., Connors, J.J., Markovitz, E.C., Howard, P.J., and Roth, G.J., "Instantaneous Profiles and Turbulence Statistics of Supersonic Free Shear Layers by Raman Excitation + Laser-Induced Electronic Fluorescence (RELIEF) Velocity Tagging of Oxygen," *Expts. in Fluids* **8** (1989) 17-24.

14. Frey, R., Lukasik, J., and Ducuing, J., "Tunable Raman Excitation and Vibrational Relaxation in Diatomic Molecules," *Chem. Phys. Ltrs.* **14** (1972) 514-517.

15. Diskin, G.S., Lempert, W.R., and Miles, R.B., "Observation of Vibrational Relaxation Dynamics in $X^3\Sigma^-_g$ Oxygen Following Stimulated Raman Excitation to the v=1 Level: Implications for the RELIEF Flow Tagging Technique," AIAA 96-0301 (1996).

16. Noullez, A., Wallace, G., Lempert, W., Miles, R.B., and Frisch, U., "Transverse Velocity Increments in Turbulent Flow Using the RELIEF Technique," (Submitted to J. Fluid Mechanics, May 1996).

17. Benzi, R., Ciliberto, S., Baudet, C., and Ruiz Chavarria, "On the Scaling of Three-Dimensional Homogeneous and Isotropic Turbulence," *Physica D* **80** (1995) 385-398.

18. Lempert, W.R., Magee, K., Ronney, P., Gee, K.R., and Haughland, R.P., "Flow Tagging Velocimetry in Incompressible Flows Using PHoto-Activated Nonintrusive Tracking Of Molecular Motion (PHANTOMM)," *Expts. in Fluids* **18** (1995) 249-257.

19. Harris, S.R., Miles, R.B., and Lempert, W.R., "PHANTOMM Flow Tagging Measurements in Complex 3-D Flows," AIAA 96-1966 (1996).

20. Escudier, M.P., "Observations of the Flow Produced in a Cylindrical Container by a Rotating Endwall," *Expts. in Fluids* **2** (1984) 189-196.

21. Brown, G.L., and Lopez, J.M., "Axisymmetric Vortex Breakdown. Part 2: Physical Mechanisms," *J. Fluid Mech.* **221** (1990) 553-576.

TABLE I: Jet Conditions

Run	Exit Dia. (cm)	Plen Press (psi)	Exit Vel. (m/s)	Tag Loc. Dia.	Time Delay (µs)	Lines Used	Line Lgth pixel	Avg. Vel. (m/s)	Turb. Int. (m/s)	Taylor Micro- scale (µ)	Kol. Scale (µ)	R_λ
A	1	10	278	38	5	5617	344	48.5	12.1	686	14.4	589
B	1	10	278	38	7	5249	344	42.7	11.9	720	14.9	605
C	1	10	278	38	5	5578	344	41.6	12.7	594	13.1	534

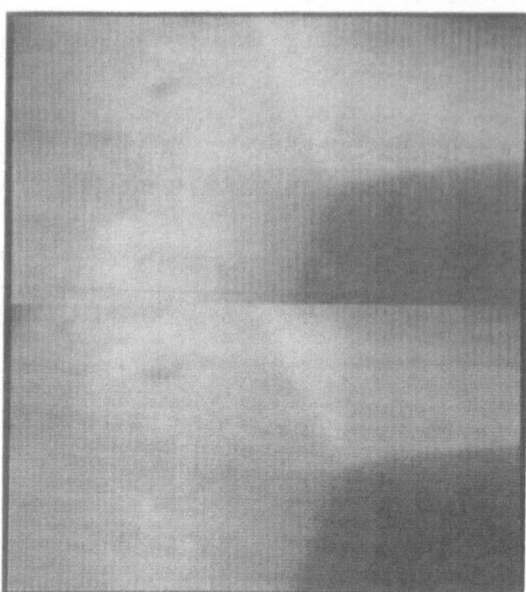

Figure 1.

Rayleigh cross sections of a 6 mm diameter underexpanded free air jet taken with argon-fluoride laser (193 nm) at the edge of the Mach disk.

Figure 2.

Sequential Rayleigh images of condensate fog highlighting the boundary layer of a Mach 3 flow. Images are separated by 20 μsec.

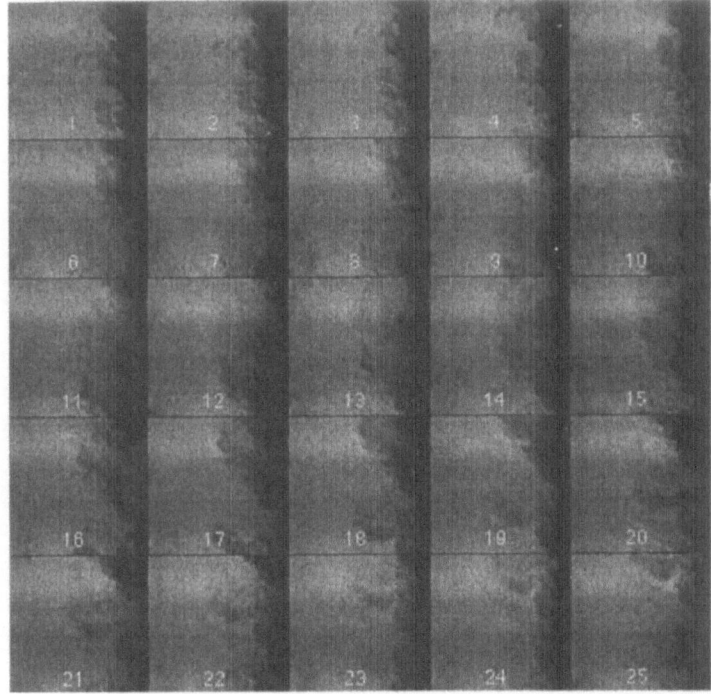

Figure 3.

Images of a Mach 2.5 turbulent boundary layer taken at 500,000 frames per second.

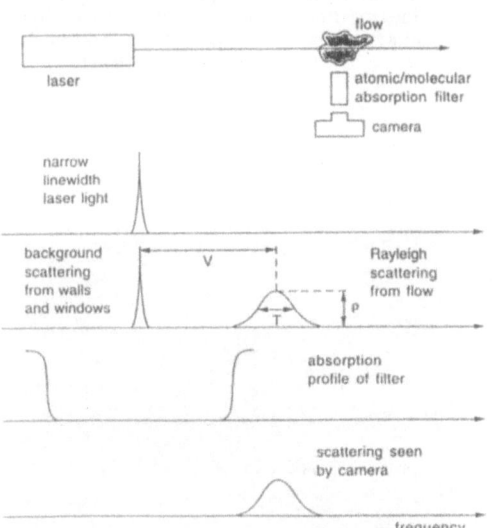

Figure 4. Diagram of background suppression, molecular filter for Rayleigh Scattering.

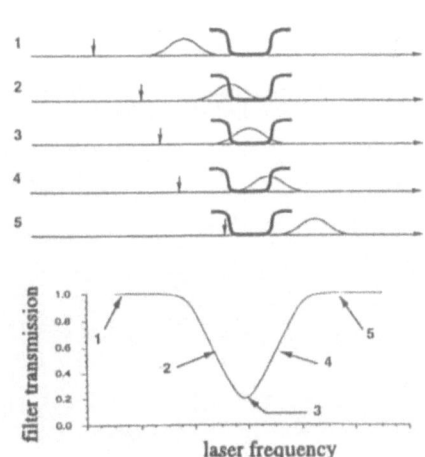

Figure 5. Diagram of velocity, temperature, and pressure measurements with molecular filter.

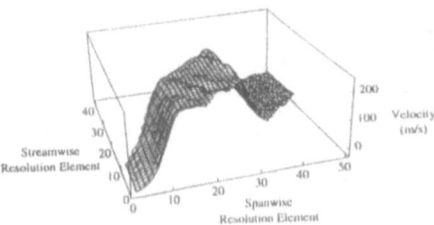

Figure 6. Velocity contour across a Mach 2 pressure-matched free jet.

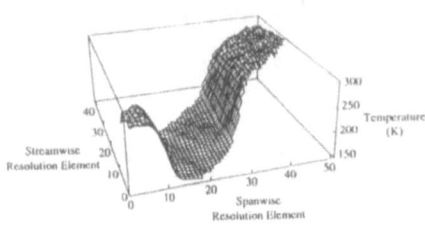

Figure 7. Temperature contour across a Mach 2 pressure-matched free jet.

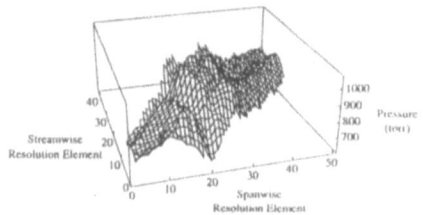

Figure 8. Pressure contour across a pressure-matched Mach 2 free jet.

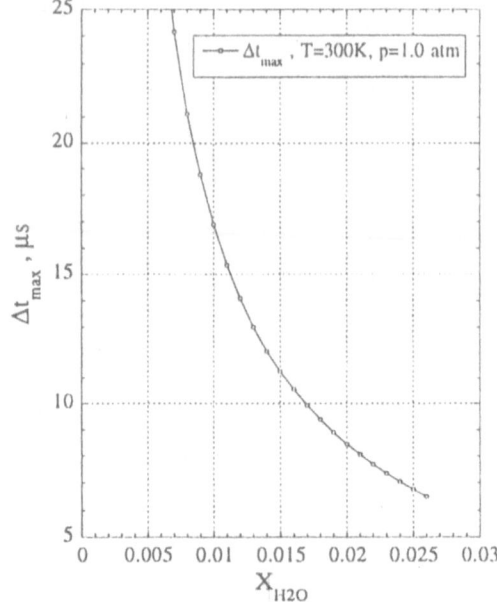

Figure 9. Maximum time between tagging and interrogation for a RELIEF line at 1 atm, 300 K, as a function of water vapor mole fraction. Time is to 10% of maximum signal.

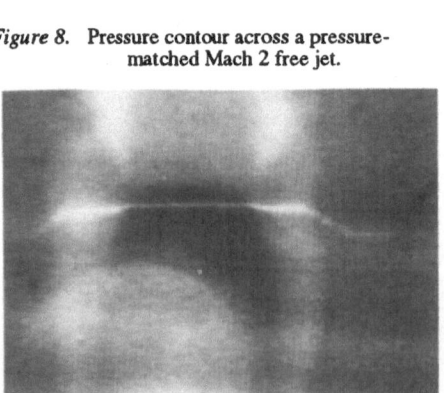

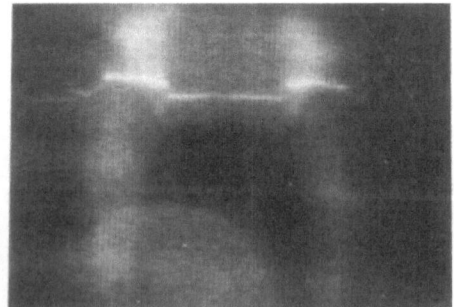

Figure 10. Six millimeter (6 mm) diameter underexpanded free air jet with simultaneous Rayleigh scattering and RELIEF lines. The lines are written into the flow 5 μsec before interrogation.

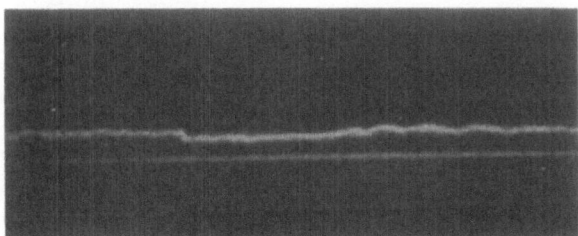

Figure 11.

Original tagged line and line
displaced after 7 μsec in a turbulent
free air jet.

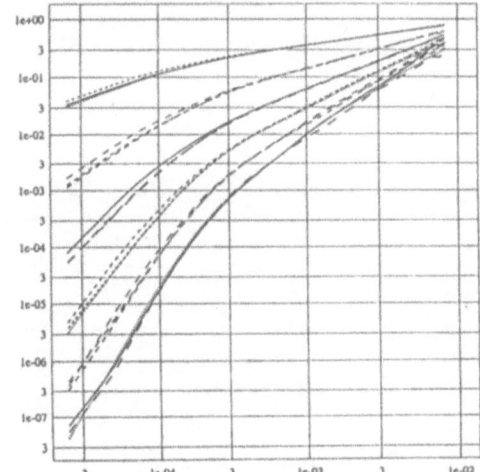

Figure 12.

First (top) through sixth (bottom)
order structure functions vs. Δx for
the turbulent free air jet.

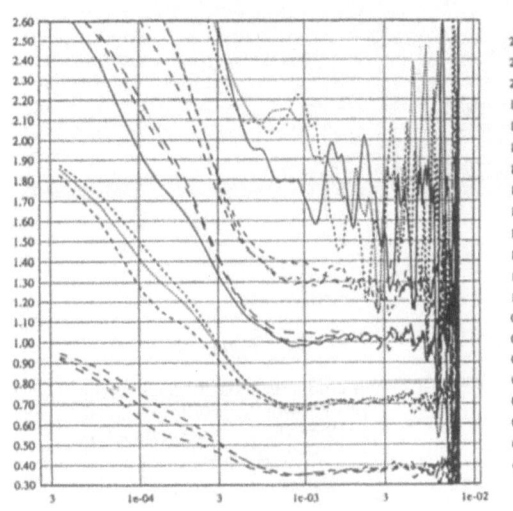

Figure 13.   Logarithmic slope of the first,
second, third, fourth, and sixth order structure
functions.

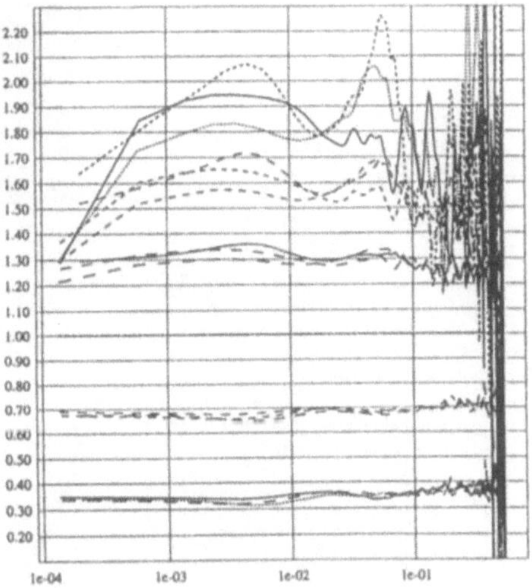

Figure 14.   Logarithmic slopes of the first,
second, fourth, fifth, and sixth order structure
functions plotted against a third order structure
function showing extended self-similarity.

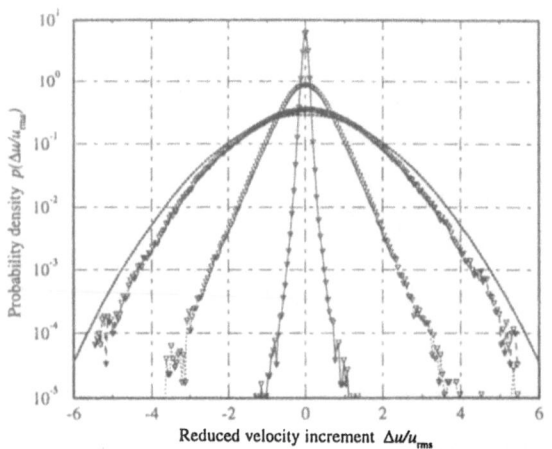

Figure 15.

PDF of the velocity difference at three different transverse spacings.

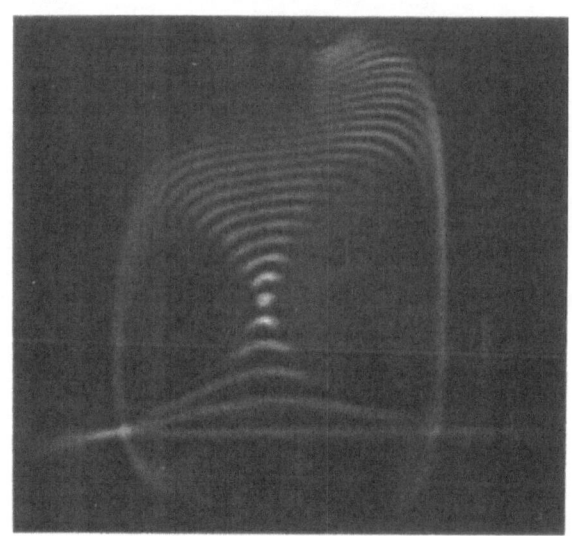

Figure 16.

Set of lines written at 1 second intervals into water contained in a stationary cylinder with a rotating top. Lines are interrogated with a pulsed dye laser focused to a sheet so they appear shorter as they rotate out-of-plane.

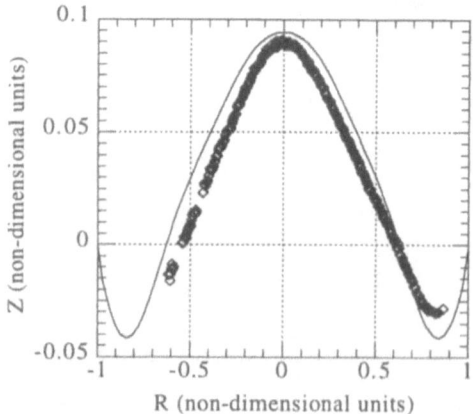

Figure 17.

A comparison of the axial velocity profile with the profile predicted for the cylinder with rotating top. Re = 996, H/R = 2.0.

LARGE SCALE CHARACTERISATION OF THE CONCENTRATION FIELD OF SUPERCRITICAL JETS OF HYDROGEN AND METHANE

E. RUFFIN, Y. MOUILLEAU and J. CHAINEAUX

Institut National de l'Environnement Industriel et des Risques
Parc Technologique Alata, BP 2, 60550 Verneuil en Halatte, FRANCE

1. Introduction

When an orifice or breach appears in the wall of a tank containing flammable gas under pressure, a jet is created which develops into an explosive cloud. Research has shown that the intensity of the explosion likely to take place in this cloud is then highly variable and depends on the cloud characteristics, such as the concentration of combustible material, the velocity field and turbulence. The experimental study developed at INERIS[1] sought to characterise the clouds formed by supercritical jets of methane and hydrogen, and the overpressures resulting from ignition of the jets at different points. Only the work on the concentration fields will be reported here. The parameters determining the composition of the explosive cloud in the experiments carried out were: the gas used (methane or hydrogen), the vent orifice diameter (25, 50, 75, 100 or 150 mm) and the time t after the commencement of venting since the tank is of finite size and does not produce steady flow conditions. The volume of the tank ($5 \ m^3$) and the pressure and temperature conditions - 40 bar and 288 K - inside it prior to the onset of venting were kept constant.

2. Theory

In the case of variable density subsonic jets, the concentration field can be described as a zone of pseudo-self-preservation. It is now clearly established that the decrease along the axis is a hyperbolic function of the distance from the orifice and that it depends strongly on the ratio (R_ρ) of the densities of the discharged gas (ρ_j) and the surrounding gas (ρ_a), and the discharge diameter D_j.

---

[1] Work carried out at INERIS as part of the European EMERGE project (Extended Modelling and Experimental Research into Gas Explosions), funded jointly by the European Economic Community and the French Ministry for the Environment

L. Fulachier et al. (eds.), IUTAM Symposium on Variable Density Low-Speed Turbulent Flows, 257–264.
© 1997 *Kluwer Academic Publishers.*

The equivalent discharge diameter D_{eq} allows a unique expression of the axial concentration profiles in these jets as follows:

$$\frac{C}{C_j} = \frac{1}{B} \frac{D_j}{X - X_C} R_\rho^{1/2} = \frac{1}{B} \frac{D_{eq}}{X - X_C} \quad with \quad D_{eq} = D_j R_\rho^{1/2} \quad and \quad C_j = 1 \quad (1a, b)$$

where X_c is the virtual abscissa of the hyperbolic decrease and X is the distance from the orifice (discharge section). This behavior was noticed by Thring and Newby (1953) and Abramovich (1963) and later confirmed by Chen and Rodi (1980), Pitts (1991) and Djeridane (1994). In fact for a given discharge velocity, the equivalent diameter D_{eq} can be interpreted as the nozzle diameter where a jet of density ρ_a should be such that its scalar flux $N_j = \rho_j U_j C_j D_j^2$ is the same as that of a jet of density ρ_j discharged through an orifice of diameter D_j.

In the case of supercritical jets, the discharge section (nozzle) is a sonic throat. At this point, the static pressure P_{NOZ} is greater than or equal to the critical pressure P_{crit} ($P_{crit} = 2 P_{atm}$ approximately). Since the static pressure in the discharge section is thus greater than atmospheric pressure, there is a sudden expansion of the jet (expansion zone) along which the pressure in the jet returns to ambient pressure. Instead of considering the supercritical jet from its actual discharge section, it is then possible to define fictional jet emission conditions just after the expansion zone, where the pressure in the jet is once again equal to the ambient pressure. Accordingly, Birch et al. (1984) define a fictional jet, starting after the expansion zone and having new characteristics of diameter D_{fic}, velocity U_{fic} and density ρ_{fic}. Birch et al. (1984) suggested the following fictional diameter[2]:

$$D_{fic} = D_j \sqrt{C_d \frac{P_{t0}}{P_{atm}} \left(\frac{2}{\gamma + 1} \right)^{\frac{\gamma+1}{2(\gamma-1)}} \left(\frac{T_{atm}}{T_{t0}} \right)^{1/4}} \qquad (2)$$

This diameter is obtained from considerations of mass flow conservation between the real discharge section and the fictional discharge section, putting the fictional velocity equal to the velocity of sound in the gas making up the jet at ambient conditions of pressure and temperature, and by putting the fictional density equal to the density of the gas under the same conditions. Notice that in the experiments carried out at INERIS, the control parameters P_{t0}, T_{t0} and also D_{fic} were variable during venting of the tank due to the finite dimension of it. Since the total pressure in the tank during venting was measured in every case, we calculated the discharge coefficient defined as follows:

$$C_d = \frac{\dot{m}}{\dot{m}_{is}} = \left(\frac{V}{RT_{t0}} \frac{dP_{t0}}{dt} \right) \Bigg/ \left(\sqrt{\gamma P_{t0} \rho_{t0}} \; \pi \frac{D_{col}^2}{4} \left(\frac{\gamma + 1}{2} \right)^{\frac{\gamma+1}{2(1-\gamma)}} \right) \qquad (3)$$

---

[2] P_{to} is the total pressure in the tank, P_{atm} is atmospheric pressure and C_d the discharge coefficient.

where $\dot{m}_{is}$ is the isentropic mass flow and $\dot{m}$ the real mass flow which is related to the pressure gradient measured in the tank.

3. Experimental rig and operating conditions

The experimental rig (figure 1) consists of a 5 m^3 test tank and a horizontal discharge pipe fitted with an orifice at its outer end the experimental nozzle diameter of which could be varied. Since the jets formed by the venting of this tank can extend axially for about 100 metres, the rig was placed on the edge of a low cliff so that the axis of the horizontal jets is 5 metres above the ground.

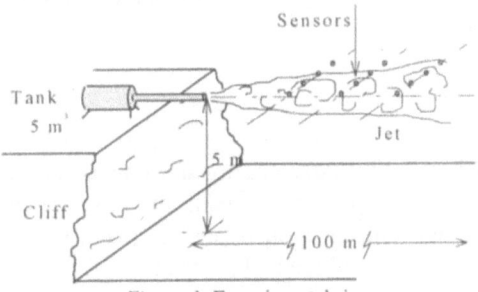

Figure 1: Experimental rig

With this arrangement it can be assumed that the ground has no effect on the development of the jets, at least for the first 50 metres. The concentration sensors are placed in the subsonic part of the jet, mounted on thin cables positioned perpendicularly to the flow.

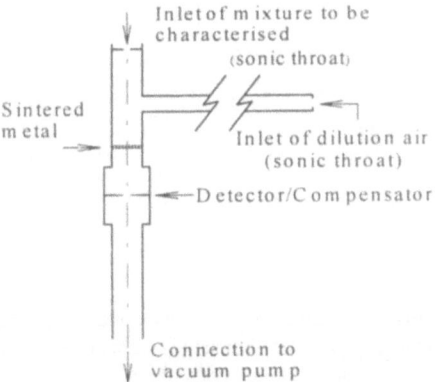

Figure 2: Diagram of a concentration sensor

Each sensor (figure 2) consists of a pellistor, a metal filament heated by an electric current and coated with a catalytic material. When a mixture of air and flammable gas comes into contact with the material, oxidazation takes place and the temperature of the filament rises. The voltage gradient needed to keep the filament at a constant temperature (about 873 K) can be easily related to the gas concentration in the mixture. In the measuring sensors developed and used at INERIS, the mixture goes through a sonic throat which prevents any flame from propagating to the outside of the sensor which can then be used in highly flammable environments without risk.

Twelve of these sensors were built and calibrated for CH_4-air and H_2-air mixtures. Validation tests on the measurement system (monitoring concentration changes) showed that these sensors were capable of following large changes in the concentration of these gases in air, without being affected by changes in the pressure of the mixture analysed, with a short response time and with no risk of igniting the mixture.

4. Results

In each test[3], which involved characterising the concentration field of the explosive cloud obtained by venting the pressurised tank, the gas concentration was measured at different points in the subsonic portion of the jet (M < 0.3), along the axis and also transversely, while venting continued. The results of the measurements shows the change in the concentration of gas (methane in this case) as a function of time as the tank empties. The measurements show that the axial attenuation of the mean mass concentration is a hyperbolic function of X (its reciprocal is linear) and that it can therefore satisfy the relation given at (1).

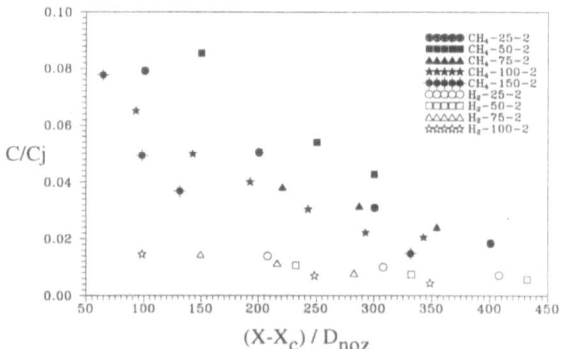

Figure 3 : Axial decrease of mass concentration as a function of distance from the discharge section normalised by the relevant diameter (D_{noz}) in supercritical jets of methane and hydrogen about two seconds after the onset of venting

---

[3] tests were made with D_{noz} equal to 25, 50, 75, 100 and 150 mm for CH_4 jets and 25, 50, 75 and 100 mm for H_2 jets

Figure 3 shows that the raw values of axial mass concentration obtained with the different jets are scattered when the concentrations are plotted as a function of distance from the nozzle normalised by the nozzle diameter. Hence this diameter is not representative of the concentration field. The results given on figure 4 show that the fictional diameter proposed by Birch *et al.* (1984), calculated for our experiments with the equation 3, does in fact group the methane jet data around one hyperbola and the hydrogen jet data around another.

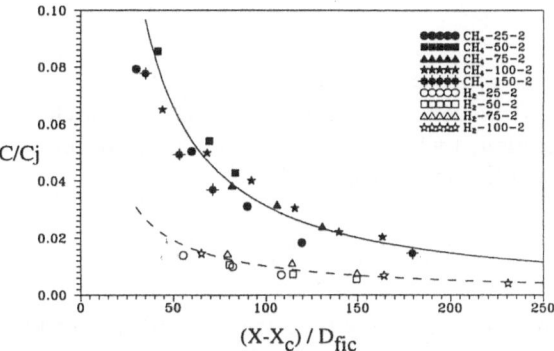

Figure 4 : Axial decrease of mass concentration as a function of distance from the discharge section normalised by the fictional diameter (D_{fic}) in supercritical jets of methane and hydrogen about two seconds after the onset of venting.

This shows that the correct allowance has been made for the effect of the pressure ratio (P_{to}/P_{atm}) and that the assumption of successive steady states which allows us the calculation of $\dot{m}$ is quite good. Moreover, if allowance is made for the ratio between

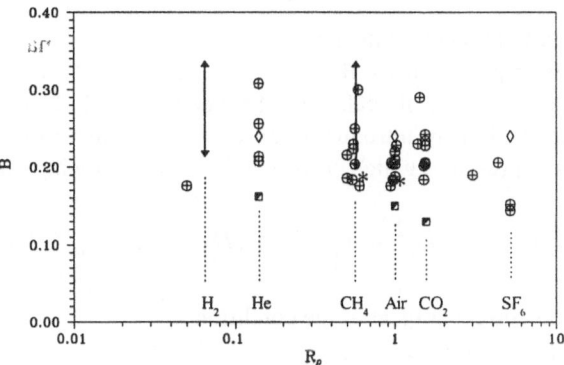

Figure 5: 'Universal' decay rate of mass concentration along the axis of subsonic jets [$B = (C_j/C) D_j/(X-X_c)(\rho_j/\rho_a)^{1/2}$] and supercritical jets [$B = (C_j/C) D_{fic}/(X-X_c)(\rho_{fic}/\rho_a)^{1/2}$]. Results for supercritical jets: ◄───►, present results; *, experiment of Birch *et al.* (1984). Results for subsonic jets: ⊕, compilation of experiments by Fulachier *et al.* (1990); ■, IMST experiment (Djeridane, 1994); ◊, calculations using a second order turbulence model (Ruffin, 1994).

the fictional densities and the density of the ambient fluid, then all the results lie along a single curve for which the decay rate B is around 0.27 (figure 5) as in the subsonic jets. These results for supercritical unsteady jets of methane and hydrogen confirm those found by Birch *et al.* (1984) for supercritical steady jets of natural gas and ethylene. They are also a good validation of the original measurement technique developed by INERIS and used in an industrial-type situation, i.e., on a very large scale.

5. Discussion

The fictional diameter proposed by Birch *et al.* (1984) is in fact based upon the hypothesis of mass flow conservation between the real discharge section and the fictional discharge section at which the ambient conditions of pressure and temperature are restored. Thus the following equation holds:

$$C_d \rho_j U_j D_j^2 = \rho_{fic} U_{fic} D_{fic}^2 \tag{4}$$

This equation does not differ formally from the one which leads to the law of pseudo-self similarity of a scalar quantity in variable density subsonic jets (equation 1), except that the underlying physical hypothesis in this case is not mass flow conservation but conservation of the scalar flux (Chen and Rodi 1980) or:

$$\rho_j U_j C_j D_j^2 = \rho_a U_j C_j D_{eq}^2 \tag{5}$$

from which the equations (1) and (2) can be established.

Thus although it was obtained on the basis of the hypothesis of mass flow conservation, the fictional diameter proposed by Birch *et al.* (1984) in the case of supercritical jets also verifies the conservation of scalar flux between the real discharge section and the fictional discharge section by introducing an additional stage to take into account the effects of changes in pressure, which corresponds to the following double equality:

$$C_d \rho_j U_j C_j D_j^2 = \rho_{fic} U_{fic} C_j D_{fic}^2 = \rho_a U_{fic} C_j D_{eq}^2 \tag{6}$$

Equation (2) is then correct subject to the condition:

$$D_{eq} = D_{fic} \left(\frac{\rho_{fic}}{\rho_a} \right)^{1/2} \tag{7}$$

Consequently the fact that a common decay rate is found for the attenuation of the scalar in supercritical jets and in variable density subsonic jets merely expresses the validity of one and the same conservation law which is that of the flux of the scalar quantity.

Finally, the works of Thring and Newby (1952) and Chen and Rodi (1980) showed that it is the momentum flux $M_j = \rho_j U_j C_j D_j^2$ which determines the actual decrease of the longitudinal velocity in variable density subsonic jets. Recently Ruffin *et al.* (1994) showed that it is also M_j which determines all the scales of characteristic length scales of turbulence in these jets. If this applies to supercritical jets and if the assumptions of Birch *et al.* (1984) concerning the density and velocity of the fictional jet are applied, we then have a fictional diameter representative of the velocity field different from that which is representative of the scalar field, i.e.:

$$C_d \rho_j U_j U_j D_j^2 = \rho_{fic} U_{fic} U_{fic} D_{fic}^{*2} \tag{8}$$

whence:

$$D_{fic}^* = D_j \sqrt{C_d \frac{P_{t0}}{P_{atm}} \left(\frac{2}{\gamma+1}\right)^{\frac{\gamma}{2(\gamma-1)}}} \tag{9}$$

Thus the hypotheses of conservation of scalar fluxes and of momentum do not lead to an unequivocal definition of the fictional diameter for the scalar quantity and the longitudinal velocity.

6. Conclusion

First of all, the large scale experimental tests carried out at INERIS confirm the results obtained by Birch *et al.* (1984) in a laboratory. In this way it was possible to validate, in an industrial-type situation and using highly reactive mixtures, a new system with short response time for measuring concentrations of flammable gases.

Secondly, it is by no means certain that the fictional diameter proposed by Birch *et al.* (1984) is representative of changes in quantities other than the mass concentration.

7. References

ABRAMOVICH, G.N. (1963) *Theory of turbulent jets*, MIT Press.

CHEN C.J. & RODI W. (1980). *Vertical turbulent buoyant jets - A review of experimental data.* The science & appl. of heat and mass transfer, Pergamon Press.

BIRCH A.D., BROWN D.R., DODSON M.G., SWAFFIELD F. (1984). The structure and concentration decay of high pressure jets of natural gas, *Combustion Science and Technology*, vol. 36, pp. 249-261.

DJERIDANE T.(1994).*Contribution à l'étude expérimentale de jets turbulents axisymétriques à densité variable*, Thèse de Doctorat, Université d'AixMarseille II.

FULACHIER L., ANSELMET F., AMIELH M. (1990). *Quelques résultats sur les écoulements subsoniques à masse volumique variable*, 27ème Colloque d'Aérodynamique Appliquée, Marseille.

PITTS W.M. (1991). Effect of global density ratio on the centerline mixing behavior of axisymmetric turbulent jets, *Experiments in Fluids*, vol. 11, pp. 125-134

RUFFIN E., SCHIESTEL R., ANSELMET F., AMIELH M. & FULACHIER L. (1994) Investigation of characteristic scales in variable density turbulent jets using a second-order model, *Physics of Fluids*, vol. 8, n°6, pp. 2785-2799.

RUFFIN E. (1994) *Etude de jets turbulents à densité variable à l'aide de modèles de transport au second ordre*, Thèse de Doctorat, Université d'AixMarseille II.

THRING M.W. & NEWBY N.P. (1953) *Combustion length of enclosed turbulent jet flames*, 4th International Symposium on Combustion, Pittsburgh.

UNSTEADY MEASUREMENT OF STATIC PRESSURE, VELOCITY AND TEMPERATURE IN THE VICINITY OF THE NOZZLE IN A VACUUM-WIND-TUNNEL

C. KIRMSE, B. SAMMLER AND H.E. FIEDLER
Technische Universität Berlin
Hermann–Föttinger–Institut für Strömungsmechanik
Fachbereich 10, Sekr. HF 1
Straße des 17. Juni 135, D–10623 Berlin, Germany

1. Introduction

Simultaneous measurements of different quantities, e.g. velocity, pressure, temperature and concentrations of mixed species in turbulent flows are difficult and often require combination of different methods. In this investigation a combination method utilizing hot–wire anemometry (HWA) and laser Doppler velocimetry (LDV) is presented which enables simultaneous measurement of temperature, velocity and static pressure. This principle provides possibilities for application also in inhomogeneous flows.

Investigations of the response of hot–wire anemometers including calibration studies have shown, that it is necessary to combine hot–wire signals with other information if unambiguous results for velocity and static pressure are asked for (Sammler *et al.*, 1993). In many cases the temperature is also a variable. So further complications can be expected. To reduce the problems we decided to combine HWA with LDV. The advantage of LDV is its near independence on static pressure and temperature.

2. Basic equations

<u>Temperature:</u> The signal voltage E_T of an unheated wire depends only on the total temperature T of the medium and may be written as a linear function, which is valid over a wide range of temperatures for very low electrical current (C_1 and C_2 are calibration constants):

$$E_T = C_1 + C_2 T \qquad (1)$$

265

L. Fulachier et al. (eds.), IUTAM Symposium on Variable Density Low-Speed Turbulent Flows, 265–268.

Velocity: The Doppler frequency f_D of the LDV output is directly proportional to the velocity component u in the main stream direction. The factor
C_3 depends only on laser wavelength and beam geometry:

$$u = C_3 \, f_D \qquad (2)$$

Static pressure: The fundamental hot–wire response–equation for the bridge
supply voltage E reads

$$E^2 = \left(\frac{T_D - T}{T_D - T_B} \right) \left[\left(0.83 + 0.17 \, \frac{p}{p_B} \right) A_B + \frac{B_B}{\sqrt{p_B}} \, \sqrt{p} \, \sqrt{u} \right] , \qquad (3)$$

with the temperature of the wire T_D, the fluid temperature during calibration T_B, the static pressure p, the static pressure during calibration p_B, the
constant in King's law A_B according to p_B and T_B and the slope in King's
law B_B under conditions of p_B and T_B.

Limitations for T, p and u:

$$
\begin{array}{rcl}
280 \, K & \leq T \leq & 300 \, K \\
50 \, kPa & \leq p \leq & 100 \, kPa \\
0.5 \, m/s & \leq u \leq & 40 \, m/s
\end{array}
$$

Eq. (3) was found to be valid for many probes of different geometry. The
pressure response changes considerably at static pressures below $20 \, kPa$.

3. Interpretation of signals

With the calibration parameters momentary values of temperature and
velocity are obtained by means of eqs. (1) and (2). By rewriting eq. (3) as
a quadratic equation of $\sqrt{p}$, the momentary value of the static pressure p
can be calculated by solving eq. (4):

$$
\begin{aligned}
0 &= \left(\frac{0.83 \, A_B \left(\frac{T_D - T}{T_D - T_B} \right) - E^2}{0.17 \, A_B \left(\frac{T_D - T}{T_D - T_B} \right)} \right) p_B + \left(\frac{B_B \, \sqrt{u} \, \sqrt{p_B}}{0.17 \, A_B} \right) \sqrt{p} + (\sqrt{p})^2 \\
&= a + b \, \sqrt{p} + (\sqrt{p})^2
\end{aligned}
$$

$$
p = \left(-\frac{b}{2} + \sqrt{\frac{b^2}{4} - a} \right)^2 , \qquad \text{where} \quad a < 0 \qquad (4)
$$

4. Preparation of the facility

To accomplish complex data aquisition in dynamic experiments simultaneous parallel digital recording of the signals of all channels must be presumed.
LDV signal bursts occur randomly depending on the seeding rate of scattering particles, which must be high enough to ensure sufficient sampling
rate. Thus it is necessary to synchronize all sampling events.

The actual experiment was done in a vacuum–wind–tunnel, which permits variation of the static pressure. The opto–mechanical and electronic
part of LDV was installed inside of the tunnel. Inside mounted hot– and
cold–wire probes were connected to their external implements by vacuum-
proof openings. The sensitive parts of LDV, temperature– and hot–wire
probe formed a sampling volume of about $1\,mm^3$ on the axis of a jet near
the nozzle.

After emptying the facility it was refilled abruptly via a large valve near
the center of the vessel. Thus, extreme fluctuations in temperature, velocity
and static pressure, could be expected.

5. Experimental results

During the first seven seconds of the experiment the measurement was carried out under steady conditions with the static pressure set at $70\,kPa$
and the temperature near $296\,K$. This provided informations about the error margins. A flush wall–mounted pressure–transducer with low frequency
characteristic was used to verify the results.

After opening the valve, air streamed into the vessel. During compression (first phase) the energy level of the gas in the vessel is increased and
consequently its temperature rises (fig. 1, top). The sudden change of the
pressure in the vessel causes a stepwise redirection of the velocity (fig. 1,
center). The temporal developement of the static pressure appears reasonable with regard to the fluctuations and the agreement between short
time averages and low frequency transducer results (fig. 1, bottom). The
fan, driving the jet, was rpm–controlled. Consequently, the fluid velocity is
slightly higher at the end of the experiment.

6. Conclusions

Dynamic measurements of velocity, temperature and static pressure are
possible by simultaneous combination of HWA and LDV readings. Assuming typical error limits it is evident that fluctuations of static pressure can
be measured by this method only via the compressibility of the medium.

According to eq.(3) the sensitivity $\partial E^2/\partial p$ decreases if p increases. Improvements may be expected at higher overheat–ratios.

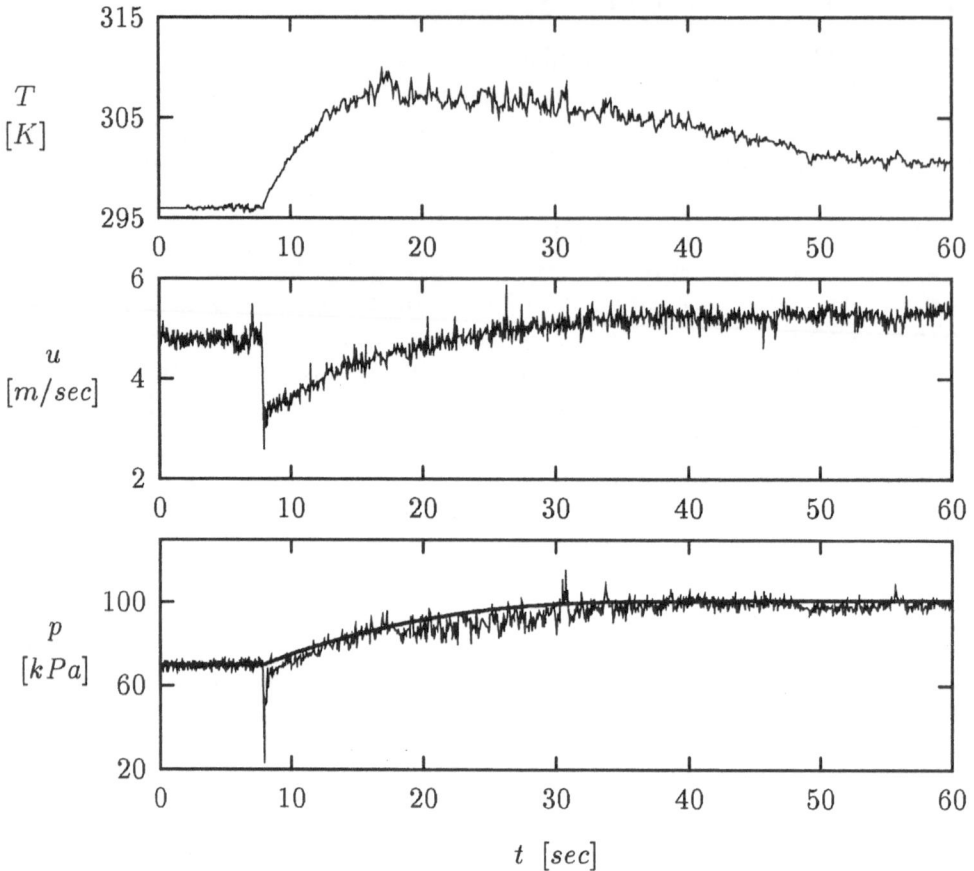

Figure 1. Time series of temperature T, velocity u and static pressure p (—) compared with transducer measurements (—)

Hot– and cold–wire probes can be exposed to very low static pressure. With reference to LDV it is advisable to use high–performance fibre–optical components to compensate for worse seeding rates and to remove all opto–mechanical elements from the interior.

Measurements by means of 2–dimensional Laser Doppler Velocimeter combined with multi hot–wire probes are under way.

References

Sammler, B., Eschenhagen, R., Graichen, K., Kitzing, H., and Seifert, G. (1993) Multi–Hot–Wire–Probes for Investigations of Turbulence, *Proceedings ASME "Thermal Anemometry 1993"*, **FED–Vol. 167**, pp. 231–240

IMPROVEMENT OF THE RESPONSE OF A FINE COLD-WIRE FOR THE MEASUREMENT OF HIGH-FREQUENCY TEMPERATURE FLUCTUATIONS

J. LEMAY AND A. BENAISSA
Laboratoire de mécanique des Fluides
Département de génie mécanique, Université Laval
Québec, Canada, G1K 7P4

1. Introduction

A number of investigations has been devoted to the determination of wire time constants (cf. the review in Brunn [1]). Nevertheless, one must notice that a limited number of contributions has been addressing this problem for very fine wires (diameter < 1 μm); LaRue *et al.* [2], Fiedler [4], Antonia *et al.* [3] and Paranthöen *et al.* [5] are among the contributors to that subject. For wire diameters larger than 1 μm, most of the results reported in the literature are in good agreement with the theoretical values. However, for the finest wires (0.63 and 0.25 μm) the forementioned authors have measured time constants larger than the theoretical ones. In the present paper, the time constant of 1 and 0.5 μm diameter wires are obtained *via*: 1) a square-wave current injection producing a Joule effect heating 2) a chopped laser beam producing a radiant square-wave heating, and 3) a tracer of heated air convected in the wake of a 5 μm wire excited with a voltage pulse. These time constants are measured as a function of the cooling velocity, the wire being exposed to the potential core of an axisymmetric air jet. Once the time constant of a wire is determined, we compensate its response by applying a numerical post-processing technique.

2. Experimental procedure

The schematics shown in figure 1 and 2 describe the dynamic calibration bench used for the determination of time constants by the 3 forementioned

L. Fulachier et al. (eds.), IUTAM Symposium on Variable Density Low-Speed Turbulent Flows, 269–272.
© *1997 Kluwer Academic Publishers.*

methods. The idea is to calibrate a given probe by the three methods during the same experiment. This ensures that we compare the methods under the same experimental conditions.

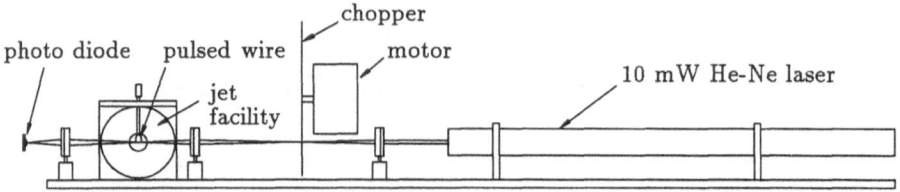

Figure 1. Dynamic calibration set-up.

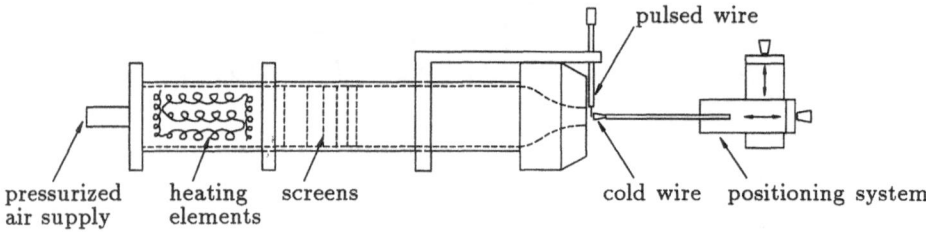

Figure 2. Jet facility (nozzle diameter of 16.5 mm).

The etching of the Wollaston wire is performed by an electrolysis process with a 5% solution of nitric acid and a dc current source which can be fine tuned. The 0.5 and 1 μm diameter wires are etched over a 0.5 and 1 mm length respectively ($l/d \simeq 1000$). A good precision on the actual diameter is necessary if one aims at comparing the time constants obtained from the experiments and the theory. The wire diameters are measured by a scanning electron microscope (JEOL 840 A) having an optimal resolution of 40 Å. The actual diameters are found to be 0.58 and 1.20 μm respectively.

3. Data acquisition and processing

The cold-wire probes are operated using in-house constant current anemometers. These CCA are characterized by the use of a mirror current source providing good dynamic performances. Hence, we achieve a high frequency modulation of the current which allows us to produce a very sharp square-wave current excitation (integrated to our CCA modules). Current levels of 0.10 and 0.15 mA are used with the 0.5 and 1 μm wires. The time constant measurements are made with a dual channel digital oscilloscope HP-54603B which has a maximum sampling rate of 20 Msample/s, an 8-bit resolution and avaraging capabilities of 8, 64 or 256 samples. Figure 3 shows an example of the oscilloscope traces. Such traces allows us to perform regression analysis in order to obtain least square estimates of the time constants. Non linear techniques are used to best fit the data.

a) current injection   b) laser excitation   c) pulsed wire

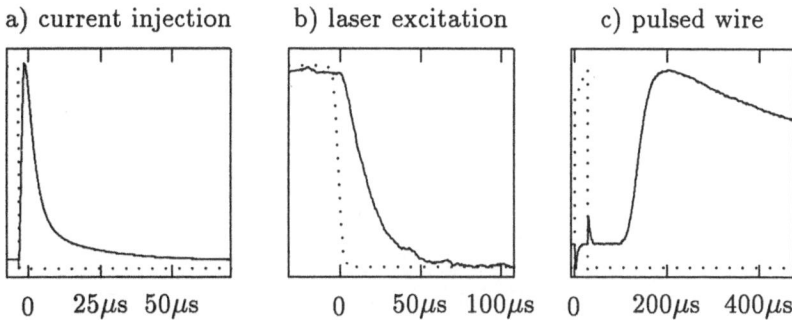

  0   25μs 50μs   0   50μs 100μs 0   200μs 400μs

Figure 3.   Typical oscilloscope traces obtained by the three methods; 0.5 μm wire diameter, $U = 8$ m/s.

For the experiments illustrating the compensation procedure, the measurements are made with a 2570 Waveform Analyser from Bakker Electronics. The measurements are performed in the following conditions: sampling rate of 50 ksample/s, low-pass antialiasing filters set at 20 kHz (6th order Butterworth), total record length per channel of 10 × 128 ksamples. This allows us to estimate autospectral density functions of temperature fluctuations over 1280 blocs of 1024 data points, giving a precision of the statistical estimate better than 2.8 %.

4. Results and discussion

The results of the cut-off frequency obtained from the time constant measurements ($f_c = 1/(2\pi\tau_w)$) are shown in figure 4. The agreement between the 3 methods and the theory is better than the one usually reported in the literature. Moreover, the current injection method is found to be the most reliable one. An example of our compensation procedure is given in figure 5. Two probes (parallel wires of 1 and 0.5 μm separated by 0.7 mm) are located on the axis of the jet at a distance of 95 mm from the nozzle exit. The local mean velocity and temperature are respectively 15 m/s and 47 °C. The 1 μm diameter wire is compensated in order to reach the dynamic capabilities of the 0.5 μm one. The compensation procedure is shown to be quite reliable, providing that the time constant of both wires is well known. The value of $\overline{\theta'^2}$ measured by the 1 μm wire is 11% lower than that measured by the 0.5 μm one. This difference is lowered to less than 1.5% when the compensation procedure is applied. The present study has been conducted in order to allow us to make accurate measurements of ϵ_θ and high-order moments in a future experiment using a strongly heated jet. It has been pointed out by Fulachier *et al.* [6] that a fine wire can be successfully used in this type of flow ($\Delta\Theta$ of the order of 800 K).

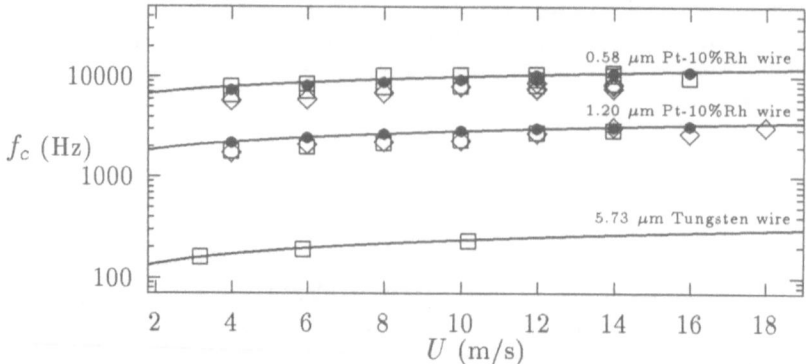

Figure 4. -3 dB cut-off frequency as a function of velocity for different wires: solid line, theoretical values; •, measurement using a square-wave current injection; □, measurement using a chopped laser beam; ◇, measurement using the pulsed wire.

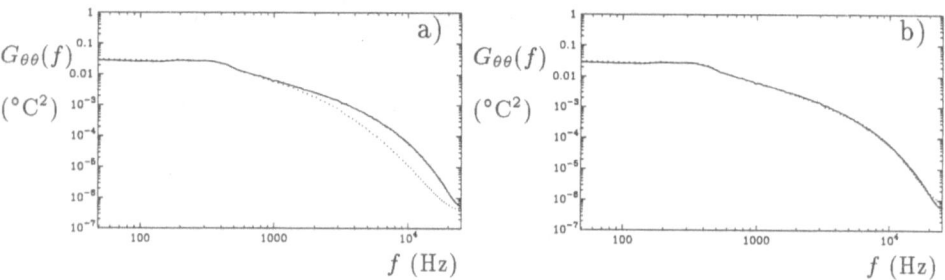

Figure 5. Example of numerical compensation: autospectral density function of temperature fluctuations; solid line, 0.5 μm wire; dotted line, 1 μm wire; a) original data; b) numerical compensation performed on the signal of the 1 μm diameter wire.

Acknowledgment

The support of the NSERC of Canada is gratefully acknowledged. We thank also Mr. Y. Jean and F. Bouak for their support and Dr. F. Anselmet (IRPHE, Marseille) for his assistance in the development of our CCA.

References

1. Bruun, H.H. (1995) *Hot-Wire Anemometry: Principle and Signal Analysis.* Oxford Science Publications, Oxford.
2. LaRue, J.C., Deaton, T. and Gibson, C.H. (1975) Measurement of High-Frequency Turbulent Temperature, *Rev. Sci. Instrum.*, **Vol. 46 No. 6**, pp. 757–764.
3. Antonia, R.A., Browne, L.W.B. and Chambers, A.J. (1981) Determination of Time Constants of Cold Wires, *Rev. Sci. Instrum.*, **Vol. 52, No. 9**, pp. 1382–1385.
4. Fiedler, H. (1978) On Data Acquisition in Heated Turbulent Flows, *Proc. of the Dynamic Flow Conf.*, Marseille, pp. 81–100.
5. Paranthöen, P., Lecordier, J.C. and Petit, C. (1982) Influence of Dust Contamination on Frequency Response of Wire Resistance Thermometer, *DISA Information*, **No. 27**, pp. 36–37.
6. Fulachier, L., Borghi, R., Anselmet, F. and Paranthöen, P. (1989) Influence of Density Variations on the Structure of Low Speed Turbulent Flows: a Report on Euromech 237, *J. Fluid Mech.*, **Vol. 203**, pp. 577–593.

MEASUREMENTS OF VELOCITY IN THE TURBULENT STAGE OF GASEOUS MIXTURES INDUCED BY SHOCK WAVES

F.POGGI, M.H. THOREMBEY, G. RODRIGUEZ and J.F. HAAS
Commissariat à l'Energie Atomique / Vaujours - Moronvilliers
BP7, F77181 Courtry, France

1. Introduction

We study the mixing of two gases of different densities arising from the shock wave-induced Richtmyer-Meshkov instability (RMI). The RMI appears when a shock wave impulsively accelerates two gases separated by an interface generally parallel to the shock plane but presenting small perturbations. The RMI is a manifestation of the baroclinic instability : the production of vorticity is proportional to the vector product of pressure gradient of the shock and the density gradient of the interface. It leads to an amplification of the interface perturbations, hence to the interpenetration of the two gases and eventually to the formation of a turbulent mixing zone. We seek to characterize the turbulent stage of the RMI in order to improve the modelization of this phenomenon. Measurements of mixing zone thickness and density profiles have already been performed in discontinuous heavy-light interfaces, where such a stage is observed [1] (see also the book of abstracts). Now, we try more direct measurements of turbulence, using laser Doppler velocimetry (LDV).

2. Experimental set-up

We study the behaviour of a discontinuous heavy-light interface between two gases, sulphur hexafluoride (SF6)-air, subjected to an acceleration by an incident upward propagating shock wave (Mach 1.45) and a deceleration by up to 3 weaker reflected shock waves in a vertical shock tube (see figure 1).

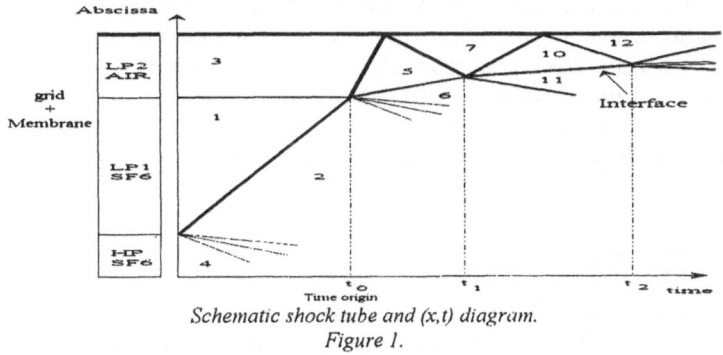

Schematic shock tube and (x,t) diagram.
Figure 1.

273

L. Fulachier et al. (eds.), IUTAM Symposium on Variable Density Low-Speed Turbulent Flows, 273–276.
© 1997 *Kluwer Academic Publishers.*

This tube has a square cross section (8 by 8 cm^2) and the distance between the initial interface position and the end plate has been set to 30 cm in the experiments described here. Discontinuous interfaces are initially created by a plastic membrane 0.3 μm thick placed directly below a thin wire mesh (wire spacing and diameter : 1010 μm and 80 μm) which does not induce any measurable turbulence by itself. The membrane is torn by the passage of the incident shock wave through the grid. Thus the initial scales of the perturbation are imposed by the mesh size only and the mixing zone is uniform across the tube. Initial scales are smaller with this grid than without it so the transition to a turbulent stage is earlier.

3. Schlieren visualization

We performed a schlieren visualization in order to measure the position and the thickness of the mixing zone during time. In these experiments, we used a Cordin high speed framing camera. Localizing the passage of the interface is necessary before measuring velocities in this flow. The images recorded show the evolution of the interface from the arrival of the first reflected shock wave until the arrival of the third reflected shock wave. We drew in figure 2 the position of the mixing zone edges and of the shock waves. The abscissa origin is the initial position of the interface and the time origin corresponds to the passage of the incident shock wave at this position.

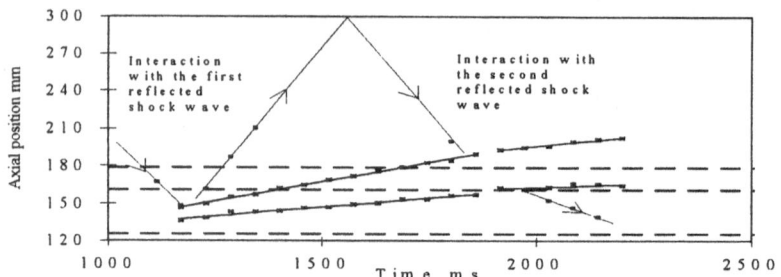

Time evolution of the mixing zone edges and of the shock waves.
Figure 2.

Remark : Other than for the LDV measurements, the Mach number is a bit weaker and we placed air instead of SF6 in the high pressure chamber, so reflected rarefaction waves interact with the mixing zone much earlier. We will have to make a schlieren visualization in the same conditions than in LDV measurements. However, the present results allow to estimate the position of the mixing zone during time, hence to choose convenient abscissa to place our LDV measurement volume.

4. Laser Doppler Velocimetry

For the laser Doppler velocimetry measurements, we use an argon laser of 300 mW maximal power and a Dantec burst spectrum analyser (BSA). We only measure the axial velocity and we assume isotropy for the interface turbulence. The diagnostic principle is classical : we measure an instantaneous and point velocity. But in our specific flows, we seed particles in the gases at rest before shock acceleration. So the concentration of particles is higher than in stationary flows. We obtain a data rate of 600 kHz. In air, we use incense cones and in SF6, where there is no combustion, we use olive oil droplets from an atomizer. In terms of turbulence rate[1], our global measurement noise is a little stronger with olive oil than with incense, (2 % versus 1.5 %), during the first velocity plateau (before the first reflected shock wave). However, with the wire mesh and the membrane, the incident shock wave is slightly pertubed so it increases the dispersion of velocity points. The measurement noise, including instrumental noise and background, is then about 2.5 % on the first plateau. Regarding our data rate and the values of the interface thickness in our flow, we needed about 30 shock tube runs to get statistically convergent data.

When the first reflected shock wave meets the interface, the flow is already turbulent, according to the RMI analysis. One of our goals is to measure the evolution of the kinetic energy level in the mixing zone through its interaction with the first reflected shock wave. Thus, we will estimate the expected increase of turbulence due to a baroclinic effect. The positions of the LDV measurement volume at 125.5 mm and 161 mm allow such an evaluation. Another goal is to quantify the dissipation of turbulence between two shocks. Some velocity measurements at 178.5 mm, compared to those at 161 mm should give such an information. These 3 abscissas are shown on Figure 2.

At 125.5 mm, we do not observe an increase of turbulence during the estimated period of the the mixing zone passage. The kinetic energy level is below the limit of our measurement noise : 16 (m/s)^2.

At 161 mm, the passage of the interface should be observed after its interaction with the first reflected shock wave and before the arrival of the second reflected shock wave. Figure 3 gives the cumulated velocity measurements at this position.

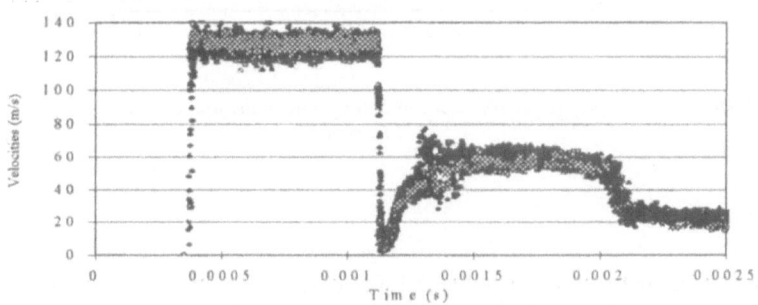

Cumulated velocity measurements at 161 mm.
Figure 3.

---

[1] The Reynolds decomposition gives the instantaneous velocity as the sum of a mean velocity and a fluctuation :

$$u = \bar{u} + u' \text{ with } \bar{u} = \frac{1}{T} \int_0^T u(t) dt .$$

The turbulence rate t is defined as : $\tau = \dfrac{\sqrt{\overline{u'^2}}}{\bar{u}}$ with $\sqrt{\overline{u'^2}} = \sqrt{\dfrac{1}{T} \int_0^T u'^2 dt}$.

In the case of isotropy, the turbulent kinetic energy is : $k = \dfrac{3}{2} \overline{u'^2}$.

The first velocity plateau at 130 m/s corresponds to air accelerated by the transmitted shock. After a short plateau at 0 m/s (air decelerated by the first reflected shock), we observe a second, perturbed plateau which includes the passage of the turbulent interface. The expected mean velocity in this zone is 47 m/s. But we observe a 59 m/s plateau, probably due to the boundary layer reversal effect [2]. We sampled the plateau duration in 30 time intervals in order to quantify the profile of k (first profile in Figure 4). With a peak at 160 $(m/s)^2$ we observe a strong increase of the kinetic energy as compared to the level measured at 125.5 mm. After the peak, k decreases before the arrival of the second reflected shock wave. This decrease is both due to dissipation (time dependent), diffusion of turbulence (both space and time dependent), and the instantaneous profile of kinetic energy (space dependent), in the interface.

At 178.5 mm, according to the schlieren results, the interface top edge should arrive before the second reflected shock wave and the passage of the bottom edge should not be observed at this location. The mean velocity level is still at 59 m/s before the second reflected shock wave arrival. With a peak at 120 $(m/s)^2$ the turbulent kinetic energy (second profile in Figure 4) is weaker than at 161 mm. Indeed, there is no production of turbulence between two shock waves while the mixing zone thickness increases.

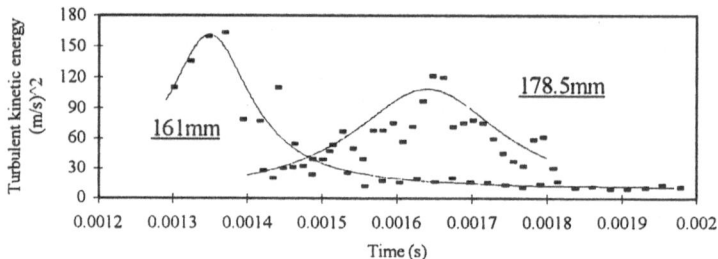

Turbulent kinetic energy profiles at 161 and 178.5 mm between the first and the second reflected shocks.
Figure 4.

5. Conclusion

We observe an increase of turbulence level essentially due to a baroclinic effect. We also obtain a decrease of turbulence between two shock waves by dissipation and diffusion effects. We still have to precisely localise the mixing zone limits by schlieren visualization in order to completely analyse our velocity measurements and to compare them to a k-ε modelization adapted to the RMI phenomenon.

6. References

[1] G. Rodriguez, I. Galametz, M.-H. Thorembey, C. Rayer and J.-F. Haas, Visualization of shocked mixing zones using differential interferometry and X-rays, 20th ISSW, Pasadena (USA), H. Hornung ed., (to be published by World Scientific Press in 1996).
[2] H. Mark, The interaction of a reflected shock wave with the boundary layer in a shock tube, J. Aero. Sciences, 24, 304, (1957).

VI. Compressible Flows

FLOWS WITH DENSITY VARIATIONS AND COMPRESSIBILITY: SIMILARITIES AND DIFFERENCES

SANJIVA K. LELE
*Department of Aeronautics and Astronautics
and Department of Mechanical Engineering,
Stanford University, Stanford, CA 94305-4035, USA*

Abstract. Similarities and differences between low speed, variable density flows and compressible flows are discussed. To aid in this comparison, flows of a binary mixture of ideal gases are considered. Under appropriate conditions significant density variations can arise in such systems, with or without compressibility. For simplicity, chemical reactions are excluded from consideration. The effects of density fluctuations and compressibility are examined in the context of the averaged transport equations with emphasis on the physical processes occuring in the flow. The roles of pressure fluctuations, differential acceleration and baroclinic vorticity are highlighted. A synopsis of the behavior of the variable density mixing layer and its compressible counterpart is given. The need for experiments and numerical simulations specifically designed to yield fundamental information about the effects of density variations and compressibility is stressed.

1. Introduction

Density variations arise in low-speed flows as well as in high-speed compressible flows. What are the similarities and differences between these two classes of flows ? For such a comparison it is useful to adopt a framework general enough to include both types of flows. A binary mixture of ideal non-reacting gases provides such a framework. At low speeds significant density fluctuations in a flow of the mixture arise due to temperature fluctuations or due to composition fluctuations. When compressibility becomes important, pressure fluctuations also contribute to the density fluctuations. Is a Lagrangian frame of reference better suited for comparing the different

L. Fulachier et al. (eds.), IUTAM Symposium on Variable Density Low-Speed Turbulent Flows, 279–301.

causes of density variation ? How should one define a 'fluid parcel' in a flow
with significant composition variation ? These questions are addressed below using the methodology common in studies of combustion (e.g. Williams
1985).

2. Binary Mixture of Ideal Gases

Consider the flow of a binary mixture of non-reacting ideal gases. Let the
mass fraction of the two species be Y_1 and Y_2. Dimensional variables are
denoted by superscript $\star$. At position $x_i^\star$ and time $t^\star$, the local density of
the two species, their local specific enthalpy, and their velocity are denoted
by $\rho_1^\star$, $\rho_2^\star$, and $h_1^\star$, $h_2^\star$, and $u_{1_i}^\star$, $u_{2_i}^\star$, respectively. It is convenient to describe
the flow in terms of the mass weighted quantities such as
$h^\star = \Sigma Y_\alpha h_\alpha^\star$, $u_i^\star = \Sigma Y_\alpha u_{\alpha_i}^\star$, where the sum over the greek indices is a sum
over the species.

The equations governing this compressible viscous flow are (e.g. Williams
1985, Kuo 1986):

$$\frac{\partial \rho^\star}{\partial t^\star} + \frac{\partial \rho^\star u_i^\star}{\partial x_i^\star} = 0 \tag{1}$$

$$\frac{\partial \rho^\star Y_\alpha}{\partial t^\star} + \frac{\partial \rho^\star u_i^\star Y_\alpha}{\partial x_i^\star} = \rho^\star \frac{DY_\alpha}{Dt^\star} = \frac{\partial}{\partial x_i^\star} \rho^\star D^\star \frac{\partial Y_\alpha}{\partial x_i^\star} \tag{2}$$

$$\frac{\partial \rho^\star u_i^\star}{\partial t^\star} + \frac{\partial \rho^\star u_i^\star u_j^\star}{\partial x_j^\star} = \rho^\star \frac{Du_i^\star}{Dt^\star} = -\frac{\partial p^\star}{\partial x_i^\star} + \frac{\partial \tau_{ij}^\star}{\partial x_j^\star} + \rho^\star f_i^\star \tag{3}$$

$$\frac{\partial \rho^\star h^\star}{\partial t^\star} + \frac{\partial \rho^\star h^\star u_i^\star}{\partial x_i^\star} = \rho^\star \frac{Dh^\star}{Dt^\star} = \frac{Dp^\star}{Dt^\star} + \Phi^\star - \frac{\partial q_i^\star}{\partial x_i^\star} \tag{4.i}$$

where $p^\star$, and $T^\star$ represent pressure and temperature, $f_i^\star$ is the external
body force (per unit mass), and $D/Dt^\star = \partial/\partial t^\star + u_i^\star \partial/\partial x_i^\star$, is the material
derivative following the mass weighted flow $u_i^\star$. The viscous stress $\tau_{ij}^\star$, the
molecular heat flux $q_i^\star$, and the rate of viscous dissipation of energy $\Phi^\star$ are
given by:

$$\tau_{ij}^\star = 2\mu^\star \{S_{ij}^\star - 1/3\delta_{ij}S_{kk}^\star\}, \qquad q_i^\star = -k^\star \frac{\partial T^\star}{\partial x_i^\star}, \qquad \Phi^\star = \tau_{ij}^\star S_{ij}^\star,$$

where the strain rate $S_{ij}^\star = 1/2\{\frac{\partial u_i^\star}{\partial x_j^\star} + \frac{\partial u_j^\star}{\partial x_i^\star}\}$, and Stokes hypothesis with zero
bulk viscosity is used to obtain the viscous stresses. Fluid properties such
as the viscosity $\mu^\star$ and thermal conductivity $k^\star$ are regarded as functions
of temperature $T^\star$. Note that species mass conservation expressed in (2)
uses Fick's law of diffusion with mass diffusivity $D^\star$, and ignores thermal
diffusion (Soret effect). Also, the heat flux associated with species diffusion

(Dufour effect) and the work done by the diffusional velocity against the body force $f_i^\star$ are neglected in (4.i).

For a mixture of ideal gases the equation of state is

$$p^\star = \rho^\star \Re^\star T^\star / W \qquad (5)$$

where $\Re^\star$ is the universal gas constant and the molecular weight W of the mixture is given by $1/W = \Sigma Y_\alpha / W_\alpha$, where W_α is the molecular weight of species α. The specific heats of the mixture at constant pressure $C_p^\star$, and at constant volume $C_v^\star$, for calorically perfect constituents are

$$C_p^\star = \Sigma Y_\alpha C_{p\alpha}^\star = \Re^\star \Sigma \frac{Y_\alpha \gamma_\alpha}{W_\alpha (\gamma_\alpha - 1)}, \qquad C_v^\star = \Sigma Y_\alpha C_{v\alpha}^\star = \Re^\star \Sigma \frac{Y_\alpha}{W_\alpha (\gamma_\alpha - 1)}.$$

It should be noted that the equation of state (5) can be rearranged to get

$$\frac{1}{\rho^\star} = \Sigma \frac{Y_\alpha}{\rho_{\alpha_f}^\star}, \qquad \rho_{\alpha_f}^\star = \frac{p^\star W_\alpha}{\Re^\star T^\star} \qquad (6)$$

where $\rho_{\alpha_f}^\star$, is the density of *pure* species α at the local pressure $p^\star$ and temperature $T^\star$. Note that $\rho_{\alpha_f}^\star \neq \rho_\alpha^\star$ and the expression (6) for the density of the mixture in terms of the density of its pure constituents (at the local conditions), is identical to the relationship commonly used for the density of a mixture of incompressible miscible liquids. Although (6) is exact for a compressible mixture of ideal gases, it stresses that the density variations are a combined result of composition variations and the variations of the density of the pure constituents in response to the temperature and pressure variations. Thus when both temperature and pressure fluctuations are negligible the density fluctuations of the gas be related to its composition fluctuations. This approximation has been adopted in modeling and in measurements of isothermal variable density turbulent flows at low speed (Shih *et al.* 1987, Panchapakesan and Lumley 1993, Chassaing *et al.* 1994, Djeridane *et al.* 1996).

For our later discussion it is convenient to record other equivalent forms of the energy equation (4.i). Since the internal energy per unit mass, $e^\star = h^\star - p^\star / \rho^\star$, it follows that

$$\frac{\partial \rho^\star e^\star}{\partial t^\star} + \frac{\partial \rho^\star e^\star u_i^\star}{\partial x_i^\star} = \rho^\star \frac{De^\star}{Dt^\star} = -p^\star \frac{\partial u_i^\star}{\partial x_i^\star} + \Phi^\star - \frac{\partial q_i^\star}{\partial x_i^\star} \qquad (4.ii)$$

Similarly the total energy per unit mass, $e_t^\star = e^\star + u_i^\star u_i^\star / 2$, satisfies

$$\frac{\partial \rho^\star e_t^\star}{\partial t^\star} + \frac{\partial \rho^\star e_t^\star u_i^\star}{\partial x_i^\star} = \rho^\star \frac{De_t^\star}{Dt^\star} = -\frac{\partial p^\star u_i^\star}{\partial x_i^\star} + \frac{\partial \tau_{ji}^\star u_i^\star}{\partial x_j^\star} - \frac{\partial q_i^\star}{\partial x_i^\star} + \rho^\star f_i^\star u_i^\star \qquad (4.iii)$$

The energy equation can also be expressed using the pressure, p^*, as the dependent variable. The equation of state (5) requires that

$$\frac{1}{p^*}\frac{Dp^*}{Dt^*} = \frac{1}{\rho^*}\frac{D\rho^*}{Dt^*} + \frac{1}{T^*}\frac{DT^*}{Dt^*} - \frac{1}{W}\frac{DW}{Dt^*}$$

Using mass balance (1) and energy balance (4.ii) in the above yields

$$\frac{Dp^*}{Dt^*} + (\frac{\Re^*}{WC_v^*} + 1)p^*\frac{\partial u_i^*}{\partial x_i^*} = \frac{\Re^*}{WC_v^*}(\Phi^* - \frac{\partial q_i^*}{\partial x_i^*}) - p^*(\frac{1}{C_v^*}\frac{DC_v^*}{Dt^*} + \frac{1}{W}\frac{DW}{Dt^*}).$$

$$(4.iv)$$

The last term is due to the diffusion of species and can be evaluated using the definitions of C_v^* and W for a gas mixture. Still another form of the energy equation is found using the Gibbs relation for the entropy per unit mass, s^* :

$$T^*ds^* = dh^* - \frac{dp^*}{\rho^*} - \Sigma\frac{\mu_\alpha^*}{W_\alpha}dY_\alpha,$$

where μ_α^* is the chemical potential of species α. Applying this to a 'fluid particle' moving with the velocity u_i^* gives the energy equation in entropy form

$$\rho^*T^*\frac{Ds^*}{Dt^*} = \Phi^* - \frac{\partial q_i^*}{\partial x_i^*} - \Sigma\frac{\mu_\alpha^*}{W_\alpha}\frac{\partial}{\partial x_i^*}\rho^*D^*\frac{\partial Y_\alpha}{\partial x_i^*} \qquad (4.v)$$

2.1. CRITERIA FOR COMPRESSIBILITY

The energy equation (4.iv) with pressure as a dependent variable yields

$$\frac{\partial u_i^*}{\partial x_i^*} = -\frac{1}{\rho^*}\frac{D\rho^*}{Dt^*} = (\frac{\Re^*}{WC_v^*} + 1)^{-1}\{-\frac{1}{p^*}\frac{Dp^*}{Dt^*} + \frac{\Re^*}{WC_v^*}\frac{1}{p^*}(\Phi^* - \frac{\partial q_i^*}{\partial x_i^*})$$

$$-(\frac{1}{C_v^*}\frac{DC_v^*}{Dt^*} + \frac{1}{W}\frac{DW}{Dt^*})\} \qquad (7)$$

This equation identifies the mechanisms responsible for changing the volume of a fluid parcel. Changes in pressure along a 'material trajectory' cause the material volume to change. Scaling the velocities as $u_i^* \sim O(U)$, spatial gradients as $\partial/\partial x_i^* \sim O(1/L)$, temporal changes as $\partial/\partial t^* \sim O(U/L)$, density as $\rho^* \sim O(\rho_r)$, mean pressure as $< p^* > \sim O(p_r)$ where ρ_r and p_r are reference density and pressure, and pressure fluctuations as $p'^* \sim O(\rho_r U^2)$ shows that (due to pressure variations alone) $\frac{\partial u_i^*}{\partial x_i^*} \sim O(U/L)M^2$, where

$M = U/C$ is a Mach number. Evidently, the compressible changes in fluid density are $O(M^2)$ and hence negligible in low-speed flows. The volume of a fluid parcel also changes due to viscous dissipation Φ, and the conductive heat-flux divergence $\frac{\partial q_i^\star}{\partial x_i^\star}$. The former effect is negligible in low-speed flows but the latter can be important when intense heat-transfer from a boundary is involved. The derivation of (7) assumes that no heat-sources are present. Also, chemical heat-release is excluded. If these were present, an additional term representing the net rate of external heat addition (per unit volume) should be added to the group of terms containing Φ in (7). In a high Peclet number (turbulent) flow, with the exception of the near-wall regions in flows involving significant heat transfer, $\frac{\partial q_i^\star}{\partial x_i^\star}$ is small relative to the convective heat flux divergence hence its effect on large-scale dilatation is also negligible. Changes in the composition of the fluid parcel due to molecular mixing also cause its volume to change (as expressed by the last group of terms in (7)). Such changes occur on the spatial scale at which the molecular mixing is active. We may conclude that the energy containing (large) eddies of a turbulent flow in a heterogeneous non-reacting gas mixture are nearly incompressible if the Mach number M is small and intense heat-exchange from a surface is not present.

In the rest of this paper the superscript $\star$ on the dimensional variables is omitted, and the Einstein summation convention is also used for compactness. Subscripts following a comma denote differentiation, and overbars or $<>$ (used interchangeably) denote averaging operations.

2.2. DENSITY VARIATIONS AND DILATATION

Even though the dilatation of fluid parcels is small in a low-speed flow, it is none-the-less critical in setting an upper bound on the density fluctuations. Mass conservation (1) yields an exact equation for the variance of density fluctuations

$$\frac{\overline{D}}{\overline{Dt}}\overline{\rho'^2} + 2\overline{u_{i,i}}\,\overline{\rho'^2} + 2\overline{\rho'u_i'}\overline{\rho}_{,i} + \overline{(\rho'^2 u_i')}_{,i} = -2\overline{\rho'u_{i,i}'}\overline{\rho} - \overline{\rho'^2 u_{i,i}'}, \qquad (8)$$

where $\frac{\overline{D}}{\overline{Dt}} \equiv \frac{\partial}{\partial t} + \overline{u}_i\frac{\partial}{\partial x_i}$, is the derivative following the Reynolds averaged mean flow. This equation shows that density fluctuations are 'produced' by the turbulence stirring up the mean density gradient and are intensified by mean compression. The terms on the right hand side (RHS), when negative, represent destruction of density fluctuations. To see this most clearly, consider a homogeneous field of density fluctuations. Only the first term is then non-zero on the left hand side. Evidently, the destruction of density fluctuations is possible only when dilatation is non-zero. This is equally true for the flow of a gas mixture at low speeds also; the fluid dilatation is linked to molecular mixing as revealed in (7).

Recognizing this role of small scale dilatation is important in the context of direct numerical simulations (DNS) and large eddy simulations (LES). As discussed in section (2.1) the dilatation fluctuations in a low speed variable density flow arise from viscous-dissipation, molecular heat-conduction and molecular mixing of different species. Since these processes are active only on small spatial scales (in a high Reynolds number flow), an accurate numerical representation of small-scale dilatation is necessary to capture the destruction mechanism of density fluctuations. In LES of compressible flows the Favre averaged form of the filtered equations (e.g. Moin *et al.* 1991) is often used. The filtered density equation in such an approach appears closed, *i.e.* it does not have any unclosed subgrid-scale terms. This is a mixed blessing. The absence of explicit subgrid-scale terms (which are often dissipative in variance budgets) requires that special care be taken with the numerical schemes to keep the density fluctuations from building up at the smallest scales.

Some studies of variable density turbulence (e.g. Sandoval 1995) use the specific volume $v = 1/\rho$ as a dependent variable. This variable is particularly convenient for studies of the mixing at constant pressure and temperature for which, in view of (6), the specific volume is a mass fraction weighted sum of the specific volumes of the constituents. The equation governing the specific volume $v = 1/\rho$ follows from (1)

$$\frac{Dv}{Dt} - vu_{i,i} = 0.$$

Since the specific volume v is a non-linear function of the density ρ, $\bar{v} \neq 1/\bar{\rho}$, and the departure $\bar{\rho}\,\bar{v} - 1 = -\overline{\rho'v'}$ is a positive definite measure of the density fluctuations (Besnard *et al.* 1992). For $(\overline{\rho'^2})^{\frac{1}{2}}/\bar{\rho} \ll 1$, it is easily shown that $-\overline{\rho'v'} \approx \overline{\rho'^2}/\bar{\rho}^2 - \overline{\rho'^3}/\bar{\rho}^3 + \ldots$. The transport equation for $\overline{\rho'v'}$ is analogous to (8) *i.e.*

$$\frac{\overline{D}}{\overline{Dt}}\overline{\rho'v'} - \bar{v}(\overline{\rho'u_i'})_{,i} - \bar{\rho}(\overline{v'u_i'})_{,i} = -2\bar{\rho}\overline{v'u_{i,i}'},$$

which once again stresses that finite dilatation fluctuations are necessary for dissipating the density variations.

In adopting the specific volume v as a dependent variable it should be remembered that since v and ρ are statistically dependent a number of identities exist which relate the statistics of v to statistics of ρ. The use of specific volume as a dependent variable simplifies the form of the balance equations for Reynolds averaged quantities. For example the mean momentum conservation is

$$\frac{D}{Dt}\overline{u}_i + (\overline{u'_i u'_j})_{,j} - \overline{u'_i u'_{j,j}} + \overline{v}\,\overline{p}_{,i} + \overline{v'p'_{,i}} = f_i + \overline{v}\,\overline{\tau_{ij,j}} + \overline{v'\tau'_{ij,j}} \qquad (9)$$

This equation should be compared to the equation obtained for $\overline{u}_i$ from (3) by a straightforward Reynolds decomposition (e.g. Hinze 1975, p. 21). The present form is free of the terms involving the mass flux $\overline{\rho'u'_i}$, but new terms involving the specific volume fluctuations, e.g. $\overline{v'p'_{,i}}$, are introduced. A combination of (9) with the Favre averaged momentum equation yields a balance equation for the mass flux, $\overline{\rho'u'_i}$,

$$\frac{D}{Dt}\overline{\rho'u'_i} + \overline{\rho'u'_i}\,\overline{u}_{i,i} + \overline{\rho'u'_j}\,\overline{u}_{i,j} + \overline{u'_i u'_j}\,\overline{\rho}_{,j} + \overline{\rho}\,\overline{u'_i u'_{j,j}} + (\overline{\rho'u'_i u'_j})_{,j}$$

$$-\{\overline{\rho}\,\overline{v} - 1\}\overline{p}_{,i} - \overline{\rho}\,\overline{v'p'_{,i}} = -\{\overline{\rho}\,\overline{v} - 1\}\overline{\tau_{ij,j}} - \overline{\rho}\,\overline{v'\tau'_{ij,j}} \qquad (10)$$

It should be noted that the body force f_i does not appear in this equation explicitly; its influence on the mass flux is indirect. Equations such as (9)-(10) were given by Besnard et al. (1992) who also proposed models for the unclosed terms in such equations.

2.3. DIFFERENT ROLES OF PRESSURE

Pressure assumes many different roles in variable density flows. The mean pressure is a thermodynamic variable. As noted earlier if the mean pressure varies appreciably relative to its reference value along a material trajectory there is, in response, a compressible change in the mean fluid density. For $M \ll 1$, the pressure variations enforce 'incompressibility' in near-field ($|x| \sim O(L)$) on the hydrodynamic time scale of $O(L/U)$. This is a hydrodynamic role of pressure, also denoted as psuedo-sound (with a related $O(M^2)$ nearly-isentropic hydrodynamic density variation, e.g. Zank and Matthaeus (1991)). The hydrodynamic pressure satisfies a Poisson equation and is responsible for non-local effects. Pressure-strain rate correlations redistribute turbulent kinetic energy among the different components. Reynolds (1995) regards the non-local effects of pressure a greater challenge to turbulence modeling than the non-linearity of the Navier-Stokes equations.

A Poisson equation for pressure is obtained by taking the divergence of (3). Its exact form is

$$\nabla^2 p = -\rho u_{j,i} u_{i,j} - (\rho f_i)_{,i} - \nabla\rho \cdot \frac{D\vec{u}}{Dt} + \tau_{ij,ij} - \rho\frac{Du_{i,i}}{Dt} \qquad (11)$$

The first term on the right hand side (RHS) of (11) is the 'source' of the usual hydrodynamic pressure fluctuations (in a uniform density flow) and the second term on the RHS gives pressure fluctuations due to density fluctuations in presence of a body force. The last two terms on the RHS are negligible in the nearly-incompressible variable-density flow. The acceleration of fluid parcels in the direction of density gradient is a significant new 'source' of pressure fluctuations in a variable density flows at low speed.

When viscous effects are negligible, *i.e.* $\rho \frac{D\vec{u}}{Dt} = -\nabla p$, this new 'source' term is revealed to be linearly dependent on pressure itself, and consequently better regarded as a modification of the elliptic operator on the left hand side of (11). Such a rearrangement is obtained by starting from (3) divided by ρ, which gives an exact equation

$$\nabla \cdot \{\frac{1}{\rho}\nabla p\} = -u_{j,i}u_{i,j} - f_{i,i} + (\tau_{ij,j}/\rho)_{,i} - \frac{Du_{i,i}}{Dt}. \tag{11a}$$

Jones (1992) discusses a Poisson equation related to (11) obtained by retaining the left hand side of (3) in conservation form and using the mass conservation, steps similar to those taken by Lighthill (1952), viz.

$$\nabla^2 p = -(\rho u_i u_j)_{,ij} - (\rho f_i)_{,i} + \frac{\partial^2 \rho}{\partial t^2} + \tau_{ij,ij}, \tag{11b}$$

and notes that due to the complexity of the equation the result sheds little light on the possible influences of density fluctuations. Pope (1987) makes a similar observation. Equations akin to (11b) are commonly used in numerical simulations of low-speed variable-density flows to find the hydrodynamic pressure (McMurtry *et al.* 1986, Sandoval 1995) but the ambiguity associated with defining the 'source' of pressure fluctuations is inherent there as well. The 'source' $\frac{\partial^2 \rho}{\partial t^2}$ in (11b) is not Galilean invariant and a further rearrangement is preferred (Sarkar 1992). The 'sources' in (11) and (11a) are Galilean invariant, but the elliptic equation (11a) appears to be less well-known. Saffman and Meiron (1989), and Pham and Meiron (1993) make use of an equation similar to (11a) in theoretical analysis and for numerical simulations of incompressible Richtmyer-Meshkov instability.

On the acoustic time scale L/C, which is $1/M$ times faster than the hydrodynamic time scale L/U, the near-field also exhibits acoustic propagation of pressure which rapidly establish the near-incompressible flow. In studies of low-speed variable-density flows these 'fast' acoustic waves are filtered out. The acoustic propagation in the far-field $|x| \sim O(L/M)$, however, occurs on the time scale L/U but as shown by Lighthill (1952) the radiated flux of acoustic power is generally only a very small fraction of the energy flux in the near-field. Acoustic propagation is associated with an acoustic energy flux (intensity) and a distribution of acoustic energy density (consisting of acoustic kinetic energy and acoustic potential energy) in the medium.

An exact transport equation for pressure fluctuations can be obtained from (4.iv), which can be simplified for the case of an ideal gas to get

$$\frac{\overline{D}}{\overline{Dt}} \frac{\overline{p'^2}}{2} = -(\overline{p'u'_j})_{,j} - \gamma\overline{p'^2}\bar{u}_{j,j} - \gamma\bar{p}\overline{p'u'_{j,j}}$$

$$-\frac{(2\gamma-1)}{2}\overline{p'^2 u'_{j,j}} - (\frac{\overline{p'^2}}{2}u'_j)_{,j} + (\gamma-1)\overline{p'u_{i,j}\tau_{ij}} + (\gamma-1)\overline{p'q'_{j,j}}. \qquad (12)$$

This diagnostic equation is not useful for general flows at low speed, since in that case at leading order it reduces to $\bar{u}_{j,j} = 0$. For compressible flows simplified forms of (12) have been used for modeling and interpreting the behavior of pressure dilatation correlation $\overline{p'u'_{j,j}}$ (Zeman 1991, Durbin and Zeman 1992, Sarkar et $al.$ 1991a). Ristorcelli (1995) develops a useful diagnostic equation for dilatation based on a low Mach number near-field approximation, and uses it to develop models for dilatational covariances.

Small amplitude (linearized) motions in a uniform mean flow with no mean property gradients satisfy a 'energy' corollary (Lee et $al.$ 1997). With neglect of dissipative effects this is

$$\{\frac{1}{2}\bar{\rho}\overline{u'_j u'_j} + \frac{\overline{p'^2}}{2\gamma\bar{p}}\}_{,t} + \{\bar{u}_i(\frac{1}{2}\bar{\rho}\overline{u'_j u'_j} + \frac{\overline{p'^2}}{2\gamma\bar{p}}) + \overline{p'u'_i}\}_{,i} = 0.$$

If the fluctuations are purely acoustic, $\frac{1}{2}\bar{\rho}\overline{u'_j u'_j}$ is the acoustic kinetic energy density, $\overline{p'^2}/(2\gamma\bar{p})$ is the acoustic potential energy density and $\overline{p'u'_i}$ is the acoustic energy flux in a medium at rest. Mean flow gradients and boundary conditions (e.g. on a solid surface near leading or trailing edges) provide coupling between the acoustic and vortical (solenoidal) disturbances.

It has been stressed by Morkovin (1992) that for an eddy to overturn over some length scale L streamwise and cross-stream acoustic communication is necessary. Similar reasoning led Bridenthal (1990) to propose that the maximum eddy-scale in a shear flow is limited by the acoustic communication to be a sonic eddy. Communication effects in shear layers with a mean velocity $U(x_2)$ have been studied in the geometric acoustic approximation by Papamoschou (1991b) and via DNS in model problems by Papamoschou and Lele (1993). The upshot of these studies is that as the Mach number across an eddy increases, the streamwise distance over which the pressure disturbance of the eddy is felt decreases rapidly. This suppresses the u'_2 disturbance which in turn reduces the $\overline{u'_1 u'_2}$. The ratio of the acoustic travel time across the eddy scale L, L/C and the mean deformation rate time scale $1/S$ gives the deformation rate Mach number $M_d = SL/C$. When M_d is small the communication delay is negligible but as M_d increases the communication delay makes the eddying process less effective. In a shear dominated flow M_d is a gradient Mach number (Sarkar, 1995) and influences the efficiency $(-\overline{u_1' u_2'}/q^2)$ with which turbulence extracts energy from the mean shear flow. In low-speed variable-density flows the acoustic communication delay is negligible and the parameter M_d should be irrelevant.

The neglect of pressure fluctuations, $p'/\overline{p}$, relative to the entropy fluctuations s'/C_v (or the related temperature fluctuation $T'/\overline{T}$) and the assumption of constant total temperature in a thin shear flows, *i.e.* flows with $\overline{U_2} \ll \overline{U_1}$, lead to the so-called Strong Reynolds Analogy (Morkovin 1961, Bradshaw 1977). This analogy is a cornerstone of compressible boundary layer studies, and is frequently used in experimental data reduction (see Gaviglio 1987 for a survey). Lele (1994) discusses the conditions necessary for the SRA assumptions to be met. Numerical simulations of shock-turbulence interaction problems (Mahesh *et al.* 1997) has shown that the unsteady motion of the shock is responsible for significant total temperature fluctuation which is carried downstream by the mean flow. The net result is that SRA is quantitatively inaccurate downstream of a shock wave.

As the Mach number associated with the turbulent fluctuations $M_t = q/C$, where q is characteristic of the velocity fluctuations, increases eddy shocklets arise in the flow (Kida and Orszag 1990, Lee *et al.* 1991, Sandham and Reynolds 1990, Papamoschou 1995). These are shock waves associated with the turbulent motion and provide another pathway for the dissipation of kinetic energy. While such eddy shocklets may have a significant impact on the acoustic noise radiated by the flow, their direct impact on the evolution of vortical turbulence appears to be small. Numerical simulations, which are restricted to $M_t < 1$, show that the contribution of dilatational dissipation with or without eddy shocklets to the turbulent kinetic energy dissipation is a negligible (Blaisdell *et al.* 1995, Huang *et al.* 1995, Freund *et al.* 1997, Vreman *et al.* 1996). In some astrophysical applications much higher values of M_t may be involved and the shocklet dissipation could be appreciable. This extra-dissipation is absent in low speed flows.

2.4. BALANCE EQUATIONS FOR MEAN FLOW

The balance equations for the mean and fluctuating quantities can be expressed in several alternative forms. The Reynolds averaged equations (given in Hinze 1975) and the density weighted equations, often called Favre averaged equations (Favre 1969), are the two most commonly adopted sets. In the mean equations for mass, momentum and species balance, additional terms involving density fluctuations arise in the Reynolds averaged form but not in the Favre averaged form. This is because there is no mass flux crossing the stream-surface based on the Favre-averaged velocity, a property not shared by the Reynolds-averaged velocity. The mean mass, momentum and species conservation equations are

$$\bar{\rho}_{,t} + (\bar{\rho}\tilde{u}_i)_{,i} = 0, \tag{13}$$

$$(\bar{\rho}\tilde{Y}_\alpha)_{,t} + (\bar{\rho}\tilde{u}_i\tilde{Y}_\alpha + <\rho u_i'' Y_\alpha'' >)_{,i} = 0, \tag{14}$$

$$(\bar{\rho}\tilde{u}_i)_{,t} + (\bar{\rho}\tilde{u}_i\tilde{u}_j + <\rho u_i'' u_j'' >)_{,j} + \bar{P}_{,i} = \bar{\rho}f_i + \bar{\mathcal{T}}_{ij,j}, \tag{15}$$

where tilde over a variable denotes Favre (density-weighted) average, and overbar denotes the usual Reynolds (volume-weighted) average. A variable with a single prime as a superscript represents the deviation relative to the Reynolds average, while a double prime as superscript represents the deviation from the Favre average.

The equation for mean total energy

$$\{\bar{\rho}(\tilde{e} + \frac{1}{2}\tilde{u}_k\tilde{u}_k + \frac{1}{2} <\rho u_k'' u_k'' >)\}_{,t} + \{\bar{\rho}\tilde{u}_j(\tilde{h} + \frac{1}{2}\tilde{u}_k\tilde{u}_k) + \frac{1}{2} <\rho u_k'' u_k'' >\}_{,j}$$

$$+\{<\rho h'' u_j'' > + <\rho u_k'' u_j'' > \tilde{u}_k + \frac{1}{2} <\rho u_k'' u_k'' u_j'' >\}_{,j} = \bar{\rho}f_i\tilde{u}_i + (\overline{u_i\mathcal{T}_{ij}})_{,j} + \bar{q}_{j,j}, \tag{16}$$

is also more compact with Favre averaging. An equation for Favre-averaged internal energy (IE) follows from (16), once the equation governing the turbulent kinetic energy (19) is subtracted

$$(\bar{\rho}\tilde{e})_{,t} + \{\bar{\rho}\tilde{u}_i\tilde{e} + <\rho u_i'' h'' > - \overline{p'u_i'} - \overline{u_i'' \mathcal{T}_{ij}} + \bar{q}_i\}_{,i} =$$

$$-\bar{P}\tilde{u}_{i,i} + \bar{\mathcal{T}}_{ij}\tilde{u}_{i,j} + \overline{u_i''P_{,i}} - \overline{p'u_{i,i}'} - \overline{u_i''\mathcal{T}_{ij,j}} + \overline{u_{i,j}'\mathcal{T}_{ij}'}. \tag{17}$$

This can be converted into an equation for the enthalpy $\tilde{h}$, since $\bar{\rho}\tilde{h} = \bar{\rho}\tilde{e} + \bar{p}$. Similarly the identity $<\rho h'' u_j'' > = <\rho e'' u_j'' > + \overline{p'u_j'} + \bar{p}\,\overline{u_j''}$ can be used to rearrange the spatial redistribution terms.

To solve for the mean quantities $\bar{p}$, $\tilde{Y}_\alpha$, $\tilde{u}_i$, and $\tilde{h}$ closure must be provided for the turbulent fluxes arising in these equations. Since the equations (13)-(17) are formally identical to their counterparts for low-speed variable-density flow it is often suggested that the closure developed for one could also be applied to the other. This commonly adopted hypothesis lacks rigor: the terms being modeled need to behave similarly in the two classes of flows for the same closure model to be used in both types of flows. Some calculation methods provide closure by modeling the unclosed terms in second moment equations and this closure is generally based on uniform density flow. The transport equations for second moments (e.g. Jones 1992, Bray 1995), however, also involve additional unclosed terms containing $\overline{u_i''} = -\overline{\rho'u_i'}/\bar{\rho}$, and $\overline{Y_\alpha''} = -\overline{\rho'Y_\alpha'}/\bar{\rho}$, etc. Bilger (1989) gives an account of flows where the role of these additional terms is as important as the conventional shear production of turbulence. Variable density flows subjected to streamwise acceleration (Starner and Bilger 1980), countergradient diffusion in premixed flames (Libby and Bray 1981) are two specific examples of this.

2.5. ENERGY BALANCE OF TURBULENCE

An equation for mean mechanical energy (ME) follows from the Favre-averaged momentum equation

$$(\bar{\rho}\frac{\tilde{u}_j\tilde{u}_j}{2})_{,t} + \{\bar{\rho}\tilde{u}_i\frac{\tilde{u}_j\tilde{u}_j}{2} + \tilde{u}_j <\rho u_i''u_j''> +\tilde{u}_i\bar{p} - \tilde{u}_j\bar{\tau}_{ij}\}_{,i} =$$

$$\bar{\rho}f_i\tilde{u}_i + <\rho u_i''u_j''> \tilde{u}_{i,j} + \bar{p}\tilde{u}_{i,i} - \bar{\tau}_{ij}\tilde{u}_{i,j}, \qquad (18)$$

and similarly a transport equation for turbulent kinetic energy (TKE) can be obtained as

$$<\frac{1}{2}\rho u_j''u_j''>_{,t} + \{\tilde{u}_i <\frac{1}{2}\rho u_j''u_j''> + <\frac{1}{2}\rho u_i''u_j''u_j''> +\overline{u_i'p'} - \overline{u_j'\tau_{ij}'}\}_{,i} =$$

$$- <\rho u_i''u_j''> \tilde{u}_{i,j} - \overline{u_i''}\bar{p}_{,i} + \overline{p'u_{i,i}'} + \overline{u_i''\bar{\tau}_{ij,j}} - \overline{u_{i,j}'\tau_{ij}'} \qquad (19)$$

Note that the body force f_i does not appear explicitly in the TKE equation. It is easily verified that the sum of the evolution equations for ME, TKE, and IE reduces identically to the total energy equation (16). This is convenient for the energy transfer paths to be drawn. The classical diagram of Favre (1969) is discussed by Lele (1994). It should, however, be stressed that such energy exchange diagrams are not unique. Huang et al. (1995) adopt the view that Favre averaging should be used only with the convective terms. This gives an alternative representation of the energy exchanges. Specifically the exchange between ME and IE occurs via $\bar{p}\,\tilde{u}_{i,i}$ and $-\bar{\tau}_{ij}\tilde{u}_{i,j}$. This causes the interactions $-\overline{u_i''}\bar{p}_{,i}$ and $\overline{u_i''\bar{\tau}_{ij,j}}$ as exchanges between ME and TKE rather than as exchanges between IE and TKE as in the classical diagram.

2.6. DIFFERENTIAL ACCELERATION

Suppose that a gas mixture with density fluctuations is subjected to an externally applied body force, or is subjected to a pressure gradient. On account of their inertia the lighter parcels of gas accelerate more rapidly than the denser parcels. This differential acceleration generates turbulent kinetic energy. From (19) the rate of TKE growth depends on the net (mass) flux $\overline{u_i''} = -\overline{\rho' u_i'}/\bar{\rho}$ and on the driving pressure gradient $\frac{\partial \bar{p}}{\partial x_i}$. The added mass of the fluid surrounding the accelerating fluid parcels places a limit on the peak differential acceleration. Since the added mass depends on the geometry of the fluid lumps being accelerated, slender streamlined streaks of light fluid can achieve accelerations which exceed the applied body force (per unit mass). The differential acceleration is responsible for countergradient transport of species in premixed flames (Libby and Bray 1981).

Recently Sandoval (1995) has studied buoyancy driven homogeneous turbulence with large density fluctuations by numerical simulations and compared his results with those obtained with the Boussinesq approximation (Batchelor et $al.$ 1992). The buoyancy driven mixing of two miscible incompressible fluids which begins when a homogeneous field of density fluctuations (and no initial velocity) is suddenly subjected to a uniform gravitational acceleration g (along the positive x_3 direction) is considered. The differential acceleration generates turbulence which initially grows but eventually decays as the molecular mixing reduces the density fluctuations. Molecular diffusion of the species is also responsible for the dilatation (as discused in section 2.1) and its correlation with density fluctuations provide the destruction mechanism for density fluctuations. The need to resolve these small-scale dilatation variations in a direct numerical simulation, restricts the study to low Reynolds numbers. However, detailed information about the turbulence evolution and specifically on the impact of the density fluctuations is obtained.

Formulating this low-speed variable-density mixing problem as a compressible, gas-phase binary-mixing problem is quite instructive. In the compressible case a strict homogeneity in x_3 is not possible; when a hydrostatic pressure variation is introduced a density gradient in x_3 must also be allowed. A quasi-homogeneous approximation is possible when the turbulence length scales are significantly smaller than the density scale-height. Full specification of the (quasi-homogeneous) mixing problem, however, requires an additional constraint. The situation is similar to the classical mixing problem in a box. Suppose two different gases at different thermodynamic states are initially separated by a partition in a box. The partition is removed and the gases mix. What is the final thermodynamic state of the mixture ? Questions such as, is the box rigid and insulated, or is it maintained at constant pressure with its volume allowed to change, need to be first answered.

Returning to the compressible buoyancy driven mixing problem, if the mean density $\bar{\rho}$ is not allowed to change with time, it follows from mass conservation and homogeneity assumptions that $\overline{\rho \tilde{u}_3} = \bar{\rho}\,\bar{u}_3 + \overline{\rho' u_3'} = constant$. Suppose that a rigid (impenetrable) surface supports the fluid mixture at some large distance from the approximately homogeneous turbulent region being analyzed, so that a hydrostatic pressure variation is set up soon after the body force is initiated. This requires that the constant in the above equation can only be zero, viz. $\tilde{u}_3 = 0$, and since $\overline{\rho' u_3'} \geq 0$, it follows that $\bar{u}_3 \neq 0$. The Favre-averaged momentum equation shows that the pressure gradient $\frac{\partial \bar{p}}{\partial x_3}$ is unchanged from its hydrostatic value of $\bar{\rho} g$. It is also evident that the mean specific volume $\bar{v}$ must evolve from its initial value $(\neq 1/\bar{\rho})$ towards its final value of $1/\bar{\rho}$.

Sandoval (1995) has studied a different problem and imposed $\bar{u}_3 = 0$ in an *inertial* frame of reference. This is rationalized on grounds that for the mixing problem of miscible incompressible fluids, the net vertical volume flux must be zero. If this condition ($\bar{u}_3 = 0$) is imposed on the compressible, binary gas-phase mixing problem, it follows that $\overline{\rho \tilde{u}_3} = \overline{\rho' u_3'} \neq 0$. The Favre-averaged vertical momentum equation then yields $\frac{\partial}{\partial t}\overline{\rho' u_3'} = -\frac{\partial \bar{p}}{\partial x_3} + \bar{\rho} g$. Since $\overline{\rho' u_3'} \geq 0$, this implies that the pressure gradient is reduced below its hydrostatic value by the unsteady mass flux. This is a non-Boussinesq effect. The turbulent kinetic energy equation $\frac{\partial}{\partial t} < \rho \frac{u_i'' u_i''}{2} >= -\overline{u_3'' \frac{\partial \bar{p}}{\partial x_3}} - \epsilon$, shows that the source of the turbulent kinetic energy is the net work done in differentially accelerating the fluid in the presence of the mean pressure gradient. Combining these equations and using $u_3'' = \bar{u}_3 - \tilde{u}_3 = -\tilde{u}_3$, gives $\frac{\partial}{\partial t}\{< \rho \frac{u_i'' u_i''}{2} > + \bar{\rho}\frac{(\tilde{u}_3)^2}{2}\} = \frac{\partial}{\partial t} < \rho \frac{u_i' u_i'}{2} >= \overline{\rho' u_3' g} - \epsilon$. It helps here to note that, in general, $< \rho u_i'' u_i'' >=< \rho u_i' u_i' > -\overline{\rho u_i''}\ \overline{u_i''}$, and $\overline{\rho \tilde{u}_i \tilde{u}_i} = \bar{\rho}\,\bar{u}_i\,\bar{u}_i - 2\bar{\rho}\,\bar{u}_i\ \overline{u_i''} + \bar{\rho}\ \overline{u_i''}\ \overline{u_i''}$. Evidently, part of the work done against gravity appears as the kinetic energy of the mean motion (associated with the net mass flux) and the remainder appears as the turbulent kinetic energy. The prediction of turbulence thus hinges on predicting the evolution of the turbulent mass flux for which a transport equation (10) was noted earlier. Using the data from numerical simulations Sandoval (1995) shows that (10) is well-approximated by $\frac{\partial}{\partial t}\overline{\rho' u_3'} = (\bar{\rho}\,\bar{v} - 1)\frac{\partial \bar{p}}{\partial x_3} + \overline{\rho v' \frac{\partial p'}{\partial x_3}}$, and in particular note the impact of large density fluctuations on the evolution of turbulence. Even for initial $(\overline{\rho'^2})^{1/2}/\bar{\rho}$ of 0.52 the changes observed in the TKE evolution (when compared to the Boussinesq case) are small.

For initial $(\overline{\rho'^2})^{1/2}/\overline{\rho}$ above 0.1 clear non-boussinesq effects are observed, however their impact on the evolution of TKE and density variance is mild. Further studies (numerical and experimental) of turbulence with large density variations are needed to guide and validate the modeling of variable density flows.

2.7. BAROCLINIC VORTICITY

The differential acceleration is also responsible for baroclinic generation of vorticity. The momentum balance (3) implies that vorticity, ω_i, is transported according to (neglecting viscous terms):

$$\frac{D}{Dt}\frac{\omega_i}{\rho} = \frac{\omega_j}{\rho}\frac{\partial u_i}{\partial x_j} + \left(\frac{\nabla\rho \times \nabla p}{\rho^3}\right)_i$$

The baroclinically generated vorticity is oriented normal to the local gradients of pressure and density and its rate of generation depends on these gradients and the angle between them. If one of the gradients is very steep the baroclinically generated vorticity can be regarded as being deposited on a vortex sheet with variable sheet strength. In many cases involving sudden acceleration, the baroclinic vorticity rolls up into vortices and induces differential motion. In turbulent flows one expects that the baroclinic vorticity is also tilted, stretched and intesified by turbulence (e.g. Sandoval 1995). Baroclinic vorticity may be useful for mixing enhancement under high-speed flow conditions, and idealized configurations have been studied in shock tube experiments (e.g. Haas and Sturtevant 1987, Jacobs 1993). Baroclinic vorticity is a central element of the Rayleigh-Taylor and Richtmeyer-Meshkov instabilities. In the problem involving the passage of turbulence through a shock wave, the baroclinic generation occurs at the distorted shock front. The mean compression in the shock wave also distorts the incident turbulence. Nonlinear behavior has been noted in the downstream development of this anisotropic turbulence (Lee et al. 1997). Recent studies (Mahesh et al. 1997) have also shown that the baroclinic generation of vorticity due to incident entropy fluctuations may be quite important in the turbulence amplification across a shock wave.

Baroclinic generation of vorticity does not require compressibility, the only requirement is that the acceleration of the fluid parcels be rotational. Stably stratified mixing layers rolling up under the Kelvin-Helmholtz instability show a secondary instability due to misaligned density gradients (perpendicular to the braids) and pressure gradients (pointing towards the stagnation points) and show secondary roll up at small-scales (Staquet, 1995). Baroclinic effects have also been noted in variable density mixing layers. Baroclinic vorticity generation in a variable density mixing layer subject to streamwise pressure gradient deserves further study. The baro-clinically generated vorticity is of the same sign as the mean flow (spanwise) vorticity, or is opposite to it, depending on whether the mean velocity gradient and the mean density gradients are parallel or antiparallel, and whether the flow is accelerated or decelerated by the mean pressure gradient. A study of these basic configurations may help develop turbulence models for the mass transport in variable density flows.

3. Variable Density Mixing Layer

Consider the mixing layer formed between two free-streams flowing in x_1 direction which are brought into contact at $x_1 = 0$. The mixing layer width $\delta(x_1)$ grows with x_1. Suppose that the conditions in the high and low speed free-streams are denoted by U_1, ρ_1, T_1, C_1, and U_2, ρ_2, T_2, C_2, respectively. Let $r = \frac{U_2}{U_1}$, and $s = \frac{\rho_2}{\rho_1}$. The static pressures in the two streams are assumed to be equal, $p_1 = p_2$. The shear across the mixing layer is, $\Delta U = U_1 - U_2$. When $\Delta U/C_1$ and $\Delta U/C_2$ are both much smaller compared to unity, the compressibility effect on the the mixing layer is negligible. Since the turbulent velocity fluctuations scale with ΔU, it follows that the 'hydrodynamic' pressure fluctuations p_i' scale as $\rho_r \Delta U^2$, where ρ_r is a representative density scale, and consequently $p_i'/p_1 \ll 1$. The low-speed mixing between the two streams thus occurs at nearly constant pressure (isobaric). Significant density fluctuations can still exist in this low-speed flow. The equation of state (5) suggests two independent ways in which significant density variations can arise: a) due to temperature difference, $\Delta T = T_1 - T_2$, between the two streams but with the same gas composition in the two streams, and b) when $T_1 = T_2$ but the composition of the two streams is different. With the same operating gas in both streams it is difficult, in practice, to obtain large density contrast between the two streams without simultaneously increasing the Mach number, so variable density effects have been explored using dissimilar gases in the two streams.

It is well known (Brown and Roshko 1974, referred to as BR here after) that the spreading rate of the low speed uniform density ($\rho_1 = \rho_2$) turbulent mixing layer, $\frac{d\delta^*}{dx^*}$, correlates linearly with the velocity ratio $\lambda = \frac{U_1-U_2}{U_1+U_2}$. This dependence can be understood as follows: the dominant eddies in the mixing layer convect at a speed $U_c = \frac{U_1+U_2}{2}$ and the mixing layer spreads due to the entrainment of free-stream fluid at a rate proportional to $\Delta U/2$ into the moving eddies. The entrainment is, however, asymmetric, and slightly more high-speed (Dimotakis 1986) fluid is entrained. When $\rho_1 \neq \rho_2$ two dominant changes occur. The convection speed U_c shifts towards the denser stream and assuming that the entrainment velocity is not impacted by density variations, it follows that (for given U_1 and U_2) the spreading rate is enhanced when the heavier fluid is carried in the slower stream. This is borne out by the BR experiments and is also consistent with the rapid initial spreading of Helium jets in air (e.g. Djeridane et al. 1996). It is worth noting that a factor of 50 change in the density ratio (at the same velocity ratio) in a mixing layer changes the growth rate by only a factor of 2, an observation which stresses that the growth rate reduction observed in high-speed mixing layers is a compressibility effect (BR). A formula for the convection speed U_c can be obtained to account for the density effects. In the moving frame the stagnation pressure of the two streams must balance at the stagnation point which exists between any pair of large eddies, i.e. $\rho_1(U_1 - U_c)^2 = \rho_2(U_c - U_2)^2$ or $U_c = \frac{U_1 + \sqrt{s}U_2}{1 + \sqrt{s}}$. Since $U_1 - U_c \neq U_c - U_2$ the mixing layer entrains more volume of the lighter fluid than the heavier fluid. A broadening of the mean density profile relative to the mean velocity profile into the dense fluid stream is also observed in the BR experiments. This broadening is most dramatic when the heavy stream is also the slower stream. Brown and Roshko interpret this as follows: the density interface between the two streams is convoluted by the engulfment associated with large eddies, but continuity of pressure implies that velocities induced in the denser stream are smaller. This is suggested as the reason why in traversing away from a low speed dense stream the mean composition changes appreciably well before the mean velocity shows any change. Fluctuating velocities were not measured in the BR experiments. Quantitative assessment of the effect of the density ratio on the profiles of Reynolds stresses in a low speed mixing layer are not available. Recent data in supersonic mixing layers shows some asymmetries (Bonnet et al. 1994). It would be interesting to know whether the profiles of $< \rho u'_i u'_j >$ are nearly symmetric. Density weighted scaling is effective in accounting for the effects due to density variations which are observed in supersonic boundary layers on adiabatic walls (Morkovin 1961, Bradshaw 1977). A recent numerical study of supersonic channel flow (Coleman et al. 1995, Huang et al. 1995) confirms that in non-hypersonic wall-bounded flow the direct effects of compressibility are minimal; the principal deviations from a uniform density flow are explained as the effects of variable density.

Fiedler et al. (1991) report on co-gradient and counter-gradient mixing layers between air-CO_2 streams. In the co-gradient configuration the mean velocity and mean density gradients are parallel, these are antiparallel for the counter-gradient case. Some limited data on the correlation between the velocity fluctuations and density fluctuations is given. Significant qualitative differences in the profiles of $\overline{u'_1 \rho'}$ were reported for the two cases. A comprehensive set of measurements taken in a variable density mixing layer would be very valuable for testing the turbulence models for variable density flows in greater detail. Studies of variable density mixing layer at density ratios outside the 1/7 to 7 range are needed to resolve some questions which have arisen from recent compressible mixing layer data which involve more extreme density ratio (Lu and Lele 1994). In particular the growth rate of low speed variable density mixing layers at extreme density ratios do not appear to have been measured.

4. Compressible Mixing Layer

Brown and Roshko (1974) showed that the reduction in the growth rate observed for a supersonic mixing layer (one stream supersonic and the other at rest) cannot be explained as a density ratio effect. By analyzing the governing equations vis-a-vis the variable density low speed case, they concluded that if the profiles of $-\overline{u_1' u_2'}$ and $\overline{\rho' u_2'}$ are not influenced by Mach number, this *compressibility effect* should be associated with an enhanced role played by the energy flux due to pressure disturbances $\overline{p' u_2'}$ in a compressible flow. Much recent work has been undertaken on compressible mixing layers since this pioneering study. Experiments by Papamoschou and Roshko (1988) show that the intrinsic compressibility effect on the growth rate can be isolated by plotting the growth rate of the compressible mixing layer normalized by the growth rate of the corresponding incompressible mixing layer at the same velocity and density ratio, against the convective Mach number M_c. The convective Mach number is defined as the Mach number of the free-streams relative to the large eddies of the flow. In the moving frame an isentropic model is used to relate the stagnation pressure obtained from the two free-streams. For the case when the specific heat ratio $\gamma_1 = \gamma_2$ this model gives the convection velocity of the large eddies as, $U_c = \frac{C_2 U_1 + C_1 U_2}{C_1 + C_2}$, and $M_{c_1} = M_{c_2} = \frac{\Delta U}{C_1 + C_2}$. When $\gamma_1 \neq \gamma_2$, the expression for U_c is more involved and $M_{c_1} \neq M_{c_2}$.

Experiments (compiled by Lele (1994)) have supported the usefulness of the convective Mach number as a parameter to describe the compressibility effects but attempts to collapse the growth rate data on a single curve also show significant 'scatter'. Different definitions of mixing layer thickness, different degree of self-similarity and splitter-plate boundary layer effects are some of the causes for the scatter. Using only the data based on the same thickness measure (e.g. vorticity thickness) appears to significantly reduce the scatter. The experiments have also provided more information on the structure of the flow. The images of the flow obtained with non-intrusive diagnostics (Clemens and Mungal 1992, 1995) show that as M_c increases the quasi-two-dimensional Brown-Roshko roller eddies which span across the flow are no longer discernable. Instead highly three-dimensional turbulent eddies are observed. Whether the eddies dominating the momentum and species transport span the width of the shear layer at high M_c is not clear.

Some differences between the two edges of the shear layer, in the three-dimensional activity and intermittancy, have also been reported (Bonnet *et al.* 1993). Measurements of velocity fluctuations suggest a decrease of the maximum values of $\overline{u'_1 u'_1}/\Delta U^2$, $\overline{u'_2 u'_2}/\Delta U^2$, and $-\overline{u'_1 u'_2}/\Delta U^2$ with M_c (Elliot and Samimy 1990, Goebel and Dutton 1991, Bonnet *et al.* 1994). It also appears that the maximum shear stress correlation coefficient and the maximum of $-\overline{u'_1 u'_2}/q^2$ do not vary significantly with M_c.

Measurements of the convection velocity obtained with different experimental techniques have been compared to the isentropic model for U_c. The measurements based on some form of optical imaging of the flow (Papamoschou 1991a, Hall *et al.* 1993, Mahadevan and Loth 1994, Ramaswamy *et al* 1996) agree with the isentropic model only for small M_c, and for $M_c \geq 0.4$ a convection velocity which is close to either of the two free-streams is obtained. Dimotakis (1991), and Papamoschou (1991a) explain these by appealing to a non-isentropic model (with shocklets in one of the two streams). Barre (1994) claims that the shocklet notion is consistent with the data from confined mixing layers only, *i.e.* those with δ/D of the order 0.1, where D is the test section width. Scalar field based measurements are indirectly related to the fluctuating velocity, making it difficult to judge whether the measured U_c represents the motion of an actively entraining eddy or not. This question is important because two-point correlation data taken with hot-wire probes or fast response pressure probes gives U_c (in the middle of the shear layer) close to the isentropic formula for all M_c values (Samimy *et al.* 1992, Barre *et al.* 1994). Two-dimensional numerical simulations (Lele, 1989) also showed that the pressure minima and maxima move at a speed given by the isentropic formula. The measured values of U_c obtained with different techniques are in apparent conflict. The importance of U_c in determining the entrainment ratio makes such measurements fundamentally important. Further laboratory and numerical experiments should help to resolve this issue.

Even with the scatter in the plot of normalized growth rate versus M_c, there is consensus on the magnitude of the reduction in the growth rate; for M_c near unity the normalized growth rate of vorticity thickness is only about 40% (of the incompressible value), the visual thickness growth rate shows a larger suppression and is about 20% (of the incompressible value). Many explanations have been proposed to explain this large reduction. Notions of dilatational dissipation (Sarkar *et al.* 1991b) or shocklet dissipation (Zeman 1990), suppression of linear instability growth rates (Sandham and Reynolds 1990), reduction of shear-stress anisotropy due to high gradient Mach number (Sarkar 1995) and reduction of inter-component energy transfer due to suppressed pressure fluctuations (Vreman *et al.* 1996) are some of the proposed explanations. Some comments on these were made in Lele

(1994) and only the newer proposals will be discussed further. The numerical simulations of compressible plane mixing layers by Vreman *et al.* (1996) and compressible round mixing layers by Freund *et al.* (1997) have shed new light on the mechanism of growth rate suppression. The conclusions reached from these studies are quite similar. The latter simulations yield a turbulent flow with a Kolmogorov-like spectrum and the large-scales also show adequate decorrelation over the computational domain. The large suppression of growth rate for supersonic shear layers is readily evident in the simulations. It is found that dilatational effects on the evolution of turbulent kinetic energy are negligible even at a convective Mach number of 1 and no shocklets were observed. The growth rate suppression is accompanied by a reductions in $\overline{u_2' u_2'}/\Delta U^2$, $-\overline{u_1' u_2'}/\Delta U^2$, and $\overline{u_3' u_3'}/\Delta U^2$, while $\overline{u_1' u_1'}/\Delta U^2$ is hardly affected and the shear stress anisotropy also appears to not change much. Budgets of the individual Reynolds stress components were studied by Freund *et al.* and showed that the shear stress and hence growth rate suppression is linked to the reduction of inter-component energy transfer via the pressure-strain-rate terms. Pressure fluctuations were also found to be suppressed in the supersonic case. It is expected that further analysis of this data base will shed light on the change in the turbulence structure with Mach number, and help explain the behavior of convection velocity obtained by different methods of visualization. The experimental data on the compressible mixing layer growth rates for M_c well above unity is quite limited. The Papamoschou and Roshko (1989) data suggests that the normalized growth rate saturates. The physical mechanism leading to this saturation of growth under hypersonic conditions deserves further study.

5. Concluding Remarks

Many similarities between low speed variable density flows and compressible flows were noted throughout this paper. Phenomena which are intrinsically compressible were noted too. The existing data on turbulent flows from laboratory experiments and numerical simulations allows some of the compressibility effects to be identified, for example the well-known spreading rate suppression in a supersonic shear layer. Likewise when density variations have a global effect on the flow such as the rapid mixing of a helium jet in air, these are well noted in the literature. Detailed understanding of how such behavior should be modeled, in a way which is both physically accurate and useful for engineering predictions, continues to evolve. New experiments (laboratory and numerical) which are specifically designed to shed light on physical and modeling issues, and executed in conjunction with theoretical and modeling studies are critical for further advances.

Acknowledgements

The author is grateful to the organizers of the IUTAM symposium on "Low Speed Variable Density Flows", July 8-10, 1996 at Marseille, for their invitation to give this lecture at the symposium and for their kind hospitality. Discussions with Professors Cantwell, Linan, Ferziger and Mungal on some of the material are gratefully acknowledged. The author is also grateful to Dr. Krishnan Mahesh for his detailed comments on a draft of this paper which led to significant improvements.

References

Barre, S., Quine, C. and Dussauge, J. P. 1994 Compressibility effects on the structure of supersonic mixing layers: experimental results. *J. Fluid Mech.* **259**, 47-78.

Barre, S. 1994 Estimate of convective velocity in a supersonic turbulent mixing layer. *AIAA J.* **32**, 211-213.

Batchelor, G. K., Canuto, V. M. and Chasnov, J. R. 1992 Homogeneous buoyancy-generated turbulence. *J. Fluid Mech.* **235**, 349-378.

Besnard, D., Harlow, F. H., Rauenzahn, R. M. and Zemach, A. C. 1992 Turbulence transport equations for variable-density flows and their relationship to two-fluid models. Los Alamos National Laboratory Report LA-12303-MS.

Bilger, R. W. 1989 Turbulent diffusion flames. *Annual Rev. Fluid Mech.* **21** 101-135.

Blaisdell, G. A., Mansour, N. N., and Reynolds, W. C. 1993 Compressibility effects on the growth and structure of homogeneous turbulent shear flow. *J. Fluid Mech.* **256**, 443-485.

Bonnet, J. P., Debisschop, J. R. and Chambres, O. 1993 Experimental studies of the turbulent structure of supersonic mixing layers. *AIAA* Paper 93-0217.

Bonnet, J. P., Chambres, O., Lammari, M., Barre, S. and Braud, P. 1994 Couches de melange turbulentes supersoniques. Rapport final, Contract DRET 91/172, Aout 1994, Univ. Poitiers, CEAT.

Bradshaw, P. 1977 Compressible turbulent shear layers. *Ann. Rev. Fluid Mech.* **9** 33-54.

Bray, K. N. C. 1995 Turbulent transport in flames. *Proc. Royal Soc. Lond.* Ser. A **451** 231-256.

Breidenthal, R. 1990 The sonic eddy - a model for compressible turbulence. *AIAA* Paper 90-0495.

Brown, G. and Roshko, A. 1974 On density effects and large structure in turbulent mixing layers. *J. Fluid Mech.* **64**, 775-816.

Chassaing, P., Harran, G. and Joly, L. 1994 Density fluctuation correlations in free turbulent binary mixing. *J. Fluid Mech.* **279**, 239-278.

Clemens, N. T. and Mungal, M. G. 1992 Two- and three-dimensional effects in the supersonic mixing layer. *AIAA J.* **30**, 973-981.

Clemens, N. T. and Mungal, M. G. 1995 Large-scale structure and entrainment in the supersonic mixing layer. *J. Fluid Mech.* **284**, 171-216.

Coleman, G. N., Kim, J. and Moser, R. D. 1995 A numerical study of turbulent supersonic isothermal-wall channel flow. *J. Fluid Mech.* **305**, 159-183.

Dimotakis, P. E. 1986 Two-dimensional shear-layer entrainment. *AIAA J.* **24**, 1791-1796.

Dimotakis, P. A. 1991 Turbulent free shear layer mixing and combustion. In *High-speed Flight Propulsion Systems*, Prog. in Astronaut. Aeronaut., ed. S. N. B. Murthy, E. T. Curran, AIAA-Washington D. C., 265-340.

Djeridane, T., Amielh, M., Anselmet, F. and Fulachier, L. 1996 Velocity turbulence properties in the near-field region of axisymmetric variable density jets. *Phys. Fluids* **8**, 1614-1630.

Durbin, P. A. and Zeman, O. 1992 Rapid distortion theory for homogeneous compressed turbulence with application to modeling. *J. Fluid Mech.* **242**, 349-370.

Elliot, G. S. and Samimy, M. 1990 Compressibility effects in free shear layers. *Phys. Fluids A* **2**, 1231-1240.

Favre, A. 1969 Statistical equations of turbulent gases. In *Problems of hydrodynamics and continuum mechanics*, SIAM Philadelphia, 231-266.

Fiedler, H. E., Lummer, M. and Nottmeyer, K. 1991. Plane mixing layer between parallel streams of different velocities and different densities. In *Advances in Turbulence*, eds. H. Branover, Y. Unger, AIAA Series on Progress in Astronautics and Aeronautics **1993**, 40-52.

Freund, J. B., Lele, S. K. and Moin, P. 1997 Direct simulation of a supersonic round turbulent shear layer. *AIAA* Paper 97-0760.

Gaviglio, J. 1987 Reynolds analogies and experimental study of heat transfer in the supersonic boundary layer. *Int. J. Heat Mass Transfer* **30**, 911-926.

Goebel, S. G. and Dutton, J. C. 1991 Experimental study of compressible turbulent mixing layers. *AIAA J.* **29**, 538-546.

Haas, J. F. and Sturtevant, B. 1987 Interaction of weak shock waves with cylindrical and spherical gas inhomogeneities, *J. Fluid Mech.* **181**, 41-76.

Hall, J. L., Dimotakis, P. E. and Rosemann, H. 1993 Experiments in nonreacting compressible shear layers. *AIAA J.* **31**, 2247-2254.

Hinze, J. O. 1975 *Turbulence*, 2nd. edn., McGraw Hill.

Huang, P. G., Coleman, G. N. and Bradshaw, P. 1995 Compressible turbulent channel flows: DNS results and modeling. *J. Fluid Mech.* **305**, 185-218.

Jacobs, J. W. 1993 The dynamics of shock accelerated light and heavy gas cylinders. *Phys. Fluids A* **5**, 2239-2247.

Jones, W. P. 1992 Turbulence modeling for combustion flows. In *von Karman Institute for Fluid Dynamics* Lecture Series 1992-03.

Kida, S. and Orszag, S. 1990 Energy and spectral dynamics in forced compressible turbulence. *J. Sci. Comput.* **5** 85-125.

Kuo, K. K. 1986 *Principles of Combustion*, pp. 161-205. John Wiley.

Lee, S., Lele, S. K. and Moin, P. 1991 Eddy shocklets in decaying compressible turbulence. *Phys. Fluids A* **3** 657-664.

Lee, S., Lele, S. K. and Moin, P. 1997 Interaction of isotropic turbulence with shock waves: effect of shock strength. to appear in *J. Fluid Mech.*

Lele, S. K. 1994 Compressibility effects on turbulence. *Ann. Rev. Fluid Mech.* **26**, 211-254.

Libby, P. A. and Bray, K. N. C. 1981 Countergradient diffusion in premixed flames. *AIAA J.* **19** 205-213.

Lighthill, M. J. 1952 On sound generated aerodynamically: I. General theory. *Proc. Roy. Soc. London*, Ser. A **221** 564-587.

Lu, G. and Lele, S. K. 1994 On the density ratio effect on the growth rate of a compressible mixing layer. *Phys. Fluids* **6**, 1073-1075.

Mahadevan, R. and Loth, E. 1994 High-speed cinematography of compressible mixing layers. *Experiments in Fluids* **17**, 179-189.

Mahesh, K., Lele, S. K. and Moin, P. 1997 The influence of entropy fluctuations on the interaction of turbulence with a shock wave. *J. Fluid Mech.* **334**, 353-379.

McMurtry, P. A., Jou, W. H., Riley, J. J. and Metcalfe, R. W. 1986 Direct numerical simulation of a reacting mixing layer with chemical heat release. *AIAA J.* **24**, 962-970.

Moin, P., Squires, K., Cabot, W. and Lee, S. 1991 A dynamic subgrid scale model for compressible turbulence and scalar transport. *Phys. Fluids A* **3**, 2746-2757.

Morkovin, M. V. 1961 Effects of compressibility on turbulent flows. In *Mecanique de la Turbulence*, ed. A. Favre, Paris:CNRS, 367-380.

Morkovin, M. V. 1992 Mach number effects on free and wall turbulent structures in light of instability flow interactions. In *Studies in Turbulence*, ed. Gatski, T. B., Sarkar, S. and Speziale, C. G. Springer-Verlag, 269-284.

Panchapakesan, N. and Lumley, J. L. 1993 Turbulence measurements in axisymmetric jets of air and helium. Part 1 and 2. *J. Fluid Mech.* **246**, 197-247.

Papamoschou, D. 1991a Structure of the compressible turbulent shear layer. *AIAA J.* **29** 680-681.

Papamoschou, D. 1991b Effect of Mach number on communication between regions of a shear layer. *Proc. Eighth Symp. on Turbulent Shear Flows* 21-5-1 to 21-5-6.

Papamoschou, D. and Roshko, A. 1988 The compressible turbulent mixing layer: an experimental study. *J. Fluid Mech.* **197**, 453-477.

Papamoschou, D. and Lele, S. K. 1993 Vortex-induced disturbance field in a compressible shear layer. *Phys. Fluids A* **5** 1412-1419.

Papamoschou, D. 1995 Evidence of shocklets in a counterflow supersonic shear layer. *Phys. Fluids* **7**, 233-235.

Pope, S. B. 1987 Turbulent premixed flames. *Ann. Rev. Fluid Mech.* **19** 237-270.

Ramaswamy, M., Loth, E. and Dutton, J. C. 1996 Free shear layer interaction with an expansion-compression wave pair. *AIAA J.* **34**, 565-571.

Reynolds, W. C. and Kassinos S. C. 1995 One-point modeling of rapidly deformed homogeneous turbulence. *Proc. Roy. Soc. Lond. A* **451**, 87-104.

Ristorcelli, J. R. 1995 A pseudo-sound constitutive relationship and closure for the dilatational covariances in compressible turbulence: an analytical theory. *ICASE Report* 95-22, submitted to *J. Fluid Mech.*

Samimy, M., Elliot, G. S. and Reeder, M. F. 1992 Compressibility effects on large structures in free shear flows. *Phys. Fluids A* **4**, 1251-1258.

Sandham, N. D. and Reynolds, W. C. 1990 Compressible mixing layer: linear theory and direct simulation. *AIAA J.* **28** 618-624.

Sandoval, D. L. 1995 The dynamics of variable-density turbulence. Thesis, LA-13037-T, Los Alamos National Laboratory.

Sarkar, S., Erlebacher, G., Hussaini, M. Y. 1991a Direct simulation of compressible turbulence in a shear flow. *Theor. Comput. Fluid Dyn.* **2**, 291-305.

Sarkar, S., Erlebacher, G., Hussaini, M. Y. and Kreiss, H. O. 1991b The analysis and modeling of dilatational terms in compressible turbulence. *J. Fluid Mech.* **227**, 473-493.

Sarkar, S. 1992 The pressure-dilatation correlation in compressible flows. *Phys. Fluids A* **4**, 2674-2682.

Sarkar, S. 1995 The stabilizing effect of compressibility in turbulent shear flow. *J. Fluid Mech.* **282**, 163-186.

Shih, T. S., Lumley, J. L. and Janicka, J. L. 1987 Second order modeling of a variable density mixing layer. *J. Fluid Mech.* **180**, 93-116.

Starner, S. H. and Bilger, R. W. 1980 LDA measurements in a turbulent diffusion flame with axial pressure gradients. *Combust. Sci. Technol.* **21**, 259-276.

Staquet, C. 1995 Two-dimensional secondary instabilities in a strongly stratified shear layer. *J. Fluid Mech.* **296**, 73-126.

Vreman, A. W., Sandham, N. D. and Luo, K. H. 1996 Compressible mixing layer growth rate and turbulence characteristics. *J. Fluid Mech.* **320**, 235-258.

Williams, F. A. 1985 *Combustion Theory*, 2nd. edn., pp. 604-627. Addison Wesley.

Zank, G. P. and Matthaeus, W. H. 1991 The equations of nearly incompressible fluids. I. Hydrodynamics, turbulence, and waves. *Phys. Fluids A* **3**, 69-82.

Zeman, O. 1990 Dilatation dissipation: the concept and application in modeling compressible mixing layers. *Phys. Fluids A* **3** 951-955.

Zeman, O. 1991 On the decay of compressible isotropic turbulence. *Phys. Fluids A* **3**, 951-955.

BALANCE OF KINETIC ENERGY IN A SUPERSONIC MIXING LAYER COMPARED TO SUBSONIC MIXING LAYER AND SUBSONIC JETS WITH VARIABLE DENSITY

O. CHAMBRES, S. BARRE AND J.P. BONNET
C.E.A.T./L.E.A. URA CNRS 191
43, route de l'Aérodrome
86036 Poitiers Cedex - FRANCE

1. Introduction

In recent years, a lot of experimental and computational work was done to study the effect of compressibility on turbulent free flows. In particular, the supersonic mixing layer was extensively studied (see (Lele, 1994) for a review). Despite all these efforts it seems that, at this time, nobody knows yet what is the real mechanism which creates the compressiblity effects observed on supersonic free flows like mixing layers or jets. It has been observed by many experimentalists that the turbulent intensity is decreased while increasing compressibility. Different authors tried to explain this fact in order to be able to take in account these effects in modelling such flows. For example Zeman (1990) and Sarkar *et al.* (1989) both proposed a model based on an extra dissipation due to dilatation to explain the observed decrease of turbulent activity in high speed flows. Applications of such models to mixing layer computations described qualitatively the flow but the results are not accurate enough to make these models available for practical applications (Sarkar & Balakrishnan, 1990). Trying to understand what the real differences are between compressible and incompressible turbulence seems to be an interesting first step to increase our knowledge.
So, we decided to measure with 2D Laser Doppler Velocimetry, a preliminary turbulent kinetic energy budget in a highly compressible mixing layer (convective Mach number close to 1) with assumptions derived from the work of different authors ((Panchapakesan & Lumley, 1993), (Wygnanski & Fiedler, 1970) and (Gruber *et al.* , 1993)). Then we compare it to the balance obtained in subsonic jets or mixing layers with and without density gradients.

L. Fulachier et al. (eds.), IUTAM Symposium on Variable Density Low-Speed Turbulent Flows, 303–308.
© 1997 *Kluwer Academic Publishers.*

2. Experimental apparatus

The tests were carried out in a blow-down, high pressure wind-tunnel (maximum duration less than one minute) producing Reynolds numbers of the order of 10^6 per cm. The cross-sectional lengths of the test section are 150 mm × 150 mm. The assembly is composed, for the supersonic part, of a two-dimensional nozzle (see *Figure* 1). The subsonic part is fed by air issuing from a settling chamber on the supersonic side. A sonic throat controls the secondary stream flow rate.

The studied mixing layer is created with a Mach 3.2 supersonic flow merging with a Mach 0.2 subsonic flow. The stagnation temperature is the same for both flows. The resulting convective Mach number is 1., the velocity ratio is 0.14 and the density ratio is 0.35. So, the density of the subsonic flow is about one third of the supersonic one.

The mean and fluctuating velocity fields are obtained with a 2D Laser Doppler Velocimeter system processed by a Fast Fourier Transform (Aerometrics DSA). Several seedings have been tested: $Si\,O_2$, ($1\mu m$ average diameter) in the supersonic part alone, oil droplets in the subsonic stream alone or $Si\,O_2$ seeding in both streams. This last solution has been accepted as the better one under the present conditions (Bonnet *et al.* , 1994). Sample of 4000 to 5000 data are collected for each component at each position, y is the transverse direction, x the streamwise one.

3. Equation of the turbulent kinetic energy

With compressible flows, one uses the Favre averaging ($\Phi = \tilde{\Phi} + \phi"$ with $\overline{\rho\phi"} = 0$). Then one obtain a similarity with equations in incompressible flows.

3.1. ASSUMPTIONS IN MIXING LAYER

The mixing layer is supposed isobaric $\dfrac{\partial \overline{P}}{\partial x_i} = 0$. The flow is supposed two-dimensional in average so $\overline{W} = 0$ and $\dfrac{\overline{\partial(\ )}}{\partial z} = 0$ and stationary $\dfrac{\overline{\partial(\ )}}{\partial t} = 0$. Moreover the flow is homogeneous in z direction so that all correlations where w' is in an odd order, are null ($\overline{u'w'} = 0, \overline{v'w'^3} = 0, ...$).

Only for the pressure p and the density ρ, one uses the Reynolds decomposition. Two relations between Reynolds ($\Phi = \overline{\phi} + \phi'$ with $\overline{\phi'} = 0$) and Favre decompositions are: $\tilde{\Phi} = \overline{\Phi} + \dfrac{\overline{\rho'\phi'}}{\overline{\rho}}$ and $\overline{\phi"} = -\dfrac{\overline{\rho'\phi"}}{\overline{\rho}} = -\dfrac{\overline{\rho'\phi'}}{\overline{\rho}} \neq 0$.

We applied also the "Very Strong Reynolds Analogy" linking ρ' and u':

$\dfrac{\rho'}{\bar{\rho}} \simeq -\dfrac{T'}{\bar{T}} = (\gamma - 1)M^2\dfrac{u'}{\bar{U}}$ knowing that this assumption is very crude in our case. But from this equation, density-velocity correlation can be transformed into velocity correlation; this is very important for the present measurements because LDV gives all the velocity correlations in Reynolds fluctuations. That is why we have written all the terms in Reynolds averaging. The correlations in w' are not measured so we have taken the values of Wygnanski & Fiedler for the triple-velocity correlations $\overline{u'w'^2}$ and the data of Gruber et al. for $\overline{w'^2}$. For the $\overline{v'w'^2}$ correlation, the assumption that this correlation is close to $\overline{v'^3}$, is also made (Panchapakesan & Lumley, 1993).

3.2. TURBULENT ENERGY BALANCE

We obtain the equation giving that $k = \dfrac{1}{2}\overline{u_i''u_i''}$:

$$\underbrace{\left[-\frac{\partial}{\partial x}(\overline{\rho}k\tilde{U}) - \frac{\partial}{\partial y}(\overline{\rho}k\tilde{V})\right]}_{CONVECTION} + \underbrace{\left[-\overline{\rho u''u''}\frac{\partial \tilde{U}}{\partial x} - \overline{\rho u''v''}\frac{\partial \tilde{U}}{\partial y} - \overline{\rho v''v''}\frac{\partial \tilde{V}}{\partial y}\right]}_{PRODUCTION} + \underbrace{[-\overline{\bar{p}\tilde{\epsilon}}]}_{DISSIP.} +$$

$$\underbrace{\left[\frac{\partial}{\partial y}(\overline{\tau_{xy}u''} + \overline{\tau_{yy}v''} + \overline{\tau_{zy}w''} - \overline{p'v''} - \overline{\rho k v''})\right]}_{DIFFUSION} + \underbrace{\left[\overline{p'\frac{\partial u''}{\partial x}} + \overline{p'\frac{\partial v''}{\partial y}} + \overline{p'\frac{\partial w''}{\partial z}}\right]}_{COMPRESSIBILITY} = 0$$

4. Results and concluding remarks

Figure 2 shows the obtained budget for the experimental case under examination and with the assumptions described in the previous chapter. The general shape is qualitatively very classical. Data are represented in a dimensionless coordinate $y^* = \dfrac{y - y_{\text{ref}}}{\delta_\omega}$, δ_ω is the vorticity thickness and $y_{\text{ref}} = y((U_1 + U_2)/2)$. In this system, the maximum shear point is located at $y^* = 0$. The terms are non-dimensioned as following: by ΔU for the velocity, by ΔU^2 for the shear and by $\rho\Delta U^3/\delta_\omega$ for the terms of the balance. It can be shown that the production and convection terms are almost symmetrical. In contrast the diffusion and dissipation-compressibility terms are more important on the high velocity side. Comparison with subsonic data can now be made. *Figure* 3 shows the evolution of the production terms for various experimental cases. The horizontal coordinate is now r/L_u where r is the distance from the high velocity side and L_u is half the velocity thickness. This convention is used to make direct comparison with jets results. The maximum shear point is now located at $r/L_u = 1$. Data for subsonic jets (I.M.S.T.) have been obtained by Djeridane (1994) (these terms are divided by $\rho(U_c - U_e)^3/L_u$, U_c is the axial velocity and U_e the external one). Data for subsonic mixing layer come from an Air-Air flow (no density difference). Results show a different behavior between mixing layers and

jets for the production term. Density ratio effect seems to be quite small
and the production in supersonic mixing layer is about 20% less than in the
subsonic case. *Figure* 4 shows the convection terms of the k-equation. Density effects are now large for the jet. The supersonic mixing layer exhibits
less convection than the subsonic one. *Figure* 5 shows the diffusion term.
Data for the Air-Helium jet seems questionable. The subsonic mixing layer
and the Air-Air jet have a similar behavior. The diffusion seems to be 20%
less in the supersonic mixing layer.

It can be concluded that the diffusion term in the k-equation is approximately the same in jets and mixing layers. However, the global balance
is very different in the two cases because production and convection are
higher in the jet case, whatever the density ratio is. It seems that compressibility decrease production and perturb significantly the symmetry of
the diffusion across the mixing layer.

References

Bonnet J.P., Chambres O., Lammari M., Barre S. & Braud P. (1994), Couches de
   mélange turbulentes supersoniques, *Rapport final, contrat DGA/DRET N° 91/172,
   Août 1994.*

Djeridane T. (1994), Contribution à l'étude expérimentale de jets turbulents axisymétriques à densité variable, *Thèse de Doctorat de l'Université de Marseille, juillet
   1994.*

Gruber M. R., Messersmith N. L. & Dutton J. C. (1993), Three-Dimensional Velocity
   Field in a Compressible Mixing Layer, *AIAA Journal* Vol. **31**, *N° 11, pp. 2061-2067.*

Lele S. K (1994), Compressibility Effects on Turbulence, *Annual Review of Fluid Mechanics*, **Vol. 26.**

Panchapakesan N.R. & Lumley J.L. (1993), Turbulence measurements in axisymmetric
   jets of air and helium - Part 1 and 2, *J. Fluid Mech.*, **Vol. 246**, *pp. 457-473*

Sarkar S. & Balakrishnan L. (1990), Application of a Reynolds Stress Turbulence Model
   to the Compressible Shear Layer, *ICASE Report, N° 90-18.*

Sarkar S., Erlebacher G., Hussani M.Y. & Kreiss H.O. (1989), The Analysis and Modeling
   of Dilatational Terms in Compressible Turbulence, *ICASE Report, N° 89-79.*

Wygnanski I. & Fiedler H. E. (1970), The two-dimensional mixing layer, *J. Fluid. Mech.*,
   Vol. 41, part 2 *pp. 327-361.*

Zeman O. (1990), Dilatation Dissipation : The Concept and Application Compressible
   Mixing Layers, *Phys. Fluids A*, **Vol. 2**, *N° 2, pp. 178-188.*

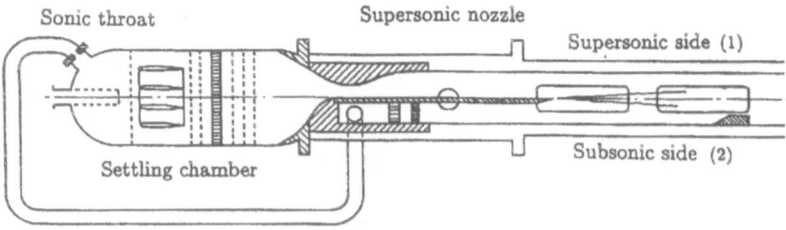

Figure 1. Description of the wind tunnel.

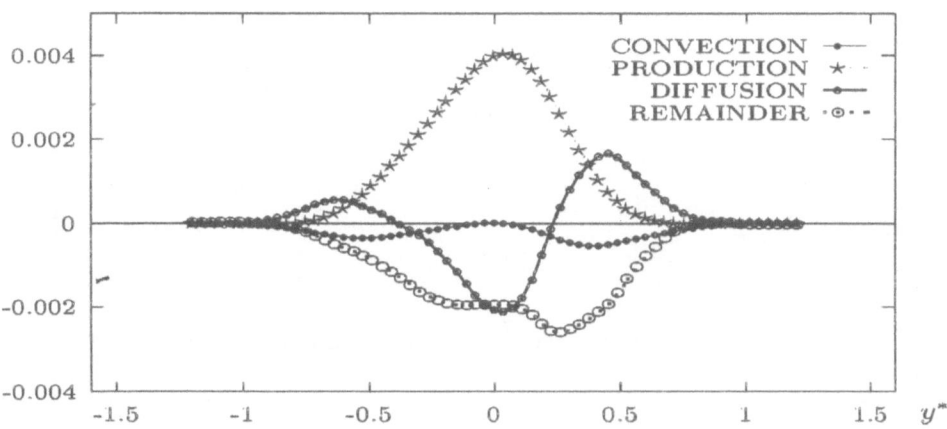

Figure 2.   Budget of turbulent kinetic energy in a supersonic mixing layer ($M_c = 1$)

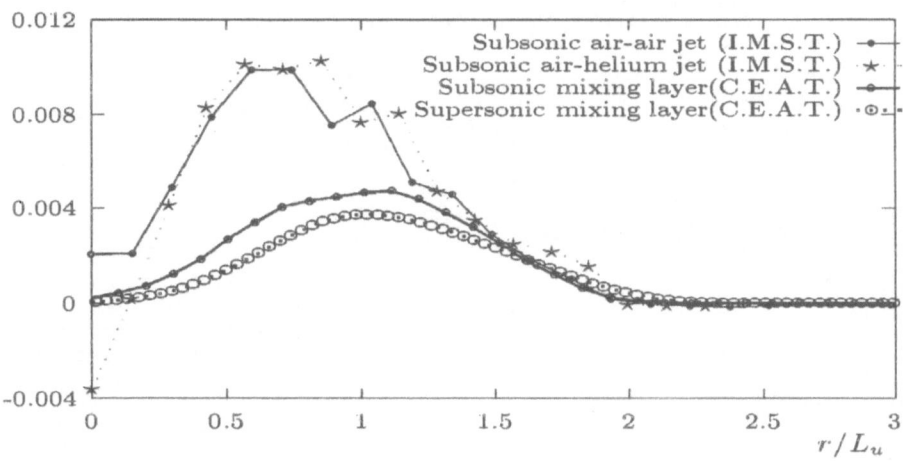

Figure 3.   Comparison of the <u>Production</u> in subsonic and supersonic mixing layers and jets.

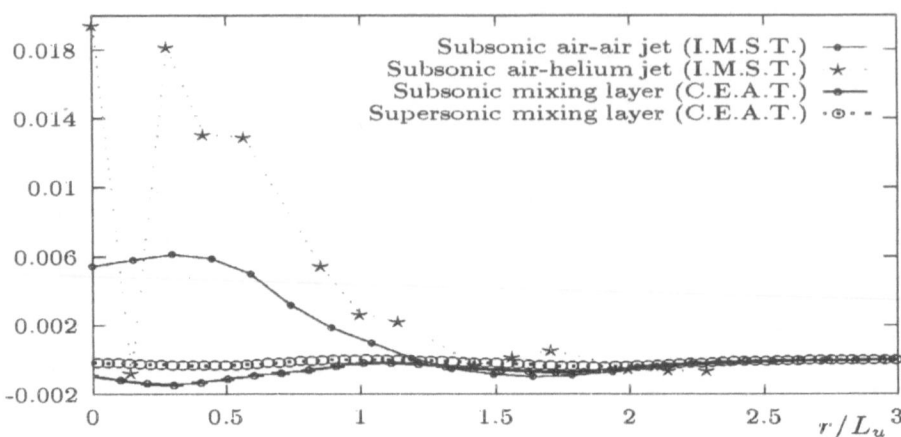

Figure 4. Comparison of the <u>Convection</u> in subsonic and supersonic mixing layers and jets.

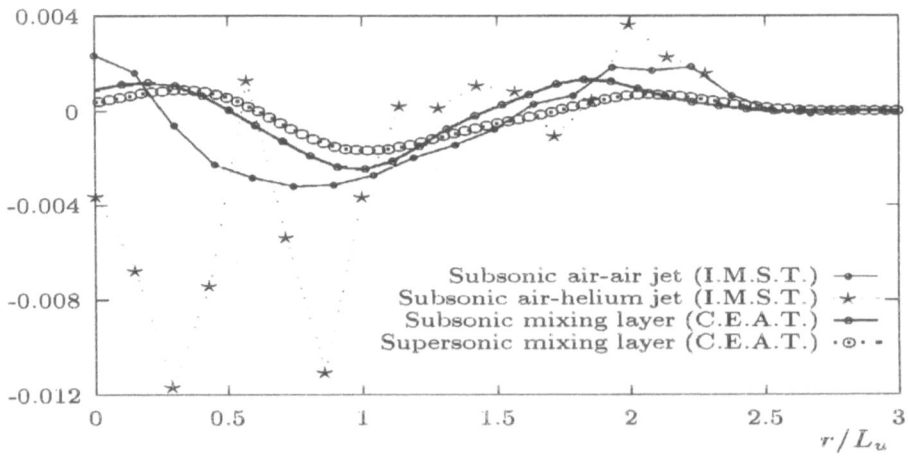

Figure 5. Comparison of the <u>Diffusion</u> in subsonic and supersonic mixing layers and jets.

COMPRESSIBILITY VS. DENSITY VARIATIONS AND THE STRUCTURE OF TURBULENCE:
A VIEWPOINT FROM EXPERIMENTS.

J.F. DEBIÈVE[+], P. DUPONT[+], A.J. SMITS[*], J.P. DUSSAUGE[+].

[+]*I.R.P.H.E., UMR CNRS Univ. Aix-Marseille I & II*
12, Av. Général Leclerc, 13003 Marseille, France
[*] *Gas Dynamics Laboratory, Princeton University*
Princeton, NJ08544, U.S.A.

1. Introduction.

The traditional description of turbulence in supersonic boundary layers by Morkovin (1962) assumes that the dynamics of energetic eddies is independent of the Mach number, as long as the flow is not hypersonic. This hypothesis leads to a description of supersonic turbulent boundary layers as flows with variable fluid properties (for example variable density and temperature), with the same underlying physical mechanisms as at low speeds. This implies in particular that the van Driest transformed velocity profile has a log-law with the same slope as in subsonic flows. A recent review by Spina, Smits and Robinson (1994) confirms this situation: the main differences in equilibrium boundary layers is found in the scale and orientation of the isocorrelations. On the other hand, it is well known that mixing layers and jets do not follow Morkovin's hypothesis since the overall properties of free shear flows are controlled by compressibility effects. The scope of this paper is to discuss from recent experimental

309

L. Fulachier et al. (eds.), IUTAM Symposium on Variable Density Low-Speed Turbulent Flows, 309–316.
© 1997 *Kluwer Academic Publishers.*

results what is known about the properties of compressible turbulence in supersonic
mixing layers and boundary layers, with consequences for simple turbulence models.

2. Supersonic mixing layers.

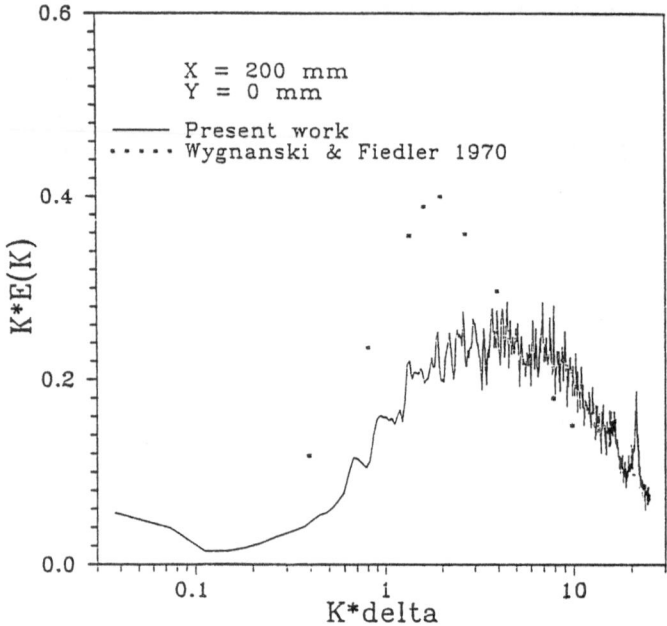

Fig. 1 : Comparisons of turbulence spectra in subsonic- and supersonic mixing layers

Turbulent diffusion is inhibited by compressibility in supersonic mixing layers. The
more popular explanation of the decrease of the rate of spread of the mixing layer is the
existence of shocklets associated with the turbulent motion. It is likely that such small
shock waves exist, but their importance in the reduction of turbulent kinetic energy by
energy dissipation and acoustic radiation has not been determined experimentally.
Recent measurements of turbulence involve visualizations, turbulence spectra and
Reynolds stress anisotropy. The visualizations made by Clemens and Mungal (1995)
suggest that at a convective Mach number of 0.62, the shape of the large scale structure
is significantly altered, with kinks in the shape of the eddies and a jagged interface. This
suggests that the energetic scales may be smaller in high speed mixing layers. An
experiment on a supersonic mixing layer at the same convective Mach number as in
Clemens & Mungal (1995) was performed by Dupont, Muscat, Dussauge (1995).

Turbulence spectra were measured by hot wire anemometry. An example is given in figure 1 where such a spectrum is compared with the measurements of Wygnanski & Fiedler (1970) in a subsonic mixing layer at about the same Reynolds number ($\Delta U\ \delta/\ \nu \sim 10^5$). The spectra E(k) are normalized to unity and the wave number k is normalized by the layer thickness δ. The premultiplied spectra k E(k) highlight the production zone where $E(k) \propto k^{-1}$; this range corresponds to the scales which extract most energy from the mean motion. It appears that the production scales extend over a much larger zone than at low speeds and that they involve larger wave numbers. Integral scales deduced from two-point correlation measurements indicate the same trend: their typical sizes are 0.2 δ while they are 0.44 δ in Wygnanski & Fiedler's experiment. The other point which was checked in this flow is the ratio $-\overline{u'v'}\ /\ \overline{u'}^2$ which is related to the anisotropy of the Reynolds stress tensor.

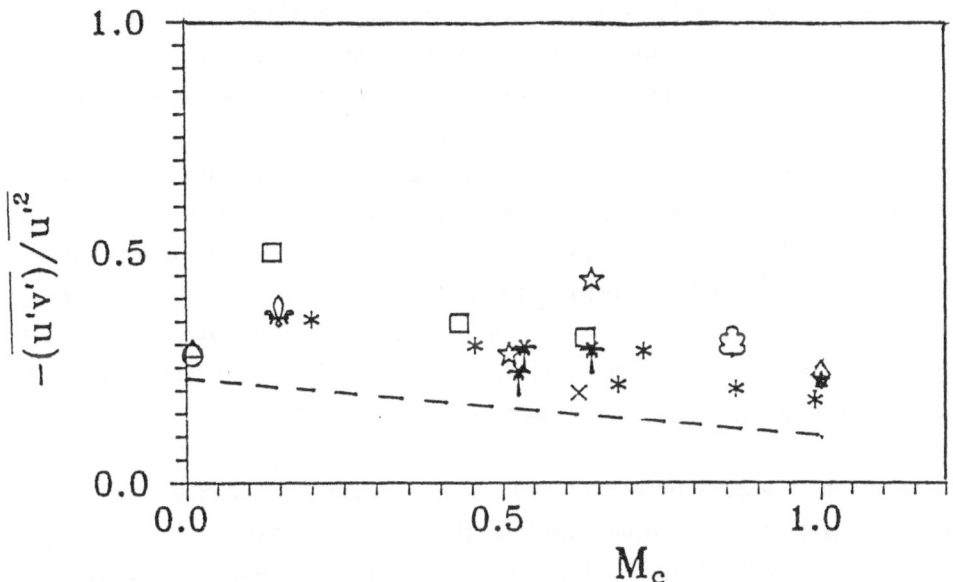

Fig. 2 : Ratio of Reynolds stresses in supersonic mixing layers (adapted from Barre et al. 1994)
Symbols : experiments ; details of the references in Barre et al. 1994.
Dashed line : results adapted from Sarkar (1995).

The experimental result are given in figure 2 where they are compared with the results of direct simulations of homogeneous compressible turbulence subjected to a mean shear with constant mean density (Sarkar (1995)). The results are scattered but the experiments suggest a decrease of this ratio of about 30%. The same trend is found in the homogeneous shear simulation, with a comparable decrease, although the turbulent Mach numbers are not the same in simulations and experiment. A possible interpretation of these results is a modification of the pressure field which leads to a change in the shape of the large eddies, and to a modification of the anisotropy of the turbulent stresses through the action of pressure strain- and pressure divergence terms. The importance of the role of possible shocklets has not yet been clearly assessed from experiments.

3. Supersonic boundary- layers.

Experimental results on supersonic boundary layer on a heated plate (T_w/T_a=2) at a Mach number of 2.3 show that the profiles of the van Driest transformed velocity account for the effect of the density change by heating, so that the same log law as on the adiabatic flat plate is found (Carvin, Debiève, Smits 1988, Dupont, Audiffren, Debiève Elena 1992). Similarly, as expected, the longitudinal velocity variance, in Morkovin's representation remains unchanged. This is in full agreement with Morkovin's hypothesis. Tomography by Rayleigh scattering has been performed in supersonic and hypersonic boundary layers and have been compared to low speed results. In the hypersonic case (M=8), the shape of the turbulent bulges involves essentially large scales. In this particular flow, the Reynolds number was small enough to produce similar effects even at low speeds, and, qualitatively speaking, no drastic effect of compressibility is observed on the shape of the eddies in such boundary layers, the principal effect apparently being due to Reynolds number.

However, some aspects of spectra of longitudinal velocity fluctuations in adiabatic boundary layers suggest possible departures from Morkovin's hypothesis. The energetic part of the spectrum is shifted to high frequencies, so that the integral- and production scales are significantly smaller than at low speed, essentially half of the subsonic value. This is illustrated in figure 3, adapted from Dussauge & Smits (1995), where a compilation of available data on production scales in the external part of the boundary layers was given. It is clear that again a reduction of the energetic scales is observed. In contrast, Laurent (1996) has shown that in the logarithmic region of the layer, the inertial range of the spectra of $\overline{u'^2}$ can be scaled as in subsonic boundary layers with a Kolmogorov representation (Fig. 4).

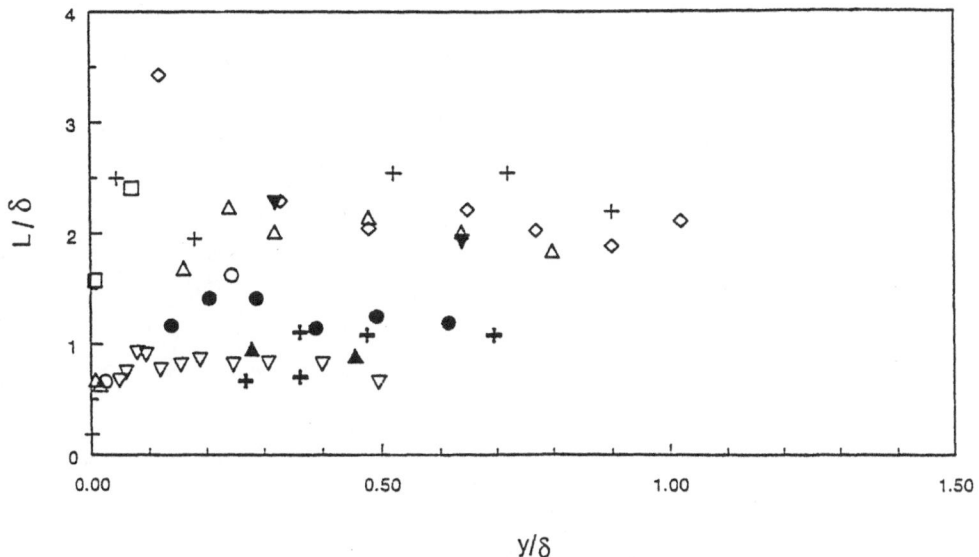

Fig. 3 : Production scales in boundary layers.
Subsonic data : + Klebanoff 1954, Δ Fulachier 1972, O, ☐ Fernholz et al. 1995 .
▲ Bestion et al., ▽ Audiffren, **+** Debiève, ● Spina & Smits, ▼ ◊ McGinley et al.
Details of the references to the experiments in Dussauge & Smits 1995.

In this figure, the Kolmogorov scales have been determined from the assumption that production balances dissipation. Note that the Kolmogorov scale is

$$\eta = \left(\frac{\rho_w}{\rho}\right)^{\frac{6m+3}{8}} \left[\frac{v_w}{u_\tau}\left(\kappa\,y^+\right)^{1/4}\right]$$

, where m is such that $\mu \propto T^m$, and $m \approx 0.75$. The quantity in square brackets is the Kolmogorov scale for constant density boundary layers with the same viscous scale, and considered at the same y^+. With $m \approx 0.75$, the exponent of the density ratio is close to 1. In adiabatic layers, $\rho_w < \rho$, which tends to reduce η. However, near the maximum of dissipation ($y^+ \approx 15$), the ratio ρ_w/ρ remains close to unity, and no major difference with the constant density case is expected.

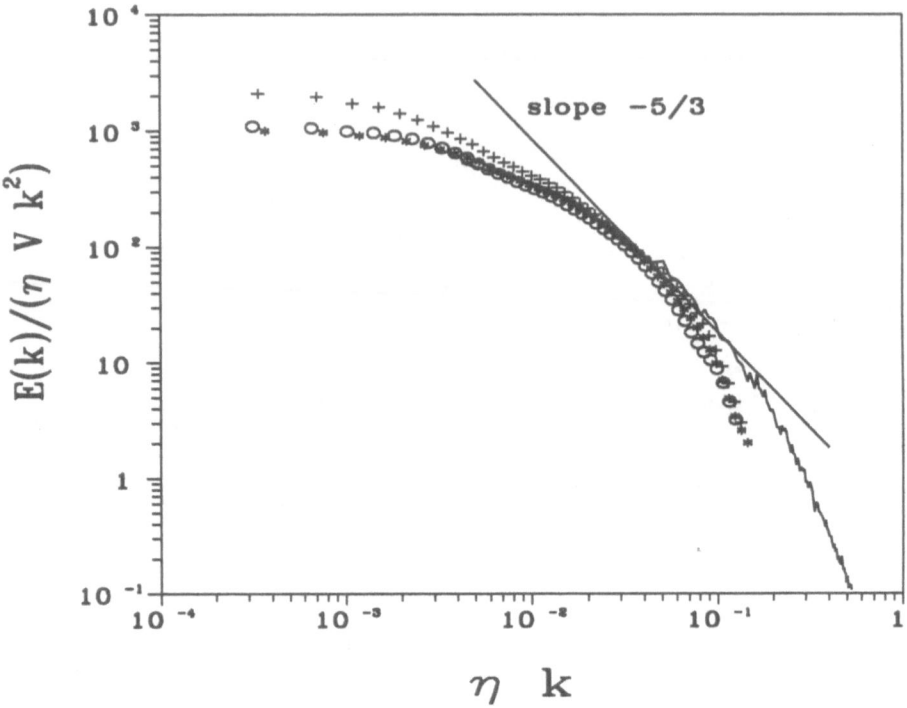

Fig. 4 : Comparisons of turbulence spectra in the log zone of subsonic and supersonic boundary layers.
Subsonic data : solid line y^+=68, from Antonia 1995
Supersonic data, M=2.3. O y^+=36 ; + y^+=61 ;* y^+=106. From Laurent (1996).

4. Discussion and conclusions.

As far as modelling issues are concerned, simple subsonic models such as mixing length hypothesis can be used to predict supersonic boundary layers with zero pressure gradient. This does not hold for mixing layers, and specific two-equation models of compressible turbulence have been developed for such flows. However, they are of limited accuracy in the case of boundary layers (Huang, Coakley, Bradshaw 1994). Several reasons can be found: i) the modelling of the equation for ε is not correct even for low speed variable density flows and ii) recent work by Sarkar (1995) indicates that the reduction of turbulence level in free shear layers could be due mainly to a decrease in the shear stress rather than to dissipation. This suggests that the mechanisms of redistribution related to the pressure fluctuations are altered by compressibility. Although there is no direct verification of this point, this mechanism is probably consistent with the trends inferred from the experimental results. In particular in boundary layers, the change in the shape of the spectra and in the production scale of $\overline{u'^2}$ occurs without changing the spatial scales for $-\overline{u'v'}$ and for turbulent transport (for example the mixing length). This can happen only if there is a modification of the level of fluctuations contributing to the energy, without changing turbulent transport. This modification implies also a change in the anisotropy of the Reynolds stress tensor. Since the inactive motions are linked to induced fluctuations, it seems that again a change in the structure of the pressure field may be inferred. These features appear even in boundary layers at moderate Mach numbers, and seem related to high speed phenomena rather than to density gradients, and can probably be considered as the first signs of the departure from Morkovin's hypothesis.

5. Acknowledgements

The authors are grateful to Prof. R. Antonia who kindly provided with the spectral data in subsonic boundary layer.

6. References.

Barre S., Quine C., Dussauge J.P. 1994, Compressibility effects on the structure of supersonic mixing layers : experimental results, *J. Fluid Mech.*, vol. 259, pp. 47-78.

Baumgartner M.L., Erbland P.J., Etz M.R., Yalin A., Muzas B.K., Smits A.J., Lempert W.R., Miles R.B. (1997), Structure of a Mach 8 turbulent boundary-layer, 35th Aerospace Science Meeting, Reno, NV, AIAA Paper 97-0765.

Carvin C., Debiève J.F., Smits A.J. (1988) The near wall temperature profile of turbulent boundary layers, AIAA Paper 88-0136.

Clemens N.T., Mungal M.G. (1995) Large scale structure and entrainment in the supersonic mixing layer, *J. Fluid Mech.*, Vol. 284, pp. 178-216.

Dupont P., Audiffren N., Elena M., Debiève J.F.(1992) Influence d'un apport pariétal de chaleur sur une couche turbulente supersonique, *Rev. Sc. et Tech. de la Défense*, Paris, N° 17, 1992-3.

Dupont P., Muscat P., Dussauge J.P. (1995) Space and Space-time statistics in a supersonic mixing layer, Symposium on Transitional and Turbulent Compressible Flow, ASME Fluids Engineering Conference, Hilton Head Island, SC.

Dussauge J.P., Smits A.J. (1995) Characteristic scales for energetic eddies in turbulent supersonic boundary layers, 10th T.S.F. Symposium, Penn State, U.S.A..

Dussauge J.P., Fernholz H.H., Finley P.J., Smith R.W., Smits A.J., Spina E. (1996) *Turbulent boundary layers in subsonic and supersonic flow*", AGARDograph 335, AGARD, Neuilly s/Seine.

Huang P. G., Bradshaw P., Coakley T.J. (1994) Turbulence models for compressible boundary layers, *AIAA Jl.*, Vol. 32, N°4, pp. 735- 740.

Laurent H. (1996) Turbulence d'une interaction onde de choc/ couche limite sur une paroi plane adiabatique ou chauffée, Thèse de Doctorat, Université de la Méditerranée (Aix-Marseille II), Marseille, France.

Morkovin M.V. (1962) Effects of compressibility on turbulent flows, *Mécanique de la Turbulence*, A. Favre ed., C.N.R.S., Paris.

Sarkar S. (1995) The stabilizing effect of compressibility in turbulent shear flow, *J. Fluid. Mech.*, vol. 282, pp. 163-186.

Spina E.F., Smits A.J. and Robinson S.K. (1994) *Ann. Rev. Fluid Mech.*, 26: 287- 319.

Wygnanski I., Fiedler H. (1970) The two-dimensional mixing region, *J. Fluid Mech.* 41, 327-361.

A NUMERICAL METHOD FOR COMPUTING FLOWS FOR ARBITRARILY SMALL AND MEDIUM MACH NUMBERS

H. BIJL AND P. WESSELING
Delft University of Technology,
Faculty of Technical Mathematics and Informatics,
PO Box 5031, 2600 GA Delft, The Netherlands

1. Introduction

We have developed a method for the numerical computation of flows with arbitrarily small and medium Mach numbers. This method is accurate and efficient for Mach numbers ranging between 0 and $\mathcal{O}(1)$. The equations are nondimensionalised such that all quantities remain finite for all Mach numbers. Discretisation is done in general coordinates on a staggered grid, using an upwind-biased density evaluation when the local Mach number comes above 0.9. The system is solved iteratively with the pressure correction method. The discretisation and solution method become identical to the classical incompressible method [4] as the Mach number tends to zero. Mach number independent accuracy and convergence is confirmed by numerical experiments.

2. Governing Equations

Adiabatic compressible flow of inviscid fluids is considered, so that the flow is governed by the Euler equations. We use the pressure as primary unknown instead of the density since in the incompressible limit the density becomes constant, whereas the pressure can still vary. In Cartesian tensor notation the equations are given by

$$\text{continuity eq.} \qquad \left(\frac{\partial \rho}{\partial p}\right)_h \frac{\partial p}{\partial t} + \left(\frac{\partial \rho}{\partial h}\right)_p \frac{\partial h}{\partial t} + (\rho u^\alpha)_{,\alpha} = 0 \qquad (1)$$

$$\text{momentum eq.} \qquad \frac{\partial \rho u^\alpha}{\partial t} + \left(\rho u^\alpha u^\beta\right)_{,\beta} = -p_{,\alpha} \qquad (2)$$

$$\text{energy eq.} \qquad \frac{\partial h}{\partial t} + (h u^\alpha)_{,\alpha} = -(\gamma - 2)\, h\, u^\alpha_{,\alpha} \qquad (3)$$

L. Fulachier et al. (eds.), IUTAM Symposium on Variable Density Low-Speed Turbulent Flows, 317–324.
© 1997 *Kluwer Academic Publishers.*

where $u^\alpha = u_\alpha$ are the Cartesian velocity components, ρ is the density, p is the pressure, and h is the enthalpy. The equation of state completes the system of equations:

$$\rho = \frac{\gamma}{\gamma - 1} \frac{p}{h}. \tag{4}$$

We will discuss internal flow in a two-dimensional channel or nozzle, the inflow and outflow boundaries are referred to by subscripts ∞ and out, respectively. When the inflow is subsonic, the boundary conditions are:

$$u_\infty^\alpha = u_\infty^\alpha(y, t) = given, \quad \alpha = 1, 2$$
$$h_\infty = h_\infty(y, t) = given,$$
$$p_{out} = p_{out}(y, t) = given.$$

For supersonic inflow, all variables should be specified on the inlet boundary, so instead of the pressure at the outflow, $p_\infty = p_\infty(y, t) = given$. At a supersonic outlet boundary all variables are linearly extrapolated from the inner region. On solid boundaries the impermeability condition is prescribed: $u_\alpha n^\alpha = 0$ with $\mathbf{n}$ the outward normal on the solid wall. The initial conditions specify p, u^α, and h.

3. Nondimensionalisation

The variables are non-dimensionalised such that they remain non-zero and finite in the limit of the characteristic Mach number $M_{ref} \to 0$. In this limit we want the non-dimensionalised Euler equations to be well defined and to reduce to the incompressible Euler equations too. This is not the case for standard compressible flow computing methods, which as a consequence break down for low Mach numbers. It is assumed that the scale of the boundary conditions is characteristic for the magnitude of the variables. Therefore the Mach number at the inlet M_∞ is chosen as reference Mach number M_{ref}. For a motivation of the units adopted see [2]. The nondimensional variables are:

$$x^1 = \frac{x^{1*}}{L^*} \qquad x^2 = \frac{x^{2*}}{L^*}.$$
$$t = \frac{t^* w_\infty^*}{L^*} \qquad p = \frac{p^* - p_{out}^*}{\rho_0^* w_\infty^{*2}}$$
$$h = \frac{h^*}{h_0^*} \qquad \rho = \frac{\rho^*}{\rho_0^*}$$
$$u^1 = \frac{u^{1*}}{w_\infty^*} \qquad u^2 = \frac{u^{2*}}{w_\infty^*},$$

where * indicates that variable is dimensional. Furthermore, w_∞^* is the speed at the inlet, L^* a reference length and p_{out} the pressure at the outlet. As reference enthalpy and density the stagnation conditions are chosen:

$$h_0^* = \left(1 + \frac{\gamma - 1}{2} M_\infty^2\right) h_\infty^*, \tag{5}$$

$$\rho_0^* = \left(1 + \frac{\gamma - 1}{2} M_\infty^2\right)^{\frac{1}{\gamma - 1}} \rho_\infty^*. \tag{6}$$

The nondimensional Euler equations are in this case the same (1)-(3). The nondimensional equation of state becomes:

$$\rho = \rho(p, h) = \frac{\gamma}{\gamma - 1} \frac{{w_\infty^*}^2}{h_0^*} \frac{p}{h} + \frac{\gamma}{\gamma - 1} \frac{p_{out}^*}{\rho_0^* h_0^*} \frac{1}{h} = \tag{7}$$

$$\frac{\gamma M_\infty^2}{1 + \frac{\gamma - 1}{2} M_\infty^2} \frac{p}{h} + \left[p_v \left(\left(1 + \frac{\gamma - 1}{2} M_\infty^2\right)^{\frac{-\gamma}{\gamma - 1}} - 1\right) + 1\right] \frac{1}{h},$$

where p_v is defined by:

$$p_v = \frac{p_{out}^* - p_0^*}{p_\infty^* - p_0^*}. \tag{8}$$

The non-dimensional subsonic boundary conditions are:

$$\begin{aligned}
(\rho u)_\infty^1 &= \left(1 + \frac{\gamma-1}{2} M_\infty^2\right)^{\frac{-1}{\gamma-1}} cos\alpha_\infty, \\
(\rho u)_\infty^2 &= \left(1 + \frac{\gamma-1}{2} M_\infty^2\right)^{\frac{-1}{\gamma-1}} sin\alpha_\infty, \\
h_\infty &= \left(1 + \frac{\gamma-1}{2} M_\infty^2\right)^{-1}, \\
p_{out} &= 0,
\end{aligned} \tag{9}$$

where α_∞ is the inlet flow angle. For supersonic inflow p_∞ must be prescribed instead of p_{out}. For the boundary condition on the momentum components M_∞ and α_∞ need to be known, while the other two boundary conditions follow directly from the scaling used. The nondimensional problem can now be solved when M_∞, α_∞, and p_v are given; M_∞ and α_∞ are necessary for the nondimensional boundary conditions, and p_v appears in the scaled equation of state. Note that the solution is independent of ρ_∞^*, but to recover the original variables from the nondimensional ones we need both h_∞^* and ρ_∞^*.

4. Invariant formulation

For computing flows in general domains the governing equations are recast in a coordinate-invariant form. The compressible Euler equations in Cartesian coordinates of (1), (2), and (3) are not tensor equations because α

occurs both as a free superscript and a free subscript, and we have a pair of equal subscripts. Raising subscripts as required by contraction with the metric tensor [1] we obtain tensor equations, which are then valid in general coordinates, according to Ricci's lemma:

$$\left(\frac{\partial \rho}{\partial p}\right)_h \frac{\partial p}{\partial t} + \left(\frac{\partial \rho}{\partial h}\right)_p \frac{\partial h}{\partial t} + (\rho U)^\alpha_{,\alpha} = 0 \tag{10}$$

$$\frac{\partial (\rho U)^\alpha}{\partial t} + \left((\rho U)^\alpha U^\beta\right)_{,\beta} = -\left(g^{\alpha\beta} p\right)_{,\beta} \tag{11}$$

$$\frac{\partial h}{\partial t} + U^\alpha h_{,\alpha} = -(\gamma - 1)h\, U^\alpha_{,\alpha} \tag{12}$$

with $(\rho U)^\alpha = \mathbf{a}^{(\alpha)} \cdot (\rho \mathbf{u})$ the contravariant momentum components, where the contravariant metric tensors are defined as

$$g^{\alpha\beta} = \mathbf{a}^{(\alpha)} \cdot \mathbf{a}^{(\beta)}, \quad \mathbf{a}^{(\alpha)} = \frac{\partial \xi^\alpha}{\partial \mathbf{x}} \tag{13}$$

the contravariant base vectors with respect to the mapping $T : \mathbf{x} = \mathbf{x}(\xi)$. Here, $\mathbf{x}$ are the Cartesian coordinates in the physical domain Ω and ξ are the boundary-conforming curvilinear coordinates in the computational domain G. The mapping is assumed to be regular, i.e. the Jacobian of the transformation, that will be denoted as $\sqrt{g}$, does not vanish. For more details see [6, 9].

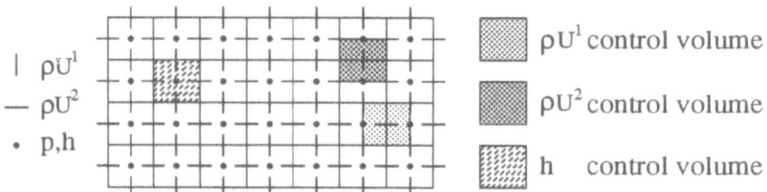

Figure 1. Cell in computational domain G and its image in the physical domain Ω

5. Discretisation in space

The compressible Euler equations will be discretised in space in boundary-fitted coordinates using a finite volume technique on a staggered grid. We have chosen for a staggered grid because of the following advantages for incompressible flow. To begin with, the differential equations can be discretised in a natural and straightforward way. Furthermore, the physical boundary conditions suffice and no measures have to be taken to avoid spurious pressure oscillations. The main difference with methods for the incompressible case is that the density is an unknown variable, which results in an extra term in the continuity equation and an extra equation:

the equation of state. Our method will be chosen such that as $M_\infty \downarrow 0$ the classical incompressible staggered grid method [4] is recovered. This may be expected to give Mach-independent convergence for small and medium Mach numbers. Figure 1 shows part of the computational grid with the staggered placement of the unknowns and the corresponding control volumes. The discretisation in general coordinates introduces mass flux components ρV^α, $V^\alpha = \sqrt{g} U^\alpha$. For more details on invariant discretisation in general coordinates, see [5].

6. Pressure Correction Method

Fully implicit discretisation in time of the continuity and momentum equations gives:

$$\left(\frac{\partial \rho}{\partial p}\right)_h^{n+1} \frac{p^{n+1} - p^n}{\delta t} + \left(\frac{\partial \rho}{\partial h}\right)_p^{n+1} \frac{h^{n+1} - h^n}{\delta t} + (\rho U^\alpha)_{,\alpha}^{n+1} = 0 \quad (14)$$

$$\frac{(\rho U^\alpha)^{n+1} - (\rho U^\alpha)^n}{\delta t} + \left((\rho U^\alpha)^{n+1} (U^\beta)^{n+1}\right)_{,\beta} = -\left(g^{\alpha\beta} p^{n+1}\right)_{,\beta}. \quad (15)$$

An efficient way to solve this system is the pressure correction method a standard workhorse for incompressible flows [4]. In this method, first, a predictor momentum field $(\rho U^\alpha)^*$ is computed from

$$\frac{(\rho U^\alpha)^* - (\rho U^\alpha)^n}{\delta t} + \left((\rho U^\alpha)^* (U^\beta)^n\right)_{,\beta} = -\left(g^{\alpha\beta} p^n\right)_{,\beta}, \quad (16)$$

where the Picard method is used for the linearisation of the convection term. Next a pressure correction $\delta p = p^{n+1} - p^n$ is computed. To find this correction equation, first (16) is subtracted from (15) neglecting the difference in the convection term:

$$\frac{(\rho U^\alpha)^{n+1} - (\rho U^\alpha)^*}{\delta t} = -\left(g^{\alpha\beta} (p^{n+1} - p^n)\right)_{,\beta}, \quad (17)$$

Van Kan [7] has shown for the incompressible case that the effect of this neglect on the accuracy is for the Crank-Nicholson method negligable, the resulting velocity at the new time level remains $\mathcal{O}(\delta t^2)$ accurate. Then the divergence of (17) is taken; with substitution of (14), taking the coefficients $\left(\frac{\partial \rho}{\partial p}\right)$ and $\left(\frac{\partial \rho}{\partial h}\right)$ at the previous time level, this gives:

$$\left(\frac{\partial \rho}{\partial p}\right)_h^n \frac{\delta p}{\delta t} + \left(\left(g^{\alpha\beta} \delta p\right)_{,\beta}\right)_{,\alpha} = \frac{(\rho U^\alpha)_{,\alpha}^*}{\delta t} - \left(\frac{\partial \rho}{\partial h}\right)_p^n \frac{\delta h}{\delta t} \quad (18)$$

When $\delta h = h^{n+1} - h^n$ is known, the pressure correction δp can be computed from (18), whereafter $(\rho U^\alpha)^{n+1}$ can be found from (17). This δh can be

computed from the following discretisation of the energy equation, which
is implicit in h but explicit in the other unknowns:

$$\frac{h^{n+1} - h^n}{\delta t} + (U^\alpha)^n h_{,\alpha}^{n+1} = -(\gamma - 1)h^{n+1}(U^\alpha)_{,\alpha}^n. \tag{19}$$

To obtain at the end of a time step the velocity components $(U^\alpha)^{n+1}$ from
$(\rho U^\alpha)^{n+1}$, these mass fluxes have to be divided by the density ρ^{n+1}, which
is computed from the equation of state $\rho^{n+1} = \rho(p^{n+1}, h^{n+1})$ see (7). In
practice the Euler equations are first discretised in space before the pressure correction method is applied, so that the equations derived in this
section are linear systems of discrete equations. These systems were solved
by a Krylov subspace iterative method for unsymmetric matrices, namely
GMRES, see [8].

7. Numerical results

First flow in a channel with a 10% bump is computed. The size of the channel is $[0, 3] \times [0, 1]$, in which a boundary fitted nonuniform grid of 63×22
cells is generated. We use time-stepping to obtain a steady solution. For this
the following termination criterion is used: $\max_{i,j} \left((u_{i,j}^\alpha)^{n+1} - (u_{i,j}^\alpha)^n \right) \le$
$\delta t 10^{-6}$. Since, the velocity components, $u_{i,j}^\alpha$, are non-dimensionalised by
the inlet speed, scaling of the termination criterion is unnecessary. As initial conditions for the time iteration, we use $\rho U^1 = \rho U^2 = 0, h = h_\infty, p_s =$
0. Computations were carried out for the following inlet Mach numbers:
$M_\infty = 10^{-6}, 0.01, 0.1, 0.5,$ and 0.675. The boundary conditions, described
in Chapter 3, depend on three parameters: the inlet Mach number, the angle of the flow at the inlet, and the pressure ratio p_v. For the channel with
bump the last two parameters were chosen to be: $\alpha_\infty = 0$, $p_v = 1$. For an
inlet Mach number of $M_\infty = 0.5$ the resulting iso-Mach lines are shown in
Figure 2. Note that the picture is symmetric, as it should be for subsonic
flow through the specified channel. The computed Mach number distributions on the upper and lower boundary of the channel are compared to
results obtained in [3] in Figure 2. In the channel with a 10% bump transonic flow arises at an inlet Mach number of 0.675. The location of the
shock was at 70% of the bump, where [3] found 72 %. The maximum Mach
number we found was 1.33 against 1.32 in [3]. At the bump, the shock is
captured in 4 cells.

The second test application was a converging/diverging nozzle with the
following contraction ratios: 5, 10, and 20. The inlet Mach number is kept
constant at 0.045. The size of the throat is kept constant at 0.4, and the
size of the outlet is kept constant at 2.5 times the size of the throat. For
all computations we have used a 49×10 nonuniform grid. The time step

was kept constant at 0.01. In the computations for flow in nozzles with contraction ratios 5 and 10, $p_v = 1$ was used, and the flow remained subsonic. The maximum Mach numbers that occurred are listed in Table 2. For a contraction ratio of 20, $p_v = 60$ was chosen so that supersonic outflow occurred. In this case upwind biased density evaluation was used to ensure thermodynamic irreversibility to satisfy the entropy condition.

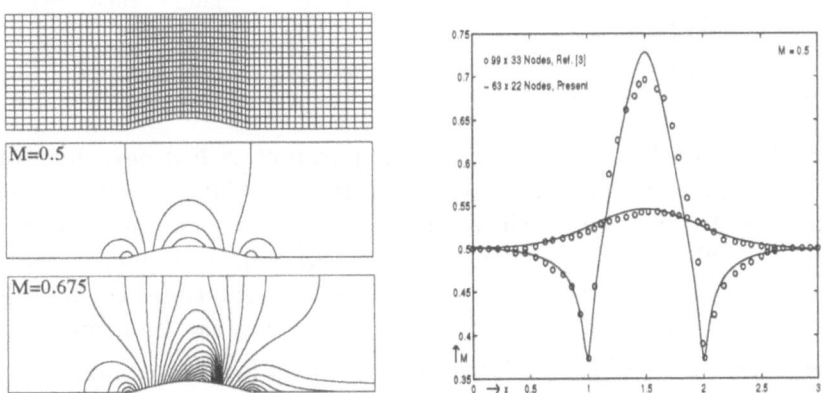

Figure 2.   Mesh and Mach number plots for $M_\infty = 0.5$, and 0.675.

The number of iterations and CPU time that were necessary in both cases are listed in Table 1. For the channel with bump neither the number of iterations nor the CPU time deteriorate for low inlet Mach numbers: also at low Mach numbers a good solution is obtained efficiently. Only for increasing Mach numbers the CPU time increases a little; this might be due to the fact that local Mach numbers become as high as .728 while we have not used any density upwinding here. In the nozzle with a contraction ratio of 5 the flow remained incompressible. The maximum Mach number that occurred was 0.24, so almost equal to 5 times the inlet Mach number. For a contraction ratio of 10 the maximum Mach number was 0.5, so the Mach number varied in this flow from 0.045 up to 0.5. In the case of a contraction ratio of 15 this range was from 0.045 up to 1.82. The CPU time and number of iterations for the nozzle do not increase significantly for lower Mach numbers, that is smaller contraction ratios, either. On the contrary, there is a larger increase of computing time for the contraction ratio 20 than for the contraction ratio 5 compared to 10. This increase of CPU time is caused by an increase of the number of time steps necessary to reach the stationary solution, which is probably due to the fact that we have kept the time step constant.

8. Concluding Remarks

A method to compute inviscid flow has been described, that has accuracy and efficiency uniform for very low and medium Mach numbers. A non-

TABLE 1. Channel with 10% bump.

Mach nr.	CPU [s]	time steps	GMRES iter/t.st.
0.5	151	509	12
0.1	73	220	18
0.01	74	200	21
10^{-6}	74	200	21

TABLE 2. Converging/diverging nozzle.

contr. ratio	max. Mach	CPU [s]	time steps	GMRES iter/t.st.
15	1.82	750	6790	14
10	.5	220	1866	14
5	.24	300	2744	14

dimensionalisation based on the reservoir parameters has been introduced such that all quantities in the scaled compressible Euler equations remain finite and $\mathcal{O}(1)$ when the Mach number goes to zero. The nondimensional equations are discretised in general coordinates on a staggered grid and solved with a pressure correction method and implicit time stepping. From the numerical experiments the following conclusions can be drawn:

- Good results are obtained for subsonic flow at low Mach numbers, even solutions at $M = 0$ are no problem.
- Mach number independent convergence is observed for subsonic flow.
- Our method is able to compute fully compressible transonic flow.
- Locations of shocks are predicted well.

References

1. R. Aris, *Vectors, tensors and the basic equations of fluid mechanics*, Prentice-Hall, Inc., Englewood Cliffs, N.J., 1962, Reprinted, Dover, New York, 1989.
2. H. Bijl and P. Wesseling, A Numerical Method for the Computation of Compressible Flows with Low Mach Number Regions, to appear in *Proceedings ECCOMAS 96*, 1996, P. Le Tallec and J. Périaux (eds.), Wiley, Chichester.
3. S. Eidelman, P. Colella and R.P. Shreeve, Application of the Godunov method and its Second-Order Extension to Cascade Flow Modelling, *AIAA J.*, 22: 1609–1615, 1984.
4. F.H. Harlow and J.E. Welch, Numerical calculation of time-dependent viscous incompressible flow of fluid with a free surface, *The Physics of Fluids*, 8:2182–2189, 1965.
5. A. Segal, P. Wesseling, J. Van Kan, C.W. Oosterlee and K. Kassels, Invariant Discretization of the Incompressible Navier-Stokes equations in Boundary Fitted Coordinates, *Int. J. Num. Meth. Fluids*, 15: 411–426, 1992.
6. P. van Beek, R.R.P. van Nooyen and P. Wesseling, Accurate discretization on nonuniform curvilinear staggered grids, *J. Comp. Phys.*, 117: 364–367, 1995.
7. J.J.I.M. Van Kan, A Second-order accurate pressure correction method for viscous incompressible flow, *SIAM J. Sci. Stat. Comp.*, 7: 870–891, 1986.
8. C. Vuik, Fast iterative solvers for the discretized incompressible Navier-Stokes equations, *Int. J. for Num. Meth. Fluids*, 22: 195–210, 1996.
9. P. Wesseling, P. van Beek and R.R.P. van Nooyen, Aspects of non-smoothness in flow computations, in *Computational Methods in Water Resources X*, 1994, A. Peters, G. Wittum, B. Herrling, U. Meissner, C.A. Brebbia, W.G. Gray and G.F. Pinder (eds.), 1263 - 1271, Kluwer, Dordrecht.

ON DENSITY AND PRESSURE FLUCTUATIONS IN UNIFORMLY SHEARED COMPRESSIBLE FLOW

S. SARKAR
Applied Mechanics and Engineering Sciences
University of California at San Diego
La Jolla, CA 92093

1. Introduction

Density fluctuations can be significant in a variety of flows such as high-speed flow of a compressible fluid, the mixing of separate constituents with different densities, reacting flow, and low-speed stratified flow. The behavior of density fluctuations in high-speed flow and their effect on the velocity is investigated here using DNS (direct numerical simulation) of compressible turbulence sustained by uniform mean shear. It is well-known [1, 2, 3] that, when convective Mach number increases, there is a large reduction in the thickness growth rate and turbulent intensities of the mixing layer. Although the shear is nonuniform in the mixing layer, a similar stabilizing effect on the turbulent kinetic energy K has been noticed in simulations [4, 5, 6] of uniformly sheared, compressible flow. We summarize the numerical procedure and our previous results [6] in section 2, investigate the evolution of thermodynamic fluctuations in section 3 and describe compressibility effects on the Reynolds stresses in section 4. The evolution in the presence of initially strong density and temperature fluctuations is discussed in section 5.

2. The evolution of turbulent kinetic energy

The numerical method used here involves sixth-order compact finite difference schemes for the spatial discretization and a third order, low storage Runge-Kutta scheme for the time advancement. The nondimensional parameters that govern the flow evolution are the initial values of gradient Mach number, $M_g = Sl/\bar{c}$, turbulent Mach number, $M_t = \sqrt{2K}/\bar{c}$, microscale Reynolds number, Re_λ, and Prandtl number Pr. Here S is the

L. Fulachier et al. (eds.), IUTAM Symposium on Variable Density Low-Speed Turbulent Flows, 325–332.
© *1997 Kluwer Academic Publishers.*

constant mean shear rate, l the integral length scale of u in the transverse shearing direction, $\bar{c}$ the mean speed of sound, and K the turbulent kinetic energy. The parameters M_g and M_t determine compressibility effects in uniformly sheared flow. In series A of the DNS, the initial value, M_{g0},

Case	M_{g0}	M_{t0}	$Re_{\lambda 0}$	Pr_0	$(SK/\epsilon)_0$
A1	0.22	0.40	14	0.7	1.8
A2	0.44	0.40	14	0.7	3.6
A3	0.66	0.40	14	0.7	5.4
A4	1.32	0.40	14	0.7	10.8

Case	M_{g0}	M_{t0}	$Re_{\lambda 0}$	Pr_0	$(SK/\epsilon)_0$
B1	0.22	0.13	14	0.7	5.4
B2	0.22	0.20	14	0.7	3.6
B3	0.22	0.40	14	0.7	1.8

TABLE 1. Parameters for series A and B of the DNS

was progressively increased, all other non-dimensional parameters remaining the same. In series B, M_{t0} was progressively increased in cases B1 to B3. The *incompressible* Navier-Stokes equations were simulated in series AI with identical relevant parameters as well as the same initial velocity and pressure fields relative to corresponding cases in series A.

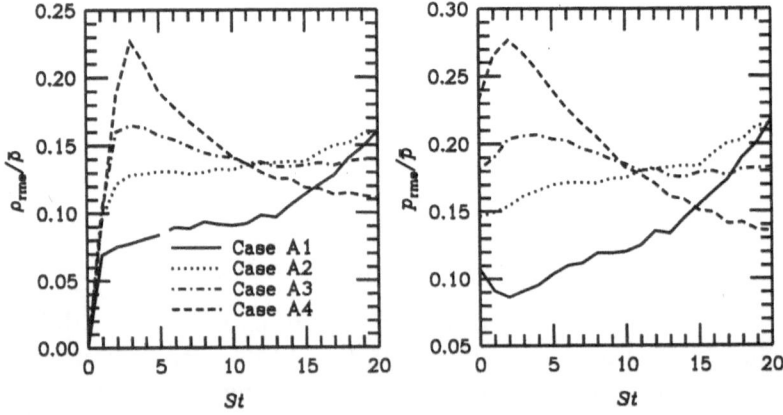

Figure 1. Evolution of (a) density, and (b) pressure fluctuations in series A.

The effects of gradient Mach number M_g and turbulent Mach number M_t were distinguished. It was established that there is a strong reduction in the asymptotic growth rate of K that occurs when gradient Mach number increases, and furthermore the reduced growth rate is primarily due to a reduced level of turbulence production. The less dramatic reduction associated with an increase in turbulent Mach number was shown to have a contribution from the pressure dilatation and compressible dissipation. If the length scale can be estimated by $l \propto K^{3/2}/\epsilon$, then $M_g \propto (SK/\epsilon)M_t$. However, the stabilizing effect of gradient Mach number observed in series A is *not* an effect of shear number SK/ϵ by itself. This was shown [6] by conducting incompressible simulations in series AI with the same relevant parameters as series A and observing that increasing SK/ϵ had only a small effect on the growth rate of K.

3. The evolution of thermodynamic fluctuations

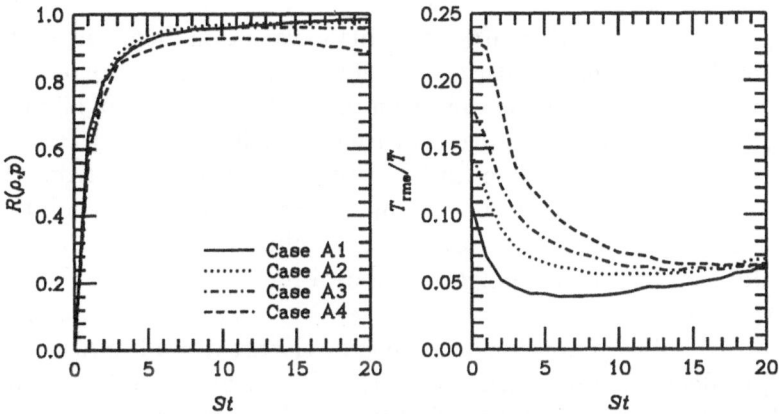

Figure 2. Evolution of (a) the correlation coefficient between density and pressure, and (b) r.m.s. temperature fluctuation in series A.

The initial conditions in series A and B correspond to uniform density, solenoidal velocity, pressure obtained by the usual Poisson equation, and temperature obtained from the ideal gas equation of state. Figs. 1-2 shows the evolution of thermodynamic fluctuations in series A. The density fluctuations can be seen in Fig. 1(a) to increase rapidly from zero to an appreciable level during the initial transient over $0 < St < 3$. The value of $\rho_{rms}/\bar{\rho}$ at $St = 3$ increases with the value of M_{g0} or equivalently from Case A1 to A4. This is not surprising because the density fluctuations are associated with pressure fluctuations in this flow and the initial value of $p_{rms}/\bar{p}$ in Fig. 1(b) increases from case A1 to case A4. The later evolution of r.m.s. density and pressure fluctuations is very similar. The close relation between ρ and p is seen in Fig. 2(a) which shows that their correlation coefficient becomes large and $R(\rho, p) \to 1$. The final value of temperature fluctuations

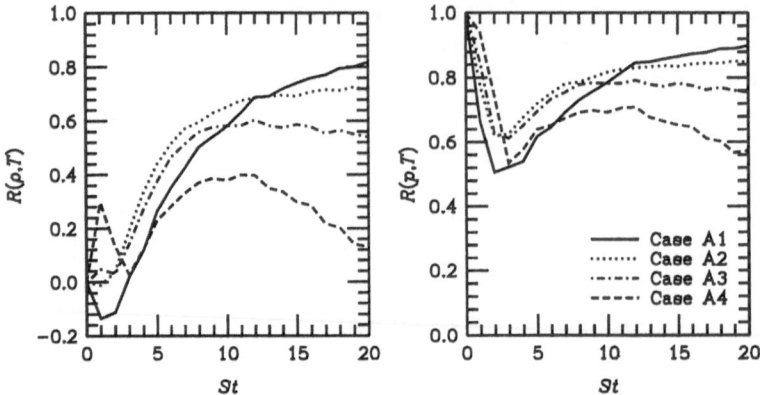

Figure 3. Evolution of the correlation coefficient between (a) temperature and density fluctuations, and (b) temperature and pressure fluctuations.

shown in Fig. (2b) is $T_{\mathrm{rms}}/\overline{T} \simeq 6\%$ with little variation between cases. Fig. 3(a) shows that the temperature field develops a positive correlation with the density field. Thus, the assumption of the Strong Reynolds Analogy that $R(\rho, T) \simeq -1$ is not supported in free shear flow. The correlation coefficient $R(p, T)$ remains positive after a preliminary decrease from the initial value $R(p, T) \simeq 1$ as shown in Fig. 3(b). In summary, the evolution of the three thermodynamic cross-correlation coefficients suggest that the 'acoustic' mode dominates the 'entropy' mode especially with respect to density fluctuations.

The asymptotic value of nondimensional growth rate $\Lambda = (1/SK)dK/dt$ was shown [6] to decrease monotonically with M_g and have a value in case A4 smaller than that in case A1 by a factor of 3. However, the final value of $\rho_{\mathrm{rms}}/\overline{\rho}$ in Fig. 1(a) does not change monotonically with M_g and, furthermore, case A4 with the largest reduction in growth rate has the smallest r.m.s. density fluctuations. Thus, the stabilizing effect of compressibility *does not* appear to increase with the magnitude of density fluctuations. Since the mean density is initially constant and remains so during the evolution the stabilizing influence is *not* a mean density effect either.

The behavior of the thermodynamic correlation coefficients in series B is similar to series A and indicates that the density and pressure fluctuations are almost perfectly correlated. However, the evolution of r.m.s pressure (and other thermodynamic variables) is quite different. The value of $p_{\mathrm{rms}}/\overline{p}$ (not shown here) increases with time for cases B1-B3 in contrast to the decrease observed for the two larger M_g cases in Fig. 1(b). In the incompressible series AI, the pressure fluctuations increase with time too unlike series A. The large difference between the evolution in series A with respect to that in series B as well as the incompressible series AI motivated

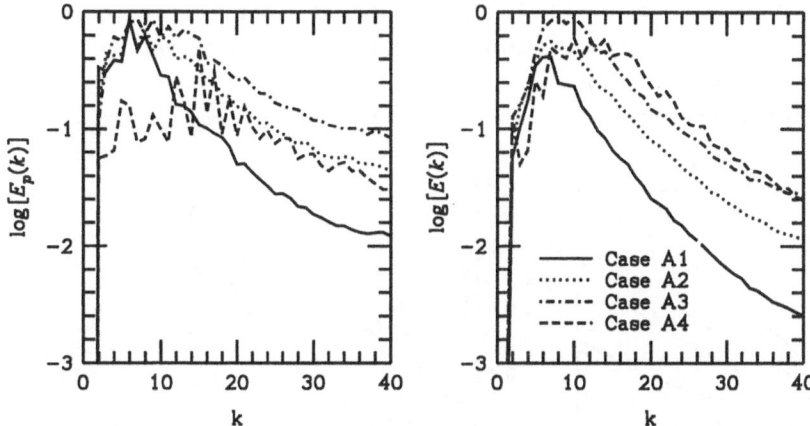

Figure 4. The spectra towards the end of simulation of (a) pressure, and (b) kinetic energy in series A. The spectra for cases A1, A2, A3 and A4 correspond to the times, $St = 19$, $St = 21$, $St = 29$ and $St = 31$, respectively.

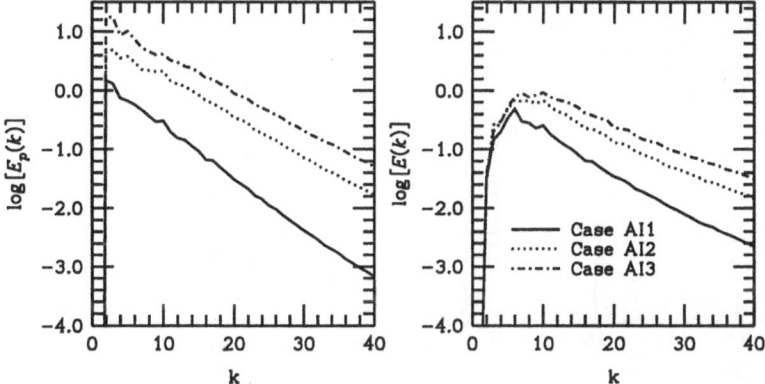

Figure 5. The spectra at the time $St = 15$ of (a) pressure, and (b) kinetic energy in the incompressible series AI.

an investigation of pressure and energy spectra. Fig. 4(a) shows the pressure spectrum $E_p(k)$ towards the end of each simulation in series A. The peak of $E_p(k)$ shifts to larger wave numbers when M_g increases and the slope becomes smaller. The energy spectra $E(k)$ in Fig. 4(b) also become flatter with gradient Mach number. Kolmogorov scaling for high Reynolds number, incompressible turbulence gives $E(k) \propto k^{-5/3}$ and $E_p(k) \propto k^{-7/3}$. Although an inertial range does not exist in the moderate Reynolds number simulations reported here, the relative shape of pressure and velocity spectra in Fig. 4 *do not* support the incompressible flow result that $E_p(k)$ has a steeper slope with respect to $E(k)$. On the other hand, the spectra shown in Fig. 5 for the corresponding incompressible series AI agree with

$E_p(k)$ being steeper than $E(k)$). The pressure and kinetic energy spectra in series B (not shown here) have similar slopes as was observed in case A1 and unlike incompressible flow.

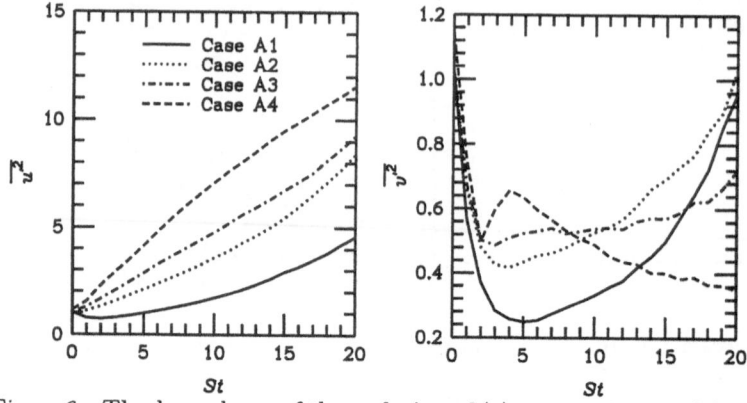

Figure 6. The dependence of the evolution of (a) streamwise and (b) transverse (shearing direction) components of Reynolds stress tensor on the gradient Mach number .

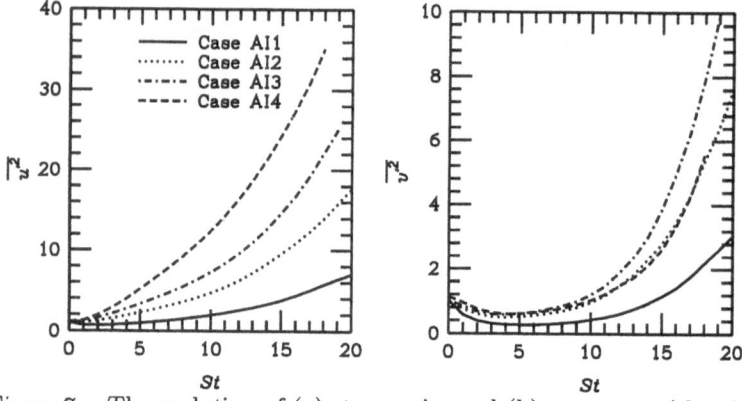

Figure 7. The evolution of (a) streamwise and (b) transverse (shearing direction) components of Reynolds stress tensor in the incompressible series AI.

4. The evolution of Reynolds stresses

The mean shear S directly sustains the streamwise Reynolds stress, $\overline{u^2}$, while the transverse and spanwise components, $\overline{v^2}$ and $\overline{w^2}$, increase due to the transfer by the corresponding components of the pressure-strain tensor. Given the large influence of M_g on the pressure field, it may be anticipated that the Reynolds stresses are strongly affected. The development of $\overline{u^2}$ and $\overline{v^2}$ in series A is shown in Fig. 6 while that in the corresponding incompressible series AI is shown in Fig. 7. Fig. 6(a) shows that $\overline{u^2}$ increases with time and there is a quantitative change in its growth rate between cases. Although the level of $\overline{u^2}$ increases with initial gradient Mach number, its normalized slope, $(1/S\overline{u^2})(d\overline{u^2}/dt)$, decreases with M_g just as is observed for the turbulent kinetic energy. The evolution of $\overline{v^2}$ in Fig. 6(b) shows a large difference between cases as well as *qualitative* differences with that shown in Fig. 7(b) for the incompressible simulations. The transverse component in Case A4 does not show eventual growth unlike the other cases. Furthermore, $\overline{v^2}$ has drastically smaller growth in series A relative to AI.

5. The influence of 'entropy' mode associated with reacting flow

In combustion applications, heat release causes large local temperature fluctuations which, at low speed, can be assumed to occur at constant thermodynamic pressure. The evolution of uniformly sheared flow in the presence of such an 'entropy' mode in the initial data is studied by performing a simulation BV1. The only difference that case BV1 has with respect to case B1 is the initially large density and temperature fluctuations, $\rho_{\mathrm{rms}}/\bar\rho = T_{\mathrm{rms}}/\bar{T} = 20\%$, which are inversely correlated, $R(\rho, T) = 1$.

Fig. 8(a) compares the evolution of r.m.s. density in cases B1 and BV1. There is a large drop in $\rho_{\mathrm{rms}}/\bar\rho$ from its initial value in case BV1 to a a level near that in case B1. There is negligible difference between the evolution of turbulent kinetic energy, Reynolds stresses and enstrophy between cases B1 and BV1. Thus, the presence of a dominant initial 'entropy' mode *does not* alter the turbulence evolution in uniformly sheared flow, at least in the absence of a distinct mechanism to sustain such a mode. Fig. 8(b) shows that the large, inverse correlation between density and temperature is lost as the 'acoustic' mode develops.

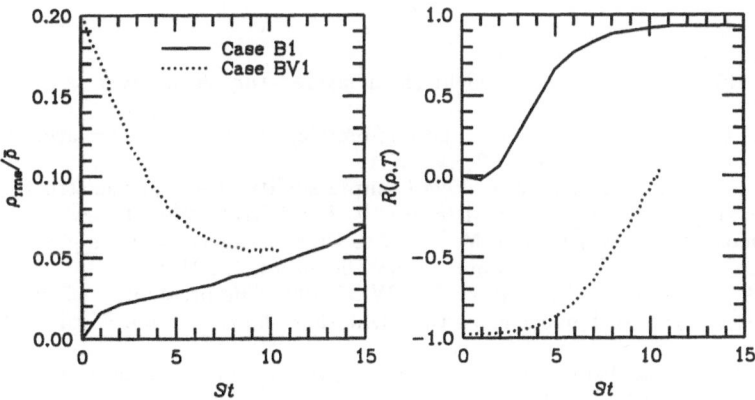

Figure 8. Evolution of (a) r.m.s. density fluctuations, and (b) the correlation coefficient between temperature and density fluctuations in cases B1 and BV1.

6. Conclusions

Our previous work [6] on uniformly sheared compressible flow had established that an increase in gradient Mach number, $M_g = Sl/\bar{c}$, is accompanied by a large reduction in the normalized growth rate of the turbulent kinetic energy that is primarily caused by a large decrease in the normalized Reynolds shear stress. The evolution of the thermodynamic fluctuations has been examined here. It appears that the 'acoustic' mode with strongly correlated density and pressure fluctuations develops in the presence of uniform shear. When gradient Mach number increases, the evolution of r.m.s. pressure and the shape of the pressure spectrum exhibit large deviations from corresponding results in incompressible flow. The consequent change in the pressure gradient term in the momentum equations (and the pressure-strain term in the Reynolds stress equations) leads to reduced levels of turbulence. Furthermore, the effect is very anisotropic with a far larger reduction in the transverse Reynolds stress relative to the streamwise component. The 'acoustic' mode is found to develop preferentially even with an initially dominant 'entropy' mode associated with combustion.

The stabilizing effect at high speeds is neither related to a variation of mean density, nor to the magnitude of density fluctuations. It is a manifestation of the differences in the evolution of pressure and its interaction with the velocity field in high-speed shear flow. Models of this stabilizing effect as a function of density variance are probably of limited use because it's origin is very different from low-speed, variable density dynamics.

Acknowledgments
The research was supported by the AFOSR grant F49620-96-1-0106.

References

1. Bradshaw, P. 1977 Compressible turbulent shear layers. *Ann. Rev. Fluid Mech.*, **9**, 33-54.
2. Papamoschou, D. & Roshko, A. 1988 The compressible turbulent shear layer: an experimental study. *J. Fluid Mech.*, **197**, 453-477.
3. Barre, S., Quine, C., & Dussauge, J. P. 1994 Compressibility effects on the structure of supersonic mixing layers: experimental results. *J. Fluid Mech.*, **259**, 47-78.
4. Sarkar, S., Erlebacher, G., & Hussaini, M. Y. 1991b Direct simulation of compressible turbulence in a shear flow. *Theor. Comput. Fluid Dynamics*, **2**, 291-305.
5. Blaisdell, G. A., Mansour, N. N., & Reynolds, W. C. 1993 Compressibility effects on the growth and structure of homogeneous turbulent shear flow. *J. Fluid Mech.*, **256**, 443-485.
6. Sarkar, S. 1995 The stabilizing effect of compressibility in turbulent shear flow. *J. Fluid Mech.*, **282**, 163-186.

A STRUCTURAL APPROACH TO THE EFFECTS
OF COMPRESSIBILITY IN TURBULENT FLOWS SUBJECTED
TO EXTERNAL COMPRESSION OR HIGH SHEAR

C. CAMBON AND A. SIMONE
Laboratoire de Mécanique des Fluides et d'Acoustique,
U.M.R CNRS n° 5509, Ecole Centrale de Lyon, BP 163,
69131 Ecully Cedex, France

1. Introduction: from 'compressed' to compressible turbulence

Flows with variable density are considered in this paper with emphasis on kinematic effects. Important variation in mean volume occurs in many engineering flows , including combustion chambers of reciprocating engines, compression corners in high speed flows, nuclear fusion by inertial confinment, not to mention astrophysics.

Depending on the value of a Mach number based on the turbulent field M_t, we will distinguish in the following *compressed* from *fully compressible* turbulence. Such a terminology has sense only when decomposing the whole field (velocity, pressure, density) into a mean (or background) contribution and a fluctuating (or disturbance) contribution. In that follows, the 'mean' field will reflect external (given a priori) compression (*e. g.* the motion of the piston in a reciprocating engine), and we will discuss only academic simplified configurations, in which the definition of the 'mean' does not yield additional problems (*e. g.* is the mean flow a laminar flow of reference, is the decomposition consistent with Reynolds or Favre averaging?). Capital letters will reffer to the mean flow, and lower case letters to the fluctuating one, for instance $U_i + u_i$ will denote the whole velocity field, except for the density field, where $\bar{\rho}$ and ρ will denote the mean and the fluctuating field, respectively. Using these preliminary definitions, the assumption of *compressed* turbulence corresponds to an incompressible fluctuating field, characterized by a *solenoidal* (divergence-free $u_{i,i} = 0$) fluctuating velocity field, in the presence of a compressible mean field, with $U_{i,i} \neq 0$. Such a

333

L. Fulachier et al. (eds.), IUTAM Symposium on Variable Density Low-Speed Turbulent Flows, 333–342.

situation seems to be relevant if $\rho/\bar{\rho} << 1$ or $M_t << 1$ which is the case looking at global statistics in a reciprocating engine, in which the mean Mach number (expected as much larger than M_t) based on the maximum piston speed is about 0.02 at $5000 tr/mn$. Some parts of the flows inside the combustion chamber, however, display more significant Mach numbers, as the valve-jet in the intake stroke.

In section 2, the effect of the external compression on 'solenoidal' turbulence will be briefly recalled and discussed as follows. As the most promising procedure, a rescaling of the fluctuating field concerning the spatial coordinates, the velocity and the time (Cambon, Mao & Jeandel, 1992) yielded removing from consideration the effect of $U_{i,i}$ on fully non-linear equations. Hence, the background equations for compressed turbulence are turned to the background equations for incompressible turbulence, but possibly with time-dependent kinematic viscosity, using this rescaling. A stringent procedure for assess any statistical closure model has been derived. Among related applications to single-point second-order closure models, we could mention a specific low-Reynolds model capable of capturing the effects of varying kinematic viscosity (Le Penven & Serres, 1994).

The second part of this paper (section 3) is devoted to investigating higher speed flows at significant Mach number, in which the assumption of *compressed* turbulence is partially released. The cases of spherical and axisymmetric compression are briefly revisited using a recent linear (RDT, Rapid Distortion Theory) approach, which was generalized towards compressible, homogeneous, and quasi-isentropic flows. New insights on the effects of intrinsic compressibility (consistent with the development of a *dilatational* part $u_{i,i}$ in the fluctuating flows) are discussed, including 'stabilizing' and 'destabilizing' effects of compressibility over a broad parameter range. Among the compressibility parameters, the most relevant was shown to be the *Distortion Mach number M_d*, which was brought to the fore by linearized theories (RDT) and DNS (Direct Numerical Simulation) results. In addition to irrotational compressing mean flow, the case of pure shear flow is extensively revisited using new RDT and DNS results.

2. Compressed turbulence

Looking at the pure kinematic description of the *given* background (or mean) field, it is useful to characterize the pure dilatational effect. Considering the mean velocity gradient $U_{i,j}$, a decomposition can be proposed in terms of a pure dilatational, pure incompressible strain and pure rotational part:

$$U_{i,j} = U_{l,l}\frac{\delta_{ij}}{3} + S_{ij} + W_{ij} \tag{1}$$

where $W_{ij} = (U_{i,j} - U_{j,i})/2 = (1/2)\epsilon_{ikj}W_k$. Regarding the Lagrangian displacement tensor $F_{ij} = \partial x_i/\partial X_j$ which characterizes the *deformation effects accumulated in time along mean trajectories* (see Cambon *et al.* 1992, Cambon 1995, and fig. 1), the following *multiplicative* decomposition is suggested:

$$F_{ij}(t,0) = J^{1/3}.S'_{il}.Q_{lj} \qquad (2)$$

where $J = Det(\mathbf{F})$ is the volumetric ratio which reflects the accumulated effects of the mean flow divergence ($\dot{J} = U_{l,l}J$), S'_{ij} is a symmetric matrix which reflects the accumulated effects of strain rate, and Q_{ij} is an orthogonal matrix which reflects the accumulated effects of rotation. For instance the application to one-dimensional axial compression (or dilatation) leads to

$$U_{i,j} = \begin{pmatrix} 0 & 0 & 0 \\ 0 & 0 & 0 \\ 0 & 0 & -c \end{pmatrix} = \begin{pmatrix} -c/3 & 0 & 0 \\ 0 & -c/3 & 0 \\ 0 & 0 & -c/3 \end{pmatrix} + \begin{pmatrix} c/3 & 0 & 0 \\ 0 & c/3 & 0 \\ 0 & 0 & -2c/3 \end{pmatrix}$$

and

$$F_{ij} = \begin{pmatrix} 1 & 0 & 0 \\ 0 & 1 & 0 \\ 0 & 0 & J \end{pmatrix} = J^{1/3}.\begin{pmatrix} J^{-1/3} & 0 & 0 \\ 0 & J^{-1/3} & 0 \\ 0 & 0 & J^{2/3} \end{pmatrix} \qquad (3)$$

with $J(t,0) = e^{\int_0^t -c(t')dt'}$, so that the 1-D axial compression is equivalent to a combination of spherical compression and incompressible axial strain.

2.1. TURBULENCE UNDERGOING SPHERICAL COMPRESSION

Mean compression is investigated in this section. According to the concept of 'compressed' turbulence (Cambon et al. 1992, Le Penven & Serres 1994), the fluctuating field is assumed to be solenoidal, $u_{i,i} = 0$, but subjected to a mean velocity gradient with $U_{i,i}(t) \neq 0$ or $J(t) \neq 1$. Such assumptions are consistent at low Mach number, provided that the ratio of fluctuating and mean density is much smaller than 1.

The case of spherical space-uniform compression characterized by

$$U_{i,j} = -\frac{c(t)}{3}\delta_{ij} \qquad F_{ij} = J^{1/3}(t)\delta_{ij} \qquad with \qquad J(t) = e^{\int_0^t c(t')dt'} \qquad (4)$$

is the special case of mean strain for which the RDT solution can be simply expressed in physical space for the fluctuating velocity field.

$$u_i(\mathbf{x},t) = J^{-1/3}(t)u_i(\mathbf{X},0) \qquad with \qquad x_i = J^{1/3}(t)X_i \qquad (5)$$

The very simple form of the RDT solution suggests the replacement of the initial data in (5) by an unknown vector function $v_i(\mathbf{X},t) = J^{1/3}u_i$ and to derive an equation for v_i from the complete non-linear equation that governs u_i, $u_{i,t} - \frac{c}{3}x_j u_{i,j} - \frac{c}{3}u_i + u_j u_{i,j} + \frac{1}{\rho}p_{,i} - \nu u_{i,jj} = 0$ with $u_{i,i} = 0$, which includes two specific terms (advection and 'production') connected with the mean compression. The new v-equation writes: $J^{2/3}v_{i,t} + v_i v_{i,J} + \frac{1}{\rho}(pJ^{2/3})_{,I} - \nu v_{i,JJ}$ with $v_{i,I} = 0$. Except for the first term which contains the time-derivative, the above equation in terms of $\mathbf{v}(\mathbf{X},t)$ (with $\partial/\partial X_i$ abridged in $_{,I}$) has the basic form of the Navier-Stokes equations (without explicit compression terms). Eventually, the factor $J^{2/3}$ is eliminated too by using a new time t^*, so that

$$dt = J^{2/3}dt^*.$$

A complete time-dependent rescaling can finally be recovered

$$\mathbf{u}^*(\mathbf{x}^*,t^*) = J^{1/3}(t)\mathbf{u}(\mathbf{x},t)$$

$$\mathbf{x}^* = J^{-1/3}\mathbf{x} \tag{6}$$

$$t^* = \int_0^t J^{-2/3}(t')dt'$$

so that the 'starred' variables satisfy the conventional Navier-Stokes equations without mean dilatational terms.

The rescaling (6) firstly was introduced by Zimont & Sabel'nikov (1975); a similar procedure is usual in the cosmological context (Frisch, private communication); in addition, it is noteworty that this rescaling belongs to a class of internal scale invariance of the fully compressible momentum equations ($u_i^* = \lambda^h u_i$, $x_i^* = \lambda x_i$, $t^* = \lambda^{1-h}t$) with $\lambda = J^{-1/3}$; $h = -1$; such a general scaling invariance could be used as a true dynamic ground for a fractal approach (Nicolleau 1994).

Another important remark is that the rescaling of velocity and space is suggested by the linear solution only (compression or dilatation of the space and conservation of the circulation), so that *it is the rescaling of time which actually takes the non-linearity into account*; if the linear dynamics were considered as 'rapid' dynamics, the rescaled time t^* could be considered as a 'slow' time: it is questionable since it is found $dt << dt^*$ for high compression (low volumetric ratio) but $dt >> dt^*$ for high dilatation. The latter comment shows that the words 'rapid' and 'slow', extensively used in the context of RDT and RSM closures, are ambiguous and cannot systematically replace 'linear' and 'nonlinear'.

2.2. SOME APPLICATIONS TO $K - \epsilon$ AND FULL RSM MODELS

The rescaling allows us to derive in a straightforward way the dynamics of spherically compressed turbulence from the one of freely decaying turbulence. Since the dynamics of freely decaying turbulence is well known, at least at high Reynolds number, and easily reflected by simple statistical models, some direct applications of the rescaling are found as follows. Starting with the law of freely decaying turbulent kinetic energy in homogeneous isotropic turbulence, or

$$q^2(t)/q^2(0) = (1 + \frac{t}{nt_0})^{-n} \quad \text{with} \quad \frac{1}{t_0} = -\frac{1}{q^2}\frac{dq^2}{dt}|_{t=0} \quad \text{and} \quad n \sim 1.2$$

and assuming that this equation holds for the 'starred variables', the law in the presence of mean spherical compression, suddenly applied at $t = 0$ is readily derived from the rescaling (6) as

$$q^2(t)/q^2(0) = J^{-2/3}(t)(1 + \frac{\int_0^t J^{-2/3}(t')dt'}{nt_0})^{-n}.$$

In the latter equation, the first factor on the left-hand-side characterizes the pure 'RDT' effect, whereas the second account for non-linearity. For instance, the case of spherical compression with a constant rate c ($J = e^{-ct} < 0$) is shown in fig. 2-b. According to the value of the initial compression rapidity ct_0, the turbulent kinetic energy is continuously increasing (pure RDT inviscid case, $ct_0 = \infty$), reaches a maximum and then decays ($ct_0 > 1/2$), or continuously decays ($ct_0 < 1/2$). The eventual decay also is strongly modified with respect to the case without compression. Similar laws are found for the dissipation rate and typical length scales, such as the integral and the Taylor length scales. For instance, in the same case of compression, it is possible to quantify the two opposite effects regarding linear (reducing the length scales) and non-linear (increasing the length scales) effects, which can be discussed from the Kelvin-circulation theorem (see Cambon 1995 and fig. 2-a).

Another direct consequence is to obtain straightforward conditions for calibrating the constants in the $k-\epsilon$ and RSM models (see Cambon, Mao & Jeandel 1992 for details). Concerning full RSM model, the rescaling removes from consideration the spherical part of the mean velocity gradient tensor $U_{l,l}\delta_{ij}/3$, so that *a model for 'compressed' turbulence can be derived from any model for incompressible turbulence in the presence of a compressible mean flow*, but possibly with variable kinematic viscosity.

3. Effects of intrinsic compressibility at finite Mach number

There is a wide consensus in the literature on the fact that the *intrinsic compressibility* (i.e. the non-zero divergence) of the turbulent field tends to reduce the amplification of turbulent kinetic energy produced by a mean velocity gradient, with respect to the pure solenoidal case. More generally, the reduction of mixing seems to be a main characteristic of compressible flows. There is no agreement, however, on how to explain and model this behaviour. Different approaches were proposed, that can be separated into two classes.

On the one hand, the 'energetic approach', mainly illustrated by second-order single-point models, try to explain and mimic the modified balance of kinetic energy through the emergence of *explicit* compressibility terms like the pressure-dilatation and the dilatation-dissipation terms.

On the other hand, the 'structural approach' investigates the effects of compressibility on the *structure* of the pressure field and/or the anisotropic structure of the velocity field; these effects being poorly reflected by the explicit correlation terms mentioned above.

In the case of homogeneous turbulence (and even in the case of the mixing layer, averaging over the cross-gradient inhomogeneous direction, Vreman 1995), both approaches can be easily discussed looking at the following equation that governs the turbulent kinetic energy:

$$\dot{\mathcal{K}} = \mathcal{P} + \Pi_d - \epsilon^{(s)} - \epsilon^{(d)}$$

in which $\mathcal{K} = \langle u_i u_i \rangle /2$, the dissipation rate ϵ is split into its solenoidal and its dilatational part, Π_d is the pressure dilatation term, and $\mathcal{P}$ stands for the production by the mean flow. The 'production' term $\mathcal{P} = -U_{i,j} < u_i u_j >$ involves the anisotropy of the turbulent flow through the deviatoric part of the Reynolds stress tensor $b_{ij} = < u_i u_j > /q^2 - \delta_{ij}/3$.

Recent works, ranging from linear analyses to full Direct Numerical Simulations, have shown that the structural approach is the only relevant one, at least for homogeneous compressed (Cambon, Coleman & Mansour, 1993) and shear flows (Sarkar, 1995), and for free mixing layers (Vreman, Sandham & Luo, 1995). The common point of these works is the reduced effect of pressure as a Distortion Mach number M_d (homogeneous distortion case) or the convective Mach number (mixing layer case) is increased. $M_d = LS/a$ compares the *sonic* timescale L/a (L is a characteristic lengthscale of energetic eddies and a is the sonic speed) with the *distortion* timescale $1/S \sim (U_{i,j} U_{i,j})^{-1/2}$, and was also referred to as *the eddy Mach number change across an eddy* (Durbin & Zeman, 1992) and *the gradient Mach number* (Sarkar, 1995). The role of M_d was initially investigated by means of a linear approach (Rapid Distortion Theory, RDT) to quasi-isentropic strongly compressible flows (Jacquin, Cambon & Blin, 1993). This procedure was made general for any background flow consistent with homogeneous turbulence (Simone & Cambon, 1995, Simone, Coleman & Cambon 1997), and compared to a full DNS approach (Blaisdell, Mansour & Reynolds, 1991). Results for spherical and one-dimensional axisymmetric compression are briefly recalled; the most salient feature is a 'destabilizing' effect of compressibility, with a monotonic increase of the kinetic energy growth rate as the distortion Mach number increases, from the solenoidal regime (or the case of *compressed* turbulence) to a *pressure-released* regime.

Restricting the study to the pure shear case (rate S), both linear and full nonlinear approaches can be reconciled over a broad parameter range. The most interesting result is the prediction by compressible RDT of a crossover in the development of the kinetic energy growth rate $\Lambda = (1/\mathcal{K})d\mathcal{K}/d(St)$ at intermediate St (fig. 3). At smallest St, the growth rate monotonically increases as the distortion Mach number increases, with the pure solenoidal and the pressure-released case as limiting, lower and upper bounds. At largest St, the inverse behaviour prevails (reduced growth rate as M_d is increased), in qualitative agreement with the behaviour emphasised in the DNS results of Sarkar (1995). It is important to point out that the 'crossover' did not occur in RDT for irrotational mean strain, so that it can be linked to the vortical part of the mean flow. Linear interactions between *solenoidal, dilatational* and pressure modes of the fluctuating fields are shown to give useful guidelines to explain all the observed results (Simone, Coleman & Cambon 1997, Friedrich 1996).

In addition to the material incorporated in Simone et al. (1997), such as, e. g., fig. 3, some new DNS computations and new data processing are in progress. DNS results about all the component of the Reynolds Stress Tensor anisotropy suggest that a one-component (in the streamwise direction) distribution is reached at large M_d, or $b_{ij} = \delta_{i1}\delta_{j1} - \delta_{ij}/3$. Such a result is consistent with a pressure-relased limit and could suggest to multiply the 'classic' pressure-strain tensor in a compressible RST model by a damping function in terms of M_d (in agreement with Cambon et al. 1993, Simone et al. 1997, Vreman 1995 and Uhlmann 1996). Different behaviours at large and small St, however, cannot be easily captured by such a simple reajustment, and a fully compressible RST model, M_d-dependent, remains a difficult challenge.

References

BLAIDELL, G. A.N., MANSOUR, N. N. AND REYNOLDS, W. C. (1991) *Stanford University, Thermosciences Division*, rep. TF-50.

CAMBON, C. (1995) Turbulence in compressed and/or swirling flows, some specific problems and application to combustion engines, *PEPIT-ERCOFTAC Euroconference*, Aachen (Germany), June 8-10 1995.

CAMBON, C., COLEMAN, G. N. AND MANSOUR, N. N. (1993) *J. Fluid Mech.* **257**, 641-665

CAMBON, C., MAO, Y. AND JEANDEL, D. (1992) *Eur. J. Mech. B/Fluids* **11**, 6, 683-703

DURBIN, P. A. AND ZEMAN, O. (1992) *J. Fluid Mech.* **242**, 349-370

FRIEDRICH, R. (1996) Compressibility effects due to turbulent fluctuations, *Invited lecture, Eur. Turb. Conf. VI*, Lausanne (switzerland), 2-5 July 1996.

JACQUIN, L., CAMBON, C. AND BLIN, E. (1993) *Phys. Fluids A* **5**, 10, 2539-2550

LE PENVEN, L. AND SERRE, G. (1994) *Ninth Symp. Turb. Shear Flow, Kyoto, Japan*
UHLMANN, M. (1996) *Phd. Thesis, Université de Lyon, to appear*
SARKAR, S. (1995) *J. Fluid Mech* **282**, 163-186
SIMONE, A. AND CAMBON, C. (1995) *10th Symp. Turb. Shear Flow, Penn. State U., USA*
SIMONE, A., COLEMAN, G. N. AND CAMBON, C. (1997) Effect of compressibility on turbulent shear flow: a RDT and DNS study, *J. Fluid Mech.* **330**, 307-339
VREMAN, A. (1995) *Phd. Thesis, University of Twente*
VREMAN, W. A., SANDHAM, N. D. AND LUO, K. H. (1995) *J. Fluid Mech.*, **320**, 235-258.

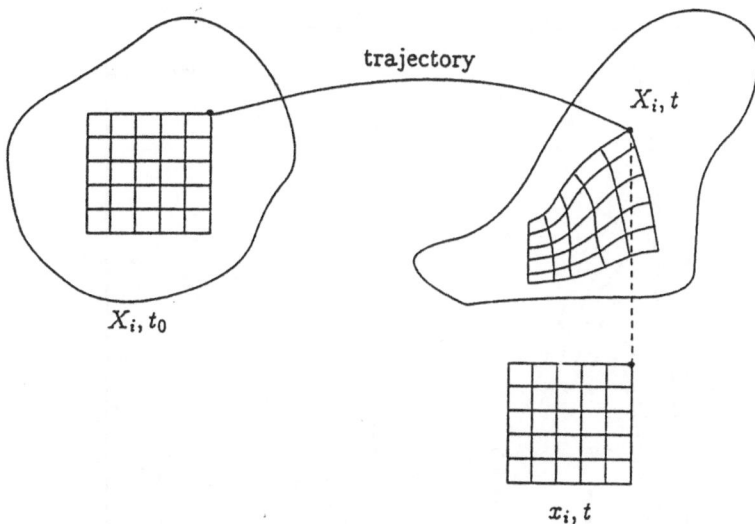

Figure 1. Lagrangian-Eulerian systems of coordinates and distortion

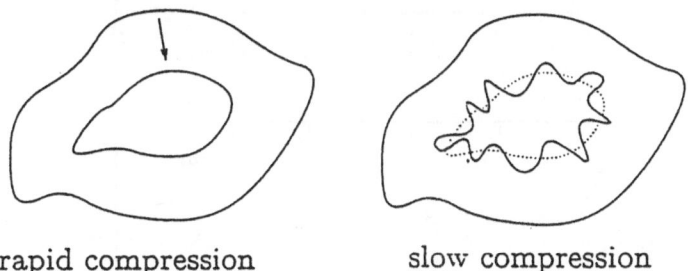

rapid compression   slow compression

Figure 2-a. Compression of a loop of fluid particle

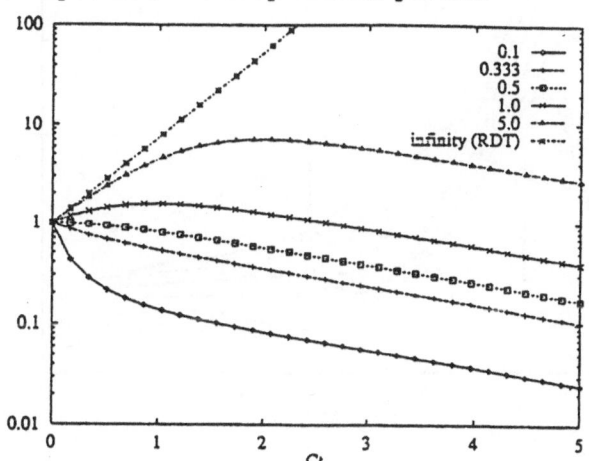

Figure 2-b. Development of the kinetic energy in isotropically compressed turbulence at high Reynolds number, for different initial compression rapidities ct_0.

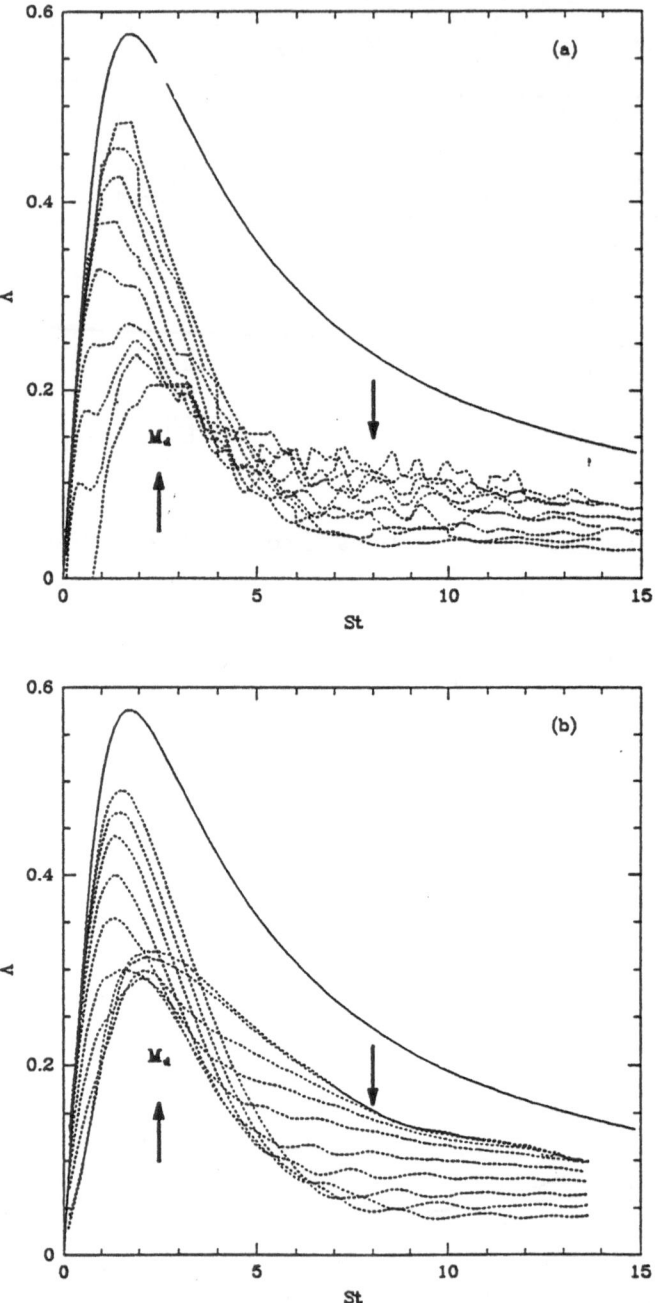

Figure 3. Histories of the temporal energy growth rates with different initial M_d: (a) DNS data, (b) RDT results. Full line: pressure-released limit. (From Simone, Coleman and Cambon, J. Fluid Mech., 1997)

Conclusion

CONCLUDING REMARKS

J. L. LUMLEY*
*Sibley School of Mechanical and Aerospace Engineering,
Cornell University, Ithaca, NY 14853*

Abstract. We suggest that a new technique for the construction of low-dimensional dynamical models be used also in flows with density fluctuations to predict levels of drag and heat transfer, as well as more subtle quantities. We give a brief introduction to the technique, which has been used in flows of uniform density. We suggest how the needed eigenfunctions can be estimated by a stability argument, equally applicable to flows with density fluctuations. We suggest a practical problem to which this approach might be applied.

*Supported in part by Contract No. F49620-92-J-0287 jointly funded by the U. S. Air Force Office of Scientific Research (Control and Aerospace Programs), and the U. S. Office of Naval Research, in part by Grant No. F49620-92-J-0038, funded by the U. S. Air Force Office of Scientific Research (Aerospace Program), and in part by the Physical Oceanography Programs of the U. S. National Science Foundation (Contract No. OCE-901 7882) and the U.S. Office of Naval Research. (Grant No. N00014-92-J-1547). Prepared for presentation at the IUTAM Symposium Variable Density Low Speed Turbulent Flows, Institut de Recherche sur les Phénomènes Hors Equilibre, July 8-10, 1996, Marseille, France.

L. Fulachier et al. (eds.), IUTAM Symposium on Variable Density Low-Speed Turbulent Flows, 345–355.
© 1997 *Kluwer Academic Publishers.*

1   Introduction

In the concluding remarks, it is traditional to give a summary and evaluation, a commentary, on the papers in the symposium. This has been an interesting three days, and there have been a number of valuable papers. However, my concluding remarks are not going to be of this traditional character. Instead, I am going to tell you about a new approach which has been applied in constant-density flows with substantial profit. I will tell you about beginnings that have been made in applying this new technique to flows with variable density. I hope I can convince you that it would be worthwhile to try this new method on other flows with variable density.

There is some precedent for concluding remarks like these. At the end of the summer of 1961 George Batchelor stood roughly where I am now and told us that the future of turbulence lay in the small scale end of the spectrum, and encouraged us all to work at that end. I told him then that I did not agree with him (which must have startled him, coming from someone who had only had his degree for four years), but I believe that history has shown that most of the progress since then has been made at the other end of the spectrum, the large scale end. In fact, five years before the 1961 meeting, Alan Townsend (George's colleague at Cambridge) had published his book on the structure of turbulent shear flows, putting forth the idea that shear flows contain what we now call coherent structures, and I think this was the first major contribution which led eventually to the work I will describe to you. To be fair, coherent structures probably go back even further, to some remarks of Liepmann around the time of the second world war (see, for example [20]). In any event, George was substantially younger than I am now when he made his remarks, so perhaps I may be permitted to say now that I believe the future belongs (at least in part) to the kind of technique I will describe below. I am prepared to be as wrong as George.

By choosing to tell you about this new technique, rather than evaluating the papers in the rest of the symposium, I am not in any sense denigrating those papers. They represent a more traditional approach, an approach which has served us well, from which we have learned much, and will continue to learn much. There is a tendency in science, as in all other fields of human endeavor, to leap to espouse new things, and abandon the old. This is not sensible: if you buy a new tool for your tool box, you do not discard all the old ones. The new tool can perhaps do some things that the old ones cannot, but the old ones should still see substantial daily use, and there will be many applications for which they are superior.

In these final few minutes of this symposium, I hope only to convince one or two of you that the new approach I will describe below is worth pursuing.

2 A Brief History of Our Approach

In the past decade our group at Cornell has developed a new approach to turbulent flows([1], [33], [2], [19], [5]). We have applied our approach primarily to the flat-plate boundary layer. The approach has become quite popular, being picked up by many research groups (c.f. [4], [6], [14], [15], [12], [16], [13], [17], [23], [24], [25], [29]), [32], [11], [10]). This is the approach that I am suggesting be applied also to the effects of density fluctuations in low-speed flows.

In the flat-plate boundary layer, we begin from the idea that the near-wall region is energetically dominated by large-scale coherent structures. It should thus be possible to construct a low-dimensional model of the wall region resolving only these coherent structures, and parameterizing the less energetic, less well-organized, smaller-scale turbulence. Such a model can be used for many purposes, if only large-scale information is needed. The model involves the following steps:

- First we need a good set of basis functions. We use the Empirical Eigenfunctions of the Proper Orthogonal Decomposition, or the Karhunen-Loève decomposition ([2]), because these are optimal from the viewpoint of convergence. In the wall region, one eigenfunction is sufficient to capture 80% of the fluctuating energy. These eigenfunctions are obtained either from experiment or from DNS.

- By Galerkin projection, we obtain a set of ordinary differential equations for the coefficients in the decomposition ([20]). The decomposition is truncated, and the modes that are not represented explicitly are parameterized by a gradient transport model. Some representation of these unresolved modes is necessary, to provide a sink for the turbulent energy; however, we believe that the precise nature of the dissipation mechanism does not affect the structure of the large scales. A unique feature of our models is the utilization of separation of scales between the coherent structures and the mean. This results in a cubic term which globally stabilizes the system and allows a very compact representation.

- Models of the near-wall region of the turbulent boundary layer have been constructed containing from four to 64 degrees of freedom ([31]); ten degrees of freedom provides a model of the wall region that is satisfactory in many respects, although even the four-dimensional model displays some of the characteristics of the wall region.

- The low-dimensional models are explored with various simulation programs constructed for this purpose ([33]), and analyzed using the techniques of dynamical systems theory.

- Note that we believe the mean time between bursts (in the laboratory or the model) to be closely related to the drag; in fact, $T_B u_\tau^2/\nu$ is thought to be approximately constant (where T_B is the mean time between bursts, u_τ is the friction velocity and ν is the kinematic viscosity), so that the drag coefficient is inversely proportional to the inter-burst time, although the data are widely scattered ([26]). This relationship coupling the interburst time and drag is appealing, but the experimental support that can be found in the literature is somewhat in conflict and disappointing; the data in [26], which is reasonably consistent, represents a careful selection of the available data. Recently, however, our own (unpublished) work with Direct Numerical Simulation of channel flow in the Minimal Flow Unit ([22]) and control of bursting (which will form part of the doctoral thesis of Peter Blossey) has given strong support to this idea. In the MFU we see each burst followed by the ejection of low-momentum fluid into the core of the flow, increasing the drag; suppression of the burst suppresses the ejection, and reduces the drag.

- The ten dimensional model displays the following properties:

  1. The model displays an intermittent phenomenon with an ejection phase and a sweep phase that strongly resembles the bursting phenomenon observed in the boundary layer. The probability distribution of inter-burst times has the observed shape ([34], [35],

[36], [21]). However, the time scales both for bursts and interburst durations are unrealistically long, a fact that was not appreciated until recently, due to the inadvertent omission of a term in [2]. We believe that the long time scales are due to the model's inclusion of only a single coherent structure, when in fact a succession of quasi-independent structures are being swept past the sensor. A simple model of this situation restores approximately the right magnitude to the time scales.

2. The low-dimensional model makes it possible to understand the dynamics of the bursting process: the instability leading to the burst, the burst itself and the reformation of streamwise vortices are determined by wall-region dynamics, and hence will scale with wall-region parameters; the time of occurrence of a burst, however, is triggered by the pressure signal from the outer part of the boundary layer ([20]). Thus, the interburst time distribution should scale jointly with both inner and outer variables.

3. In the presence of favorable and unfavorable pressure gradients, the mean time between bursts (and hence, the inverse of the drag coefficient) increases and decreases in accordance with observation ([33]).

4. In turbulent drag reduction by polymer additives, the structure of the wall region remains approximately the same, but increases in scale. It has been speculated that this is due to an increased lossiness in the turbulent part of the flow, due to the extensional viscosity of the polymer ([26]). When the ten-dimensional model is used to describe a flow with a stretched wall region, mean drag is reduced, and increased lossiness in the turbulence is required to maintain the system, consistent with the speculative explanation ([3]).

5. The low-dimensional model has been used to explore the possibility of active drag reduction. Active drag reduction corresponds to burst-suppression, that is, to keeping the system from bursting as long as possible, increasing the mean time between bursts (and consequently reducing the drag) ([8], [9], [7]). It has been shown using the model that active drag reduction is formally equivalent

to drag reduction by polymer additives, which also acts to suppress the bursting ([3]).

- Models of this general type can be used whenever it is desired to simulate the large scales of a boundary layer in response to various effects. We have just seen how such a model can be used to investigate one possible drag-reduction technique. It can also be used ([20]):

  1. To simulate the fluctuating pressure pattern on a surface under a boundary layer, either to predict panel vibration or sonar self-noise;

  2. To simulate index of refraction fluctuations occurring in the surface mixed layer of the planetary boundary layer, giving rise to multi-path interference in digital communication links;

  3. To simulate the effect of streamline curvature and flow divergence, which has a profound influence on interburst period and drag [27].

The applications are limited only by the imagination. The point here is, that through the model, we have established a relation between the time between bursts (and hence the drag or momentum mixing) and various other physical effects - polymer additives, pressure gradients, and so forth. In these concluding remarks, we suggest that the same connection can be exploited in attempting to understand the influence of low-speed density fluctuations.

3   Prediction of the Eigenfunctions

In the boundary layer, we have used the empirical orthogonal eigenfunctions obtained from the experiments of Herzog [18] or the Direct Numerical Simulations of Moin and Moser [29]. Both of these are very labor-intensive, either in terms of computational time or of experimental time. However, it is possible to predict the form of these eigenfunctions, using a non-linear stability argument [30]; we begin with a classical energy-method stability argument, but we must allow the Reynolds stress to be modified by the presence of the growing eigenfunction, as well as the mean velocity profile. Proceeding in this way, we manage to predict not only the form of the eigenfunctions, but also the peak of the eigenvalue spectrum.

This approach can also be applied to flows with density fluctuations; in [28] we present preliminary results for sheared Rayleigh-Bénard convection; these results do not include anisotropic eddy viscosities and mean-profile feedback, and do not predict directly the form of the POD eigenfunctions; however, a very satisfactory comparison can be made with the second order moments obtained from a DNS.

4   Suggestions for the Future

This is a new way of understanding the mechanics of turbulent flow, to the extent that the mechanics of interest are dominated by the large scales. In principle it can be applied with profit to any turbulent flow. Nothing is essential save the idea of constructing a low-dimensional model resolving only the coherent structures and parameterizing the rest. Everything else is technique. We have suggested one technique which is fairly rationally based and which results in moderately realistic imitations of the large scales in the boundary layer, but others will surely be discovered. Low speed flows with density fluctuations contain complex physics, and it seems likely that low-dimensional models of these flows would have analytical and predictive value, and would be useful in design.

In [28] we show how the POD approach can be extended to flows with density fluctuations. In the Boussinesq approximation the resulting equations can be reduced to a set very similar to that used by Aubry et al in [2], except in four dimensions (to accomodate the density fluctuations). There is, of course, an additional term corresponding to the buoyancy. This is a linear term, just like the additional linear term that appears due to streamline curvature or flow divergence in [27]. When the situation (density gradient, or streamline curvature) is destabilizing, it results in an increase in the unstable eigenvalue, resulting in a decrease in the interburst time, with a resulting increase in the drag and heat transfer.

This has applications in the atmospheric and oceanic surface mixed layer, of course, as well as many industrial heat transfer situations. In heat transfer in turbine blade passages, for example, if the blade is cooled to protect it from the hot gases, the higher density near the blade surface, acted on by the centrifugal field produced by the high rotational speed, will certainly contribute to the intensification of rolls at the surface, increasing the heat transfer substantially over what would be present in an isothermal flow due to streamline curvature. This effect can be analyzed using our low-dimensional models.

The eigenfunctions necessary for such an analysis need not be obtained by laborious experiment or computation. In [28] we suggest how these eigenfunctions can be estimated using a modified energy method technique in flows with density fluctuations.

References

[1] N. Aubry. *A Dynamical System/Coherent Structure Approach to the Fully Developed Turbulent Wall Layer*. PhD thesis, Cornell University., 1987.

[2] N. Aubry, P. Holmes, J. L. Lumley, and E. Stone. The dynamics of coherent structures in the wall region of a turbulent boundary layer. *Journal of Fluid Mechanics*, 192:115–173, 1988.

[3] N. Aubry, J. L. Lumley, and P. Holmes. The effect of modeled drag reduction in the wall region. *Theoret. Comput. Fluid Dynamics*, 1:229–248, 1990.

[4] K. S. Ball, L. Sirovich, and L. R. Keefe. Dynamical eigenfunction decomposition of turbulent channel flow. *Int. Jour. for Num. Meth. in Fluids*, 12:585–604, 1991.

[5] G. Berkooz, P. Holmes, and J. L. Lumley. The proper orthogonal decomposition in the analysis of turbulent flows. *Annual Review of Fluid Mechanics*, 25:539–575, 1993.

[6] D. H. Chambers, R. J. Adrian, P. Moin, D.S. Stewart, and H. J. Sung. Karhunen-Loève expansion of Burgers model of turbulence. *Phys. Fluids*, 31:2573–2582, 1988.

[7] B. D. Coller, P. Holmes, and J. L. Lumley. Control of bursting in boundary layer models. *Appl. Mech Rev.*, 47 (6), part 2:S139–S143, 1994. Mechanics USA 1994, ed. A. S. Kobayashi.

[8] B. D. Coller, P. Holmes, and J. L. Lumley. Controlling noisy heteroclinic cycles. *Physica D*, 72:135–160, 1994.

[9] B. D. Coller, P. Holmes, and J. L. Lumley. Interaction of adjacent bursts in the wall region. *Phys. Fluids*, 6 (2):954–961, 1994.

[10] A. E. Deane, I. G. Keverkidis, G. E. Karniadakis, and S. A. Orszag. Low-dimensional models for complex flows: Application to grooved channels and circular cyliders. *Physics of Fluids A*, 3(10):2337–2354, 1991.

[11] A. E. Deane and L. Sirovich. A computational study of Raleigh-Bénard convection part I. Rayleigh number scaling. *Journal of Fluid Mechanics*, 222:231–250, 1991.

[12] M. Glauser, S. J. Leib, and W. K. George. Coherent structures in the axisymmetric turbulent jet mixing layer. In *Turbulent shear flows 5*. Springer-Verlag, 1987.

[13] M. Glauser, X. Zheng, and W. K. George. The streamwise evolution of coherent structures in the axisymmetric jet mixing layer. In T. B. Gatski, S. Sarkar, and C. G. Speziale, editors, *Studies in Turbulence*, pages 207–222, New York, etc., 1992. Springer.

[14] M. N. Glauser and W. K. George. Orthogonal decomposition of the axisymmetric jet mixing layer including azimuthal dependence. In G. Comte-Bellot and J. Mathieu, editors, *Advances in turbulence*. Springer-Verlag, 1987.

[15] M. N. Glauser and W. K. George. An orthogonal decomposition of the axisymmetric jet mixing layer utilizing cross-wire velocity measurements. In *6th symposium turbulent shear flows*, 1987.

[16] M. N. Glauser, X. Zheng, and C. R. Doering. The dynamics of organized structures in the axisymmetric jet mixing layer. In M. Lesieur and O. Metais, editors, *Turbulence and coherent structures*. Kluwer, 1989.

[17] A. Glezer, A. J. Kadioglu, and A. J. Pearlstein. Development of an extended proper orthogonal decomposition and its application to a time periodically forced plane mixing layer. *Physics of Fluids A*, 1:1363–73, 1989.

[18] S. Herzog. *The Large Scale Structure in the Near Wall Region of a Turbulent Pipe Flow.* PhD thesis, Cornell Univ., 1986.

[19] P. J. Holmes. Can dynamical systems approach turbulence? In J.L. Lumley, editor, *Whither turbulence? Turbulence at the Crossroads.* pages 195–249, New York, 1990. Springer.

[20] P. J. Holmes, G. Berkooz, and J. L. Lumley. *Turbulence, Coherent Structures, Dynamical Systems and Symmetry.* Cambridge University Press, 1996.

[21] P. J. Holmes and E. Stone. Heteroclinic cycles, exponential tails and intermittency in turbulence production. In T. B. Gatski, S. Sarkar, and C. G. Speziale, editors, *Studies in Turbulence*, pages 179–189. Springer-Verlag, 1992.

[22] J. Jimenez and P. Moin. The minimal flow unit in near-wall turbulence. *Journal of Fluid Mechanics*, 225:213–240, 1991.

[23] M. Kirby, J. Boris, and L. Sirovich. An eigenfunction analysis of axisymmetric jet flow. *Journal of computational physics*, 90 no. 1:98–122. 1990.

[24] M. Kirby, J. Boris, and L. Sirovich. A proper orthogonal decomposition of a simulated supersonic shear layer. *International journal for numerical methods in fluids*, 10:411–428, 1990.

[25] Z-C. Liu, Adrian R. J., and T. J. Hanratty. Reynolds-number similarity of orthogonal decomposition of the outer layer of turbulent wall flow. Technical Report TAM 748, UILU-ENG-94-6004, University of Illinois, Department of Theoretical and Applied Mechanics, 1994.

[26] J. L. Lumley and I. Kubo. Turbulent drag reduction by polymer additives: a survey. In B. Gampert, editor, *The Influence of Polymer Additives on Velocity and Temperature Fields*, pages 3–24. Springer-Verlag, 1985.

[27] J. L. Lumley and B. Podvin. Dynamical systems theory and extra rates of strain in turbulent flows. *Journal of Experimental and Thermal Fluid Science*, 1996. Peter Bradshaw Symposium; in press.

[28] J. L. Lumley and A. Poje. Low-dimensional models for flows with density fluctuations. In *IUTAM Symposium: Flows with Variable Density*, Grenoble, September 1994.

[29] P. Moin and R. D. Moser. Characteristic-eddy decomposition of turbulence in a channel. *Journal of Fluid Mechanics*, 200:471–509, 1989.

[30] A. C. Poje and J. L. Lumley. A model for large scale structures in turbulent shear flows. *Journal of Fluid Mechanics*, 285:349–369, 1994.

[31] S. Sanghi and N. Aubry. Mode interaction models for near-wall turbulence. *J. Fluid Mech.*, 247:455–488, 1993.

[32] L. Sirovich and A. E. Deane. A computational study of Raleigh-Bénard convection part II. Dimension considerations. *Journal of Fluid Mechanics*, 222:251–265, 1991.

[33] E. Stone. *A Study of Low Dimensional Models for the Wall Region of a Turbulent Layer*. PhD thesis, Cornell University., 1989.

[34] E. Stone and P. J. Holmes. Noise induced intermittency in a model of a turbulent boundary layer. *Physica D*, 37:20–32, 1989.

[35] E. Stone and P. J. Holmes. Random perturbations of heteroclinic cycles. *SIAM J. on Appl. Math.*, 50 no. 3:726–743, 1990.

[36] E. Stone and P. J. Holmes. Unstable fixed points, heteroclinic cycles and exponential tails in turbulence production. *Phys. Lett. A*, 155:29–42, 1991.

List of Participants

K.J. ADOU
Univ. Cocody
Fac. des Sciences et Techniques
Dept de Mathématiques 22 BP
582 ABIDJAN 22
IVORY COAST
Tel.: (225) 44 90 00
Fax: (225) 44 83 97

F. ANSELMET
IRPHE - UMR 6594 - CNRS
Univ. Aix-Marseille I & II
12 Ave. du Général Leclerc
13003 MARSEILLE
FRANCE
Tel.: (33) 04 91 50 54 39
Fax: (33) 04 91 08 16 37

S.N. ARISTOV
Inst. of Continuous Media
Mechanics Ural Branch
Russian Academy of Sciences
1 Academian Korolyov Str.
614061 PERM
RUSSIA
Tel.: (7) 34 22 39 64 09
Fax: (7) 34 22 90 49 72

A. BENAISSA
Queen's University
Dept. of Mechanical Engineering

KINGSTOM K7L 3N6
CANADA
Tel.: (1) 61 35 45 72 52
Fax: (1) 61 35 45 64 89

R.W. BILGER
Univ. of Sydney
Dept. of Mechanical Engineering
NSW 2006 SYDNEY
AUSTRALIA
Tel.: (61) 2 35 129 28
Fax: (61) 2 69 291 69

M. AMIELH
IRPHE - UMR 6594 - CNRS
Univ. Aix-Marseille I & II
12 Ave. du Général Leclerc
13003 MARSEILLE
FRANCE
Tel.: (33) 04 91 50 54 39
Fax: (33) 04 91 08 16 37

J.C. ANTORANZ
UNED - Dept. Fisica Fundamental
APDO. DE CORREOS 60141

28008 MADRID
SPAIN
Tel.: (34) 1 398 71 22
Fax: (34) 1 398 66 97

S. BARRE
CEAT - LEA - URA CNRS 191
43 Route de l'Aérodrome

86036 POITIERS Cedex
FRANCE
Tel.: (33) 05 49 53 70 00
Fax: (33) 05 49 53 70 01

H. BIJL
Delft Univ. of Technology
Dept. Maths and Informatics
P.O. BOX 5031
2600 GA DELFT
THE NETHERLANDS
Tel.: (31) 15 278 72 90
Fax: (31) 15 278 72 09

J.P. BONNET
CEAT - LEA - URA CNRS 191
43 Rue de l'Aérodrome
86036 POITIERS CEDEX
FRANCE
Tel.: (33) 05 49 53 70 00
Fax: (33) 05 49 53 70 01

R. BORGHI
ESM2 - IRPHE
Univ. Aix-Marseille I & II - CNRS
Technopôle de Château Gombert
13451 MARSEILLE Cedex 20
FRANCE
Tel.: (33) 04 91 05 44 14
Fax: (33) 04 91 11 38 38

A. BOUNIF
Univ. Sciences & Technologie
Institut Génie Mécanique
BP 1505 - ELMNAOUAR
31000 ORAN
ALGERIA
Tel.:
Fax: (213) 637 16 92

C. CAMBON
Ecole Centrale de Lyon
Lab. Méca. Fluides & Acoustique
36 Ave. Guy de Collongues - BP 163
69131 ECULLY Cedex
FRANCE
Tel.: (33) 04 72 18 60 00
Fax: (33) 04 78 64 71 45

J.P. CHABARD
EDF / Lab. Nat. d'Hydraulique

6 Quai Watier - BP 49
78401 CHATOU Cedex
FRANCE
Tel.: (33) 01 30 87 72 44
Fax: (33) 01 30 87 80 86

O. CHAMBRES
CEAT - LEA - URA CNRS 191
43 Route de l'Aérodrome

86036 POITIERS Cedex
FRANCE
Tel.: (33) 05 49 53 70 00
Fax: (33) 05 49 53 70 01

V.G. CHAPIN
ENSICA
Fluid Mechanics Dept.
1 Place Emile Blouin
31056 TOULOUSE Cedex
FRANCE
Tel.: (33) 05 61 61 86 62
Fax: (33) 05 61 58 75 24

P. CHASSAING
ENSICA
Fluid Mechanics Dept.
1 Place Emile Blouin
31056 TOULOUSE Cedex
FRANCE
Tel.: (33) 05 61 61 86 56
Fax: (33) 05 61 61 86 63

G. COMTE-BELLOT
Ecole Centrale de Lyon
Lab. Méca. Fluides & Acoustique
36 Ave. Guy de Collongues - BP 163
69131 ECULLY Cedex
FRANCE
Tel.: (33) 04 78 33 81 27
Fax: (33) 04 78 64 71 45

J.F. DEBIEVE
IRPHE - UMR 6594 - CNRS
Univ. Aix-Marseille I & II
12 Ave. du Général Leclerc
13003 MARSEILLE
FRANCE
Tel.: (33) 04 91 50 54 39
Fax: (33) 04 91 08 16 37

C. DOPAZO
Univ. Zaragoza
Fluid Mechanics Dept.
C/ Maria de Luna 3.
50015 ZARAGOZA
SPAIN
Tel.: (34) 76 51 82 18
Fax: (34) 76 76 18 82

O. DUBREUIL
Quantel
17 Ave. de l'Atlantique
BP 23
91941 LES ULIS Cedex
FRANCE
Tel.: (33) 01 69 29 17 00
Fax: (33) 01 69 29 17 29

P. DUPONT
IRPHE - UMR 6594 - CNRS
Univ. Aix-Marseille I & II
12 Ave. du Général Leclerc
13003 MARSEILLE
FRANCE
Tel.: (33) 04 91 50 54 39
Fax: (33) 04 91 08 16 37

I. EAMES
St Catharines College
Trampington Street
CAMBRIDGE CB3 9EW
UK
Tel.: (44) 12 23 33 79 18
Fax: (44) 12 23 33 83 40

R. ELAMRAOUI
EDF / Lab. Nat. d'Hydraulique
6 Quai Watier - BP 49

78401 CHATOU Cedex
FRANCE
Tel.: (33) 01 30 87 72 44
Fax: (33) 01 30 87 80 86

L. FALLOT
GDF - DETN - CERTSA

361 Ave. du Président Wilson - BP 33
93211 LA PLAINE SAINT DENIS Cedex
FRANCE
Tel.: (33) 01 49 22 47 57
Fax: (33) 01 49 22 49 67

F. DUMOUCHEL
Université de Rouen
CORIA - URA 230 - CNRS
Place Emile Blondel - BP 118
76821 MONT SAINT AIGNAN Cedex
FRANCE
Tel.: (33) 02 35 14 65 75
Fax: (33) 02 35 70 83 84

J.P. DUSSAUGE
IRPHE - UMR 6594 - CNRS
Univ. Aix-Marseille I & II
12 Ave. du Général Leclerc
13003 MARSEILLE
FRANCE
Tel.: (33) 04 91 50 54 39
Fax: (33) 04 91 08 16 37

J.L. EDY
Deltalab

38340 VOREPPE
FRANCE
Tel.: (33) 04 76 50 04 54
Fax: (33) 04 76 56 74 36

A.B. EZERSKY
Russian Academy of Sciences
Institute of Applied Physics
46 Uljanov Str.
603600 NIZHNY NOVGOROD
RUSSIA
Tel.: (7) 83 12 36 72 91
Fax: (7) 83 12 36 72 91

A. FAVRE
IRPHE
Résidence Le Chambord
122, Rue du Commandant Rolland
13008 MARSEILLE
FRANCE
Tel.: (33) 04 91 77 65 86
Fax: (33) 04 91 08 16 37

M. FAVRE- MARINET
LEGI / IMG
Domaine Universitaire - BP 53X
38041 GRENOBLE Cedex
FRANCE
Tel.: (33) 04 76 82 50 49
Fax: (33) 04 76 82 52 71

H.E. FIEDLER
Technical Univ. of Berlin
Hermann Fottinger Institute
Fur Thermo - Fluiddyn
Str. des 17 Juni 135
D 10623 BERLIN
GERMANY
Tel.: (49) 30 31 42 33 59
Fax: (49) 30 31 42 11 01

V.A. FROST
Russian Academy of Sciences
Inst. for Problems in Mechanics
Lab. Maths. & Physical Modelling
in Hydrodynamics
101 Pr Vernadskogo
MOSCOW 117526
RUSSIA
Tel.: (7) 095 434 46 09
Fax: (7) 095 938 20 48

P.L. GARCIA-YBARRA
UNED
Dept. Fisica Fundamental
Apdo. de Correos 60141
28080 MADRID
SPAIN
Tel.: (34) 13 98 71 27
Fax: (34) 13 98 66 97

M. GERMANO
Politecnico di Torino
Aeronautica e Spaziale
Dipartimento di Ingegneria
Corso Duca degli Abruzzi 24
10129 TORINO
ITALY
Tel.: (39) 11 564 68 14
Fax: (39) 11 564 68 99

A.T. FEDORCHENKO
Moscow Phys. Techn. Institute
Fluid Mechanics Dept.
141700 DOLGOPRUDNYI
RUSSIA
Tel.: (7) 095 408 43 27
Fax: (7) 095 921 10 04

P. FRAUNIE
LSEET
Univ. de Toulon & du Var
BP 132

83957 LA GARDE CEDEX
FRANCE
Tel.: (33) 04 94 14 24 16
Fax: (33) 04 94 14 24 17

L. FULACHIER
IRPHE - UMR 6594 - CNRS
Univ. Aix-Marseille I & II
12 Ave. du Général Leclerc

13003 MARSEILLE
FRANCE
Tel.: (33) 04 91 50 54 39
Fax: (33) 04 91 08 16 37

D. GARRETON
EDF / Lab. Nat. d'Hydraulique
6 Quai Watier - BP 49

78401 CHATOU Cedex
FRANCE
Tel.: (33) 01 30 87 72 61
Fax: (33) 01 30 87 80 86

A. GHARBI
Université des Sciences de Tunis
Dept. de Physique
Campus Universitaire

1060 TUNIS
TUNISIA
Tel.: (216) 1 512 600
Fax: (216) 1 885 073

J.F. HAAS
CEA - DAM
Dept. Détonique - CIT
Ctre Etudes Vaujours-Moronvilliers - BP 7
77181 COURTRY
FRANCE
Tel.: (33) 01 49 36 83 77
Fax: (33) 01 49 36 75 01

C. HONORE
Ecole Polytechnique
Laboratoire PMI
91128 PALAISEAU CEDEX
FRANCE
Tel.: (33) 01 69 33 32 44
Fax: (33) 01 69 33 30 23

J.C.R. HUNT
UK Meteorological Office
London Road
Bracknell

BERKS RG12 2SZ
UK
Tel.: (44) 1 344 85 66 08
Fax: (44) 1 344 85 69 09

L. JOLY
ENSICA
Fluid Mechanics Dept.

1 Place Emile Blouin
31056 TOULOUSE CEDEX
FRANCE
Tel.: (33) 05 61 61 86 62
Fax: (33) 05 61 58 75 90

W. KOLLMANN
Univ. of California at Davis

MAME Dept.
CA 95616 DAVIS
USA
Tel.: (1) 91 67 52 14 52
Fax: (1) 91 67 52 41 58

J.M. HERARD
EDF / Lab. Nat. d'Hydraulique
6 Quai Watier - BP 49

78401 CHATOU CEDEX
FRANCE
Tel.: (33) 01 30 87 70 37
Fax: (33) 01 30 87 80 86

P. HUERRE
Ecole Polytechnique
LADHYX - Lab. Hydrodynamique
91128 PALAISEAU CEDEX
FRANCE
Tel.: (33) 01 69 33 49 89
Fax: (33) 01 69 33 30 30

O. IIDA
Nagoya Institute of Technology
Dept. of Mechanical Engineering
Gokiso-Cho
Showa-Ku
NAGOYA 466
JAPAN
Tel.: (81) 52 735 53 47
Fax: (81) 52 735 53 42

C. KIRMSE
Technical Univ. of Berlin
Hermann Fottinger Institute
Fur Thermo - Fluiddyn
Str. des 17 Juni 135
D 10623 BERLIN
GERMANY
Tel.: (49) 30 31 42 33 59
Fax: (49) 30 31 42 11 01

I. KOUDOUGOU
Univ. Evry Val Essonne CEMIF/LEST
Lab. Energétique & Struct. Therm.
40 Rue du Pelvoux
91020 EVRY CEDEX
FRANCE
Tel.: (33) 01 69 47 75 62
Fax: (33) 01 69 47 75 99

I.A. KRYUKOV
Russian Academy of Sciences
Inst. for Problems in Mechanics
101 Pr. Vernadscogo
MOSCOW 117526
RUSSIA
Tel.:
Fax: (7) 095 938 20 48

M. LARABI
USTO
BP 1505
EL MNAOUER
31000 ORAN
ALGERIA
Tel.:
Fax: (213) 637 16 92

B.E. LAUNDER
UMIST - Dept. Mechanical Engineering
PO BOX 88
Sackville Str.
MANCHESTER M60 1QD
UK
Tel.: (44) 16 12 00 37 01
Fax: (44) 16 12 00 37 23

P. LE GAL
IRPHE - UMR 6594 - CNRS
Univ. Aix-Marseille I & II
12 Ave. du Général Leclerc
13003 MARSEILLE
FRANCE
Tel.: (33) 04 91 50 54 39
Fax: (33) 04 91 08 16 37

S.K. LELE
Stanford University
Dept. Mechanical Engineering
& Aeronautics & Astronautics
STANFORD CA 94305-4035
USA
Tel.: (1) 415 723 77 21
Fax: (1) 415 725 33 77

J. LEMAY
Université Laval
Lab. Mécanique des Fluides
Dept. de Génie Mécanique
QUEBEC GIK 7P4
CANADA
Tel.: (1) 418 656 21 04
Fax: (1) 418 656 74 15

O. LEUCHTER
ONERA
8 Rue des Vertugadins

92190 MEUDON
FRANCE
Tel.: (33) 01 46 23 51 00
Fax: (33) 01 46 23 51 51

A. LOPEZ-MARTIN
UNED
Dept. Fisica Fundamental
Apdo. de Correos 60141
28008 MADRID
SPAIN
Tel.: (34) 13 98 71 27
Fax: (34) 13 98 66 97

J.L. LUMLEY
Cornell Univ.
Sibley School of Mech.
& Aerosp. Engin.
Upson and Grumman Halls
ITHACA NY 14853-7501
USA
Tel.: (1) 607 255 40 50
Fax: (1) 607 255 12 22

M. MEDALE
IUSTI
Technopôle de Château-Gombert
5, Rue Enrico Fermi

13453 MARSEILLE Cedex 13
FRANCE
Tel.: (33) 04 91 10 68 68
Fax: (33) 04 91 10 69 69

R.B. MILES
Princeton Univ.
Dept. of Mech. & Aerosp. Engin.
Room D-414-Engin Quad
Olden Street
PRINCETON NEW JERSEY 08544
USA
Tel.: (1) 609 258 51 31
Fax: (1) 609 258 11 39

J.P. MINIER
EDF / Lab. Nat. d'Hydraulique
6 Quai Watier - BP 49

78401 CHATOU Cedex
FRANCE
Tel.: (33) 01 30 87 72 44
Fax: (33) 01 30 87 80 86

R. MOREAU
EPM-MADYLAM
ENSHMG - BP 95

38402 ST MARTIN D'HERES Cedex
FRANCE
Tel.: (33) 04 76 82 52 06
Fax: (33) 04 76 82 52 49

A.A. PAVELIEV
Research Institute
of Thermal Processes
8 Onezhskaja Str.
MOSCOW 125438
RUSSIA
Tel.: (7) 095 456 46 08
Fax: (7) 095 456 82 28

F. POGGI
CEA - DAM
Dept. Détonique - CIT
Ctre Etudes Vaujours-Moronvilliers - BP 7
77181 COURTRY
FRANCE
Tel.: (33) 01 49 36 83 77
Fax: (33) 01 49 36 75 01

C. MINARD
Spectra Physics
2-3 Rue de Madrid
BP 7408

38074 ST QUENTIN-FALLAVIER
FRANCE
Tel.: (33) 04 74 94 43 77
Fax: (33) 04 74 95 50 78

H.K. MOFFATT
Cambridge Univ.
Dept. Appl. Maths & Theor. Physics
Silver Street
CAMBRIDGE CB4 1HX
UK
Tel.: (44) 12 23 33 78 56
Fax: (44) 12 23 33 79 18

P. PARANTHOEN
Université de Rouen
CORIA - URA 230 - CNRS
Place Emile Blondel - BP 118
76821 MONT SAINT AIGNAN Cedex
FRANCE
Tel.: (33) 02 35 14 65 80
Fax: (33) 02 35 70 83 84

W.M. PITTS
Nat. Inst. of Standards & Techno.
U.S. Dept. of Commerce
Fire Research Lab. - B258/224
GAITHERSBURG MD 20899-0001
USA
Tel.: (1) 301 975 64 86
Fax: (1) 301 975 40 52

S. PUGLIESE
IUSTI
Technopôle de Château-Gombert
5, Rue Enrico Fermi
13453 MARSEILLE Cedex 13
FRANCE
Tel.: (33) 04 91 10 68 68
Fax: (33) 04 91 10 69 69

J. QUINARD
IRPHE - UMR 6594 - CNRS
Univ. Aix-Marseille I & II
Campus de St Jérôme - Service 252
13397 MARSEILLE Cedex 20
FRANCE
Tel.: (33) 04 91 28 81 02
Fax: (33) 04 91 63 52 61

P. REYNIER
IMFT - URA 005 CNRS - INPT
Inst. de Mécanique des Fluides
Allée du Prof. Camille Soula
31400 TOULOUSE
FRANCE
Tel.: (33) 05 61 28 58 31
Fax: (33) 05 61 28 58 99

E. RUFFIN
INERIS
Parc Technologique ALATA-BP 2
60550 VERNEUIL EN HALATTE
FRANCE

Tel.: (33) 03 44 55 67 78
Fax: (33) 03 44 55 68 99

J.P.H. SANDERS
Lab. Comb. / Systèmes Réactifs
1C Ave. Recherche Scientifique

45071 ORLEANS CEDEX 2
FRANCE
Tel.: (33) 02 38 51 54 65
Fax: (33) 02 38 51 78 75

K. SARDI
Imperial College of
Science Technology & Medicine
Mechanical Engineering Dept.
Exhibition Road
LONDON SW7 2BX
UK
Tel.: (44) 0171 589 51 11
Fax: (44) 0171 823 88 45

C. REY
IRPHE - UMR 6594 - CNRS
Univ. Aix-Marseille I & II
12 Ave. du Général Leclerc
13003 MARSEILLE
FRANCE
Tel.: (33) 04 91 28 93 89
Fax: (33) 04 91 28 94 94

X. ROGUE
CEA - DAM
Dept. Détonique - CIT
Ctre Etudes Vaujours-Moronvilliers BP 7
77181 COURTRY
FRANCE
Tel.: (33) 01 49 36 83 77
Fax: (33) 01 49 36 75 01

V.A. SABELNIKOV
Central Aerohydrodyn. Inst.
Tsa Gi
140160 ZHUKOVSKI
RUSSIA
Also at Ecole Centrale de Lyon
c/o D. JEANDEL and S. SIMOENS
Tel.: (33) 04 72 18 60 00
Fax: (33) 04 78 64 71 45

D.L. SANDOVAL
Los Alamos National Lab.
Applied Theoretical & Computational
Physics Group XNH MS-F664
LOS ALAMOS NEW MEXICO 87544
USA
Tel.: (1) 505 665 46 14
Fax: (1) 505 665 35 61

B. SARH
Lab. Comb. / Systèmes Réactifs
1C Ave. Recherche Scientifique

45071 ORLEANS CEDEX 2
FRANCE
Tel.: (33) 02 38 51 54 65
Fax: (33) 02 38 51 78 75

S. SARKAR
Univ. of California at San Diego
Dept. of AMES
9500 Gilman Drive
Mail Code 0411
LA JOLLA-CA 92093-0411
USA
Tel.: (1) 619 534 82 43
Fax: (1) 619 534 75 99

R. SAUREL
IUSTI
Technopôle de Château-Gombert
5, Rue Enrico Fermi
13453 MARSEILLE Cedex 13
FRANCE
Tel.: (33) 04 91 10 68 68
Fax: (33) 04 91 10 69 69

A.N. SECUNDOV
Central Inst. of Aviation Motors
2, Aviamotornaya Str.
Scient. & Research Centre ECOLEN
MOSCOW 111250
RUSSIA
Tel.: (7) 095 361 66 38
Fax: (7) 095 267 13 54

X. SILVANI
Lab. Comb. / Systèmes Réactifs
1C Ave. Recherche Scientifique

45071 ORLEANS CEDEX 2
FRANCE
Tel.: (33) 02 38 51 54 65
Fax: (33) 02 38 51 78 75

O. SIMONIN
EDF / Lab. Nat. d'Hydraulique
6 Quai Watier - BP 49

78401 CHATOU CEDEX
FRANCE
Tel.: (33) 01 30 87 78 30
Fax: (33) 01 30 87 72 53

S.I. SATAKE
Kogakuin University
Dept. of Mech. System Eng.
Nishi-Shinjuku 1-24-2
Shinjuku-Ku
TOKYO 163-91
JAPAN
Tel.: (81) 33 34 21 211
Fax: (81) 33 34 00 108

R. SCHIESTEL
IRPHE - UMR 6594 - CNRS
Univ. Aix-Marseille I & II
1, Rue Honnorat
13003 MARSEILLE
FRANCE
Tel.: (33) 04 91 10 78 29
Fax: (33) 04 91 08 58 91

L. SHAO
Ecole Centrale de Lyon
Lab. Méca. Fluides & Acoustique
36 Ave. Guy de Collongues - BP 163
69131 ECULLY Cedex
FRANCE
Tel.: (33) 04 72 18 60 00
Fax: (33) 04 78 64 71 45

S. SIMOENS
Ecole Centrale de Lyon
Lab. Méca. Fluides & Acoustique
36 Ave. Guy de Collongues - BP 163
69131 ECULLY Cedex
FRANCE
Tel.: (33) 04 72 18 60 00
Fax: (33) 04 78 64 71 45

A.J. SMITS
Princeton Univ. - Gas Dynamics Lab.
PO BOX CN5263
PRINCETON
NEW JERSEY 8544-5263
USA
Tel.: (1) 609 258 51 17
Fax: (1) 609 258 22 76

B. SQUALLI
Univ. Evry Val Essonne CEMIF/LEST
Lab. Energétique & Struct. Therm.
40 Rue du Pelvoux
91020 EVRY CEDEX
FRANCE
Tel.: (33) 01 69 47 75 62
Fax: (33) 01 69 47 75 99

D. STEPOWSKI
Université de Rouen
CORIA - URA 230 - CNRS
Place Emile Blondel - BP 118
76821 MONT SAINT AIGNAN Cedex
FRANCE
Tel.: (33) 02 35 14 65 85
Fax: (33) 02 35 70 83 84

A.M.K.P. TAYLOR
Imperial College of
Science Technology & Medicine
Mechanical Engineering Dept.
Exhibition Road
LONDON SW7 2BX
UK
Tel.: (44) 0171 589 51 11
Fax: (44) 0171 823 88 45

T. TSUJI
Nagoya Institute of Technology
Dept. of Mechanical Engineering
Gokiso-Cho
Showa-Ku
NAGOYA 466
JAPAN
Tel.: (81) 52 735 53 33
Fax: (81) 52 735 53 47

M. UHLMAN
Ecole Centrale de Lyon
Lab. Méca. Fluides & Acoustique
36 Ave. Guy de Collongues - BP 163
69131 ECULLY Cedex
FRANCE
Tel.: (33) 04 72 18 60 00
Fax: (33) 04 78 64 71 45

D. VANDROMME
Université de Rouen
CORIA - URA 230 - CNRS
Place Emile Blondel - BP 118
76821 MONT SAINT AIGNAN Cedex
FRANCE
Tel.: (33) 02 35 52 84 20
Fax: (33) 02 35 52 84 21

I.I. WERTGEIM
Institute of Continuous Media
Mechanics Ural Branch
Russian Academy of Sciences
1 Academician Korolyov Str.
614061 PERM
RUSSIA
E-mail: iiw@lab15.icmm.perm.su

V. ZAKHAROV
Centre de Physique Théorique
Luminy - Case 907

13288 MARSEILLE Cedex 09
FRANCE
Tel.: (33) 04 91 26 95 00
Fax: (33) 04 91 26 95 53

S. ZALESKI
Univ. P & M Curie Paris VI
CNRS URA 229 - LMM
Tour 66 - Case 162
75252 PARIS Cedex 05
FRANCE
Tel.: (33) 01 44 27 25 58
Fax: (33) 01 44 27 52 59

In addition, about 35 scientists and students from IRPHE were also present.

Author Index

Mechanics

FLUID MECHANICS AND ITS APPLICATIONS
Series Editor: R. Moreau

Aims and Scope of the Series

The purpose of this series is to focus on subjects in which fluid mechanics plays a fundamental role. As well as the more traditional applications of aeronautics, hydraulics, heat and mass transfer etc., books will be published dealing with topics which are currently in a state of rapid development, such as turbulence, suspensions and multiphase fluids, super and hypersonic flows and numerical modelling techniques. It is a widely held view that it is the interdisciplinary subjects that will receive intense scientific attention, bringing them to the forefront of technological advancement. Fluids have the ability to transport matter and its properties as well as transmit force, therefore fluid mechanics is a subject that is particularly open to cross fertilisation with other sciences and disciplines of engineering. The subject of fluid mechanics will be highly relevant in domains such as chemical, metallurgical, biological and ecological engineering. This series is particularly open to such new multidisciplinary domains.

Kluwer Academic Publishers – Dordrecht / Boston / London

Mechanics

FLUID MECHANICS AND ITS APPLICATIONS
Series Editor: R. Moreau

21. J.P. Bonnet and M.N. Glauser (eds.): *Eddy Structure Identification in Free Turbulent Shear Flows.* 1993   ISBN 0-7923-2449-8
22. R.S. Srivastava: *Interaction of Shock Waves.* 1994   ISBN 0-7923-2920-1
23. J.R. Blake, J.M. Boulton-Stone and N.H. Thomas (eds.): *Bubble Dynamics and Interface Phenomena.* 1994   ISBN 0-7923-3008-0
24. R. Benzi (ed.): *Advances in Turbulence V.* 1995   ISBN 0-7923-3032-3
25. B.I. Rabinovich, V.G. Lebedev and A.I. Mytarev: *Vortex Processes and Solid Body Dynamics. The Dynamic Problems of Spacecrafts and Magnetic Levitation Systems.* 1994   ISBN 0-7923-3092-7
26. P.R. Voke, L. Kleiser and J.-P. Chollet (eds.): *Direct and Large-Eddy Simulation I.* Selected papers from the First ERCOFTAC Workshop on Direct and Large-Eddy Simulation. 1994   ISBN 0-7923-3106-0
27. J.A. Sparenberg: *Hydrodynamic Propulsion and its Optimization.* Analytic Theory. 1995   ISBN 0-7923-3201-6
28. J.F. Dijksman and G.D.C. Kuiken (eds.): *IUTAM Symposium on Numerical Simulation of Non-Isothermal Flow of Viscoelastic Liquids.* Proceedings of an IUTAM Symposium held in Kerkrade, The Netherlands. 1995   ISBN 0-7923-3262-8
29. B.M. Boubnov and G.S. Golitsyn: *Convection in Rotating Fluids.* 1995   ISBN 0-7923-3371-3
30. S.I. Green (ed.): *Fluid Vortices.* 1995   ISBN 0-7923-3376-4
31. S. Morioka and L. van Wijngaarden (eds.): *IUTAM Symposium on Waves in Liquid/Gas and Liquid/Vapour Two-Phase Systems.* 1995   ISBN 0-7923-3424-8
32. A. Gyr and H.-W. Bewersdorff: *Drag Reduction of Turbulent Flows by Additives.* 1995   ISBN 0-7923-3485-X
33. Y.P. Golovachov: *Numerical Simulation of Viscous Shock Layer Flows.* 1995   ISBN 0-7923-3626-7
34. J. Grue, B. Gjevik and J.E. Weber (eds.): *Waves and Nonlinear Processes in Hydrodynamics.* 1996   ISBN 0-7923-4031-0
35. P.W. Duck and P. Hall (eds.): *IUTAM Symposium on Nonlinear Instability and Transition in Three-Dimensional Boundary Layers.* 1996   ISBN 0-7923-4079-5
36. S. Gavrilakis, L. Machiels and P.A. Monkewitz (eds.): *Advances in Turbulence VI.* Proceedings of the 6th European Turbulence Conference. 1996   ISBN 0-7923-4132-5
37. K. Gersten (ed.): *IUTAM Symposium on Asymptotic Methods for Turbulent Shear Flows at High Reynolds Numbers.* Proceedings of the IUTAM Symposium held in Bochum, Germany. 1996   ISBN 0-7923-4138-4
38. J. Verhás: *Thermodynamics and Rheology.* 1997   ISBN 0-7923-4251-8
39. M. Champion and B. Deshaies (eds.): *IUTAM Symposium on Combustion in Supersonic Flows.* Proceedings of the IUTAM Symposium held in Poitiers, France. 1997   ISBN 0-7923-4313-1
40. M. Lesieur: *Turbulence in Fluids.* Third Revised and Enlarged Edition. 1997   ISBN 0-7923-4415-4; Pb: 0-7923-4416-2

Kluwer Academic Publishers – Dordrecht / Boston / London

Mechanics

FLUID **MECHANICS AND ITS APPLICATIONS**
  Series Editor: R. Moreau

41. L. Fulachier, J.L. Lumley and F. Anselmet (eds.): *IUTAM Symposium on Variable Density Low-Speed Turbulent Flows*. Proceedings of the IUTAM Symposium held in Marseille, France. 1997   ISBN 0-7923-4602-5

Kluwer Academic Publishers – Dordrecht / Boston / London

ERCOFTAC SERIES

KLUWER ACADEMIC PUBLISHERS – DORDRECHT / BOSTON / LONDON